287
Current Topics in Microbiology and Immunology

Editors

R.W. Compans, Atlanta/Georgia
M.D. Cooper, Birmingham/Alabama
T. Honjo, Kyoto · H. Koprowski, Philadelphia/Pennsylvania
F. Melchers, Basel · M.B.A. Oldstone, La Jolla/California
S. Olsnes, Oslo · M. Potter, Bethesda/Maryland
P.K. Vogt, La Jolla/California · H. Wagner, Munich

L. Enjuanes (Ed.)

Coronavirus Replication and Reverse Genetics

With 49 Figures

Professor Dr. Luis Enjuanes
Centro Nacional de Biotecnología
Department of Molecular and Cell Biology
Campus Universidad Autónoma, Cantoblanco
38049 Madrid
Spain

e-mail: L.Enjuanes@cnb.uam.es

Cover illustration by Luis Enjuanes
Simplified structure of a coronavirus prototype (transmissible gastroenteritis virus, TGEV) showing the envelope with several structural proteins: spike (S), two topologies of the membrane (M or M') proteins, envelope (E) protein, and nucleoprotein (N). A nucleocapsid can be observed inside the envelope virus, containing the genomic RNA, the N protein, and the carboxiterminus of the M protein.

Library of Congress Catalog Card Number 72-152360

ISSN 0070-217X
ISBN 3-540-21494-1 Springer Berlin Heidelberg New York

This work is subject to copyright. All rights are reserved, whether the whole or part of the material is concerned, specifically the rights of translation, reprinting, reuse of illustrations, recitation, broadcasting, reproduction on microfilms or in any other way, and storage in data banks. Duplication of this publication or parts thereof is permitted only under the provisions of the German Copyright Law of September 9, 1965, in its current version, and permission for use must always be obtained from Springer-Verlag. Violations are liable for prosecution under the German Copyright Law.

Springer is a part of Springer Science+Business Media
springeronline.com
© Springer-Verlag Berlin Heidelberg 2005
Printed in Germany

The use of general descriptive names, registered names, trademarks, etc. in this publication does not imply, even in the absence of a specific statement, that such names are exempt from the relevant protective laws and regulations and therefore free for general use.
Product liability: The publishers cannot quarantee that accuracy of any information about dosage and application contained in this book. In every individual case the user must check such information by consulting the relevant literature.

Editor: Dr. Rolf Lange, Heidelberg
Desk editor: Anne Clauss, Heidelberg
Production editor: Andreas Gösling, Heidelberg
Cover design: design & production GmbH, Heidelberg
Typesetting: Stürtz AG, Würzburg
Printed on acid-free paper 27/3150/ag – 5 4 3 2 1 0

Preface

The *Coronaviridae* family is included in the *Nidovirales* order together with the *Arteriviridae* and *Roniviridae*. Possibly the first recorded coronavirus-related disease was feline infectious peritonitis in 1912. However, until the late 1960s the coronaviruses were not recognized as pathogens responsible for human diseases (common cold), and it was in 2003 when human coronaviruses (HCoVs) received worldwide attention with the emergence of the severe and acute respiratory syndrome (SARS), produced by a coronavirus (SARS-CoV), that has infected more than 8,000 people in 32 countries, killing about 10%. The increase in research on coronaviruses soon led to the discovery of another human coronavirus (HCoV-NL63), which is prevalent in 7% of hospital patients and has been associated with bronchiolitis and, possibly, conjunctivitis.

Coronaviruses have been identified in mice, rats, chickens, turkeys, pigs, dogs, cats, rabbits, horses, cows, and humans. Coronaviruses are associated mainly with respiratory, enteric, hepatic, and central nervous system diseases. In humans and fowl, coronaviruses primarily cause upper respiratory tract infections, while porcine and bovine coronaviruses establish enteric infections that result in severe economic losses. HCoVs are responsible for 10%–20% of common colds, and have been implicated in gastroenteritis, high and low respiratory tract infections, and rare cases of encephalitis. HCoVs have also been associated with infant necrotizing enterocolitis and are tentative candidates for multiple sclerosis.

In some coronavirus members, such as transmissible gastroenteritis virus (TGEV), three levels can be distinguished in the virion structure: the envelope, the core, and the nucleocapsid formed by the genome and the nucleoprotein (N). The CoV genome is a single-stranded positive-sense RNA genome of 27–32 kb that is infectious. Coronaviruses have the largest genome known for an RNA virus and probably one of the longest stable RNAs in nature. The genome of all coronaviruses contains a basic set of genes: the replicase (Rep 1a and 1b), the spike (S), envelope (E), membrane (M), and nucleoprotein (N) arranged in the order 5'-Rep1a-1b-S-E-M-N-3' and a variable number of genes encoding nonstructural proteins intercalated between these genes in a position characteristic of each virus group. The production of coronavirus subgenomic mRNAs involves the fusion of sequences that are noncontiguous in the viral genome. Several models have been proposed to explain this discontinuous synthesis. Nevertheless, the model of Sawicki and Sawicki (1995), which proposes a discontinuous step during minus-strand RNA synthesis, is best supported by the available biochemical and genetic studies.

Coronavirus reverse genetics was performed in the two last decades using defective genomes, as cDNAs encoding full-length genomes were not available due to the large size of the coronavirus RNAs, posing relevant limitations. During the 1990s, reverse genetics in coronaviruses was possible by targeted recombination developed by Paul Master's group. This useful technology enabled the modification of the coronavirus genome by recombination between a replicating coronavirus genome and nonreplicating or replicating RNAs introduced into the same cell. This technology still remains a very useful tool for modifying the coronavirus genome. Significant progress was made in 2000 with the construction of infectious cDNAs encoding coronavirus genomes using a variety of technologies. Historically, the construction of infectious cDNA clones started with the assembly of Qβ phage (4.5 kb) cDNA, and was followed by the construction of cDNA of RNA viruses with increasing complexity such as brome mosaic virus (with three RNA segments with sizes between 2.1 and 3.2 kb), poliovirus (7.5 kb), closteroviruses (19 kb) and, finally, coronaviruses (27–32 kb). The two main problems associated with the construction of infectious cDNA clones—the fidelity of reverse transcriptase and the toxicity of sequences derived from eukaryotic viruses in bacteria—were aggravated for coronavirus genomes due to their extremely large size. These problems have been overcome by following different strategies. The first infectious cDNA clone was constructed for TGEV in bacterial artificial chromosomes (BACs). These plasmids presented a reduced toxicity in bacteria since only a single copy, maximum two per cell, is produced. Other approaches were the assembly of a full-length cDNA in vitro, or the use of poxviruses as cloning vectors.

The construction of infectious cDNAs for different coronaviruses [TGEV, HCoV-229E, infectious bronchitis virus (IBV), mouse hepatitis virus (MHV), and SARS-CoV] enables nowadays deep insights into coronavirus replication and transcription mechanisms, virus–host interactions, and development of strategies to protect against coronavirus-induced diseases. This volume consists of eight chapters. The first two provide a perspective on coronavirus genome structure, replication, and transcription. The third chapter concentrates on the replicase, the most complex coronavirus gene. The fourth chapter reviews the viral and cellular proteins involved in coronavirus replication. This chapter includes cellular factors involved in virus–host interaction, a new avenue that is attracting the attention of many scientists. The fifth chapter reviews the design of targeted recombination in coronaviruses and its application to the analysis of the replication and morphogenesis of this virus family. The construction of virus vectors derived from RNA viruses is a comprehensive process that requires for optimum performance the availability of an infectious cDNA clone, knowledge of virus transcription mechanisms to optimize mRNA levels, determination of the essential and nonessential genes to create room for heterologous genes, and a strategy for vector safety. These aspects are reviewed in the sixth chapter, using TGEV as a model, and also in the last two chapters, using HCoV-229E and MHV genomes. The content of these chapters clearly shows that a new family of virus vectors based on coronavirus genomes has emerged with high potential, due to the variety of tissue and species tropism of these vectors, their large size (providing them with a large cloning capacity), and the possibility of multigenic expression using the extensive repertoire of subgenomic mRNAs that are transcribed in coronaviruses. In addition, in the last chapter the construction of the first infectious cDNA clone of SARS-CoV is described, allowing the study of the molecular biology of this virus in order to understand the molecular basis of its virulence and the development of recombinant vaccines to prevent coronavirus-induced diseases.

I certainly hope that this volume will be useful to academic researchers, scientists involved in human and animal health, and enterprises involved in the fight against coronavirus-induced diseases. I would like to thank all the authors, the staff of Springer, and my colleagues at the CNB, CSIC (Madrid), for their help in the preparation of this book.

L. Enjuanes

List of Contents

Coronavirus Genome Structure and Replication
D.A. Brian and R.S. Baric..................................... 1

Coronavirus Transcription: A Perspective
S.G. Sawicki and D.L. Sawicki 31

The Coronavirus Replicase
J. Ziebuhr 57

Viral and Cellular Proteins Involved in Coronavirus Replication
S.T. Shi and M.M.C. Lai 95

Coronavirus Reverse Genetics by Targeted RNA Recombination
P.S. Masters and P.J.M. Rottier.............................. 133

Coronavirus Reverse Genetics and Development
of Vectors for Gene Expression
L. Enjuanes · I. Sola · S. Alonso · D. Escors · S. Zúñiga 161

Reverse Genetics of Coronaviruses Using Vaccinia Virus Vectors
V. Thiel and S.G. Siddell.................................... 199

Development of Mouse Hepatitis Virus
and SARS-CoV Infectious cDNA Constructs
R.S. Baric and A.C. Sims 229

Subject Index 253

List of Contributors

(Their addresses can be found at the beginning of their respective chapters.)

Alonso, S.   161

Baric, R.S.   1, 229

Brian, D.A.   1

Enjuanes, L.   161

Escors, D.   161

Lai, M.M.C.   95

Masters, P.S.   133

Rottier, P.J.M.   133

Sawicki, D.L.   31

Sawicki, S.G.   31

Shi, S.T.   95

Siddell, S.G.   199

Sims, A.C.   229

Sola, I.   161

Thiel, V.   199

Ziebuhr, J.   57

Zúñiga, S.   161

Coronavirus Genome Structure and Replication

D. A. Brian[1] (✉) · R. S. Baric[2,3]

[1] Departments of Microbiology and Pathobiology, University of Tennessee, College of Veterinary Medicine, Knoxville, TN 37996-0845, USA
dbrian@utk.edu
[2] Department of Microbiology and Immunology, School of Medicine, University of North Carolina at Chapel Hill, Chapel Hill, NV 27599-7400, USA
[3] Department of Epidemiology, Program of Infectious Diseases, School of Public Health, University of North Carolina at Chapel Hill, Chapel Hill, NC 27599-7400, USA

1	Introduction	2
2	Common Features of Genome Structure Among Coronaviruses	3
3	Cis-Acting RNA Elements in Coronavirus Genome Replication	8
3.1	The 5′ UTR and the Translation Step(s) Preceding Genome Replication.	8
3.2	The Pseudoknot and Slippery Sequence Involved in the −1 Ribosomal Frameshifting at the ORF 1a/1b Junction	10
3.3	Cis-Acting Elements Required for Membrane Association of the RNA with the Replication Complex	10
3.4	5′ and 3′-Proximal RNA Cis-Acting Elements for DI RNA (and Presumably Genome) Replication	12
3.5	Internal Cis-Acting Signals for DI RNA (and Possibly Also for Genome) Replication	17
4	Packaging Signals	17
5	Minimum Sequence Requirements for (Autonomous) Genome Replication	18
6	Importance of Gene Order for Genome Replication	19
7	Future Directions	21
	References	22

Abstract In addition to the SARS coronavirus (treated separately elsewhere in this volume), the complete genome sequences of six species in the coronavirus genus of the coronavirus family [avian infectious bronchitis virus-Beaudette strain (IBV-Beaudette), bovine coronavirus-ENT strain (BCoV-ENT), human coronavirus-229E strain (HCoV-229E), murine hepatitis virus-A59 strain (MHV-A59), porcine transmissible gastroenteritis-Purdue 115 strain (TGEV-Purdue 115), and porcine epidemic diarrhea virus-CV777 strain (PEDV-CV777)] have now been reported. Their lengths range from 27,317 nt for HCoV-229E to 31,357 nt for the murine hepatitis virus-A59, establishing the coronavirus genome as the largest known among RNA viruses. The basic organization of the coronavirus genome is shared with other members of the Nidovirus order (the torovirus genus, also in the family *Coronaviridae*, and members of the family *Arteriviridae*) in that the nonstructural proteins involved in proteolytic processing, genome replication, and subgenomic mRNA synthesis (transcription) (an estimated 14–16 end products for coronaviruses) are encoded within the 5′-proximal two-thirds of the genome on gene 1 and the (mostly) structural proteins are encoded within the 3′-proximal one-third of the genome (8–9 genes for coronaviruses). Genes for the major structural proteins in all coronaviruses occur in the 5′ to 3′ order as S, E, M, and N. The precise strategy used by coronaviruses for genome replication is not yet known, but many features have been established. This chapter focuses on some of the known features and presents some current questions regarding genome replication strategy, the *cis*-acting elements necessary for genome replication [as inferred from defective interfering (DI) RNA molecules], the minimum sequence requirements for autonomous replication of an RNA replicon, and the importance of gene order in genome replication.

1
Introduction

Despite its unique property as the largest of the known plus-strand RNA genomes, the coronavirus genome shares with those of other plus-strand RNA viruses (excepting retroviruses) the properties of (1) infectiousness [and not using a packaged RNA-dependent RNA polymerase (RdRp)] (Brian et al. 1980; Schochetman et al. 1977) and (2) replication in the cytoplasm in close association with cellular membranes (Denison et al. 1999; Dennis and Brian 1982; Gosert et al. 2002; Sethna and Brian 1997; Shi et al. 1999; van der Meer et al. 1999). Many of the basic features of coronavirus genome structure and replication have been described in recent reviews (Cavanagh et al. 1997; Enjuanes et al. 2000a, 2000b; Lai and Cavanagh 1997; Lai and Holmes 2001; Luytjes 1995; van der Most and Spaan 1995). With the advent of reverse genetics enabling site-directed mutagenesis of any part of the genome (Almazan et al. 2000; Casais et al. 2001; Masters 1999; Thiel et al. 2001; Yount et al. 2000, 2002), many of the mechanistic features of coronavirus genome replication that could previously be learned only from direct manipulation of defective interfering (DI) RNA can now be examined in the context of the whole virus genome. In this chapter, we review the current knowledge of coronavirus genome structure and organization and the *cis*-acting elements in coronavirus replication and raise selected questions that we believe are important for approaching a better understanding of coronavirus genome replication.

2
Common Features of Genome Structure Among Coronaviruses

In addition to the SARS coronavirus (treated separately elsewhere in this volume), the genomes of six species of coronaviruses have now been fully sequenced and reported in GenBank (as of November 2002): IBV-Beaudette (NC 001451, Boursnell et al. 1987), BCoV-ENT (NC 003045, Chouljenko et al. 2001), MHV-A59 (NC 001846, Leparc-Goffart et al. 1997), HCoV-229E (NC 002645, Herold et al. 1993; Thiel et al. 2001), TGEV-Purdue (NC 002306, Almazan et al. 2000; Eleouet et al. 1995; Penzes et al. 2001), and PEDV-CV777 2001 (NC 003436, Kocherhans et al. 2001). These, representing all three coronavirus serogroups (Siddell 1995), are schematically depicted in Fig. 1. Additional strains of BCoV [BCoV-LUN (AF391542, Chouljenko et al. 2001)], BCoV-Mebus (U00735, Nixon and Brian, unpublished data) and BCoV-Quebec (AF220295, Yoo and Pei 2001), and MHV [MHV-2 (AF201929, Sarma et al. 1999)] have also been reported. The genome sizes range from 27,317 nt for HCoV-229E to 31,357 nt for MHV-A59, establishing them as the largest known among RNA viruses (Enjuanes et al. 2000a; Lai and Cavanagh 1997). The following similarities in genome structure among the six can be noted:

1. The 5' UTRs ranging in length from 209 to 528 nt contain a similarly positioned short, AUG-initiated open reading frame (ORF) relative to the 5' end [Table 1; a situation that, by current terminology, is problematic because the "untranslated region" now becomes in part potentially translatable and thus should preferably be called a "leader" (Morris and Geballe 2000). The term "leader," however, has an established meaning in the nidovirus lexicon (Lai and Cavanagh 1997; see subsequent chapters, this volume) of a 5'-terminal, genome-encoded sequence of 65–98 nt appearing on the 5' terminus of each subgenomic mRNA species]. For purposes of this review, "5' UTR" will refer to the sequence upstream of ORF 1 (gene 1) despite the internally positioned short ORF. The short AUG-initiated ORFs (except for HCoV-229E) begin in a suboptimal Kozak context for translation (Table 1) (Kozak 1991) and potentially encode peptides of 3–11 amino acids.
2. The 3' UTRs range from 288 to 506 nt [although some strains of IBV have 3' UTRs of greater length because of internal sequence duplications (Williams et al. 1993)], all possess an octameric sequence of GGAAGAGC beginning at base 73 to 80 upstream from the poly(A) tail, and all possess a 3'-terminal poly(A) tail (Table 1).

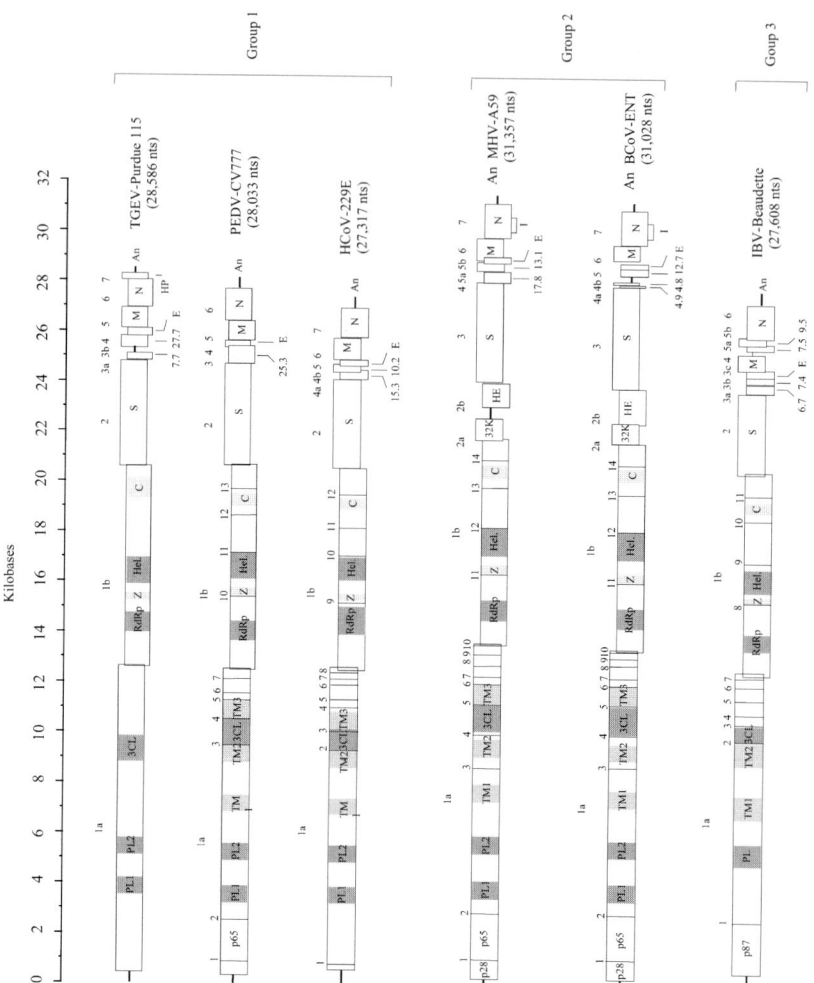

Fig. 1. Genomes of the six sequenced species of coronaviruses known prior to the discovery of the SARS coronavirus. Maps are drawn to approximate scale, and species are shown in decreasing order of size within each of the three groups. The representations are derived from data in the GenBank as of November 2002. For gene 1 (ORFs 1a and 1b) the predicted protease cleavage sites are indicated by numbers and domains of known or predicted function are shaded and identified (*PL*, papain-like protease; *3CL*, poliovirus 3C-like protease; *TM*, transmembrane domain; *RdRp*, RNA-dependent RNA polymerase; *Z*, zinc finger (metal-binding) domain; *Hel*, helicase domain; *C*, conserved sequence domain). Genes 2–8 (or 9) are identified by their transcript name (1a, 1b, etc.) or their abbreviated name of the protein product (*S*, spike; *E*, envelope; *M*, membrane; *N*, nucleocapsid; *HP*, hydrophobic protein; *HE*, hemagglutinin-esterase; *I*, internal). Literature references are described with the GenBank information (see text)

Table 1. Properties of the coronavirus 5' UTR, intra-5' UTR short ORF, and 3' UTR

Coronavirus	Length of 5' UTR (number of nt upstream of gene 1)	Position and Kozak context[a] of the intra-5' UTR short ORF start codon	Number of amino acids encoded by the 5' UTR short ORF	Amino acid sequence of the 5' UTR short ORF product	Length of 3' UTR (number of nt)	Position of the first nt in the octamer GGAAGAGC upstream from the 3' poly(A) tail
TGEV-Purdue	314	117UCUaugA	3	MKS	279	76
PEDV-CV777	296	105GUUaugC	10	MLLEAGVEFH	334	73
HCoV-229E	292	86GCUaugG	11	MAGIFDAGVVV[b]	462	74
MHV-A59	209	99UCCaugC	8	MPAGIVLS	324	81
BCoV-ENT	210	100UCUaugC	8	MPVGVDFS[c]	288	78
IBV-Beaudette	528	131UGGaugG	11	MAPGHLSGFCY	506	80

[a] The optimal Kozak context for translation initiation is GCCaugG (Kozak 1991).
[b] A second ORF beginning 16 nt downstream from the first and in the plus 1 reading frame relative to the first encodes the amino acids MLES.
[c] The second amino acid in the BCoV-Mebus strain is L.

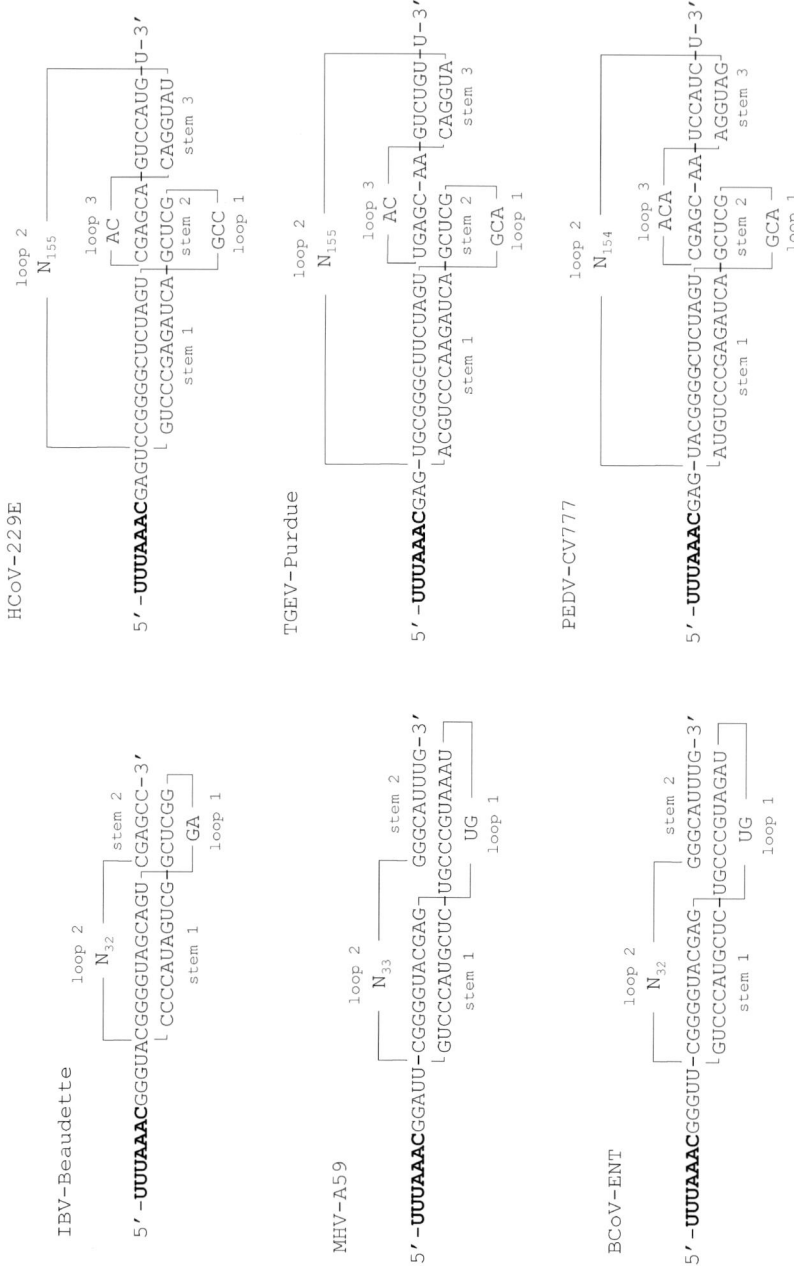

3. All have an extremely large gene 1 (separated into ORFs 1a and 1b and extending over approximately two-thirds of the genome) encoding nonstructural proteins involved in proteolytic processing of the gene 1 polyprotein products, virus genome replication, and sgmRNA synthesis (transcription). In each, gene 1 is translated as ORFs 1a and 1ab, with 1ab resulting from a pseudoknot-induced −1 ribosomal frame shifting event at a slippery sequence of UUUAAAC at the ORF 1a/1b junction (Fig. 2) (Brown and Brierley 1995).
4. All encode the structural spike (S) glycoprotein, small envelope (E) protein, membrane (M) glycoprotein, and nucleocapsid (N) protein, in that order, 5′→3′ within the 3′-proximal one-third of the genome. A variable number of other ORFs appearing to be virus- or group-specific, many apparently encoding nonstructural proteins, are also found here. These (and their potential products) include ORF 3a (7.7-kDa protein), ORF 3b (27.7-kDa protein), and ORF 7 [0.7-kDa hydrophobic protein (HP)] in TGEV; ORF 3 (25.3-kDa protein) in PEDV; ORF 4a (15.3-kDa protein) and ORF 4b (10.2-kDa protein) in HCoV-229E; ORF 2a (32-kDa protein), ORF 2b [65-kDa complete or 34.6-kDa truncated hemagglutinin-esterase (HE) protein, depending on the strain], ORF 4 (17.8-kDa protein), ORF 5a (13.1-kDa protein), and an ORF internal to gene 7 [23-kDa internal (I) protein] in MHV; ORF 2a (32-kDa protein), ORF 2b (65-kDa HE protein), ORF 4a (4.9-kDa protein), ORF 4b (4.8-kDa protein), ORF 5 (12.7-kDa protein), and an ORF internal to gene 7 (23-kDa I protein) in BCoV; and ORF 3a (6.7-kDa protein), ORF 3b (7.4-kDa protein), ORF 5a (7.5-kDa protein), and ORF 5b (9.5-kDa protein) in IBV (Fig. 1; Brown and Brierly 1995, and references listed in the GenBank information noted above). Some of these, such as ORFs 3a and 3b in TGEV (McGoldrick 1999; Wesley et al. 1991) and ORFs 2a

Fig. 2. Pseudoknotted structures and slippery sequences responsible for highly efficient (25%–30%) −1 ribosomal frameshifting at the ORF 1a and 1b junction in gene 1 of the six coronaviruses shown in Fig. 1. The slippery sequence UUUAAAC, identified in bold, is the same in all sequenced genomes. The IBV pseudoknot-induced frameshifting was the first nonretroviral example of ribosomal frameshifting in higher eukaryotes (Brierley et al. 1987, 1989). The pseudoknots in MHV (Bredenbeek et al. 1990) and BCoV (Yoo and Pei 2001) are nearly identical and are similar to the structure in IBV. In HCoV-229E an elaborated pseudoknot with three stems was shown by mutation analysis to be the functional frameshifting structure (Harold and Siddell 1993). In TGEV (Eleouet et al. 1995) and in PEDV (Kocherhans et al. 2001) an elaborated pseudoknot was also predicted based on similarities to HCoV-229E

(Schwarz et al. 1990), 2b (HE) (Luytjes et al. 1988), 4 (Weiss et al. 1993; Yokomori and Lai 1991), 5a (Yokomori and Lai 1991), and I (Fischer et al. 1997) in MHV, have been shown to be nonessential for replication in cell culture, and their function in virus replication remains undetermined (de Haan et al. 2002).

Presumably all coronavirus genomes are capped with a 5′ methylated nucleotide, but so far this has been demonstrated only in MHV (Lai et al. 1982).

3
Cis-Acting RNA Elements in Coronavirus Genome Replication

3.1
The 5′ UTR and the Translation Step(s) Preceding Genome Replication

As with all nonretroviral plus-strand RNA viruses, a necessary early step in genome replication is translation of the genome for production of the RdRp and other proteins required for viral genome replication. The presence of a 5′ terminal methylated cap on MHV genomic and subgenomic mRNAs (Lai et al. 1982) would suggest that coronaviruses use a cap-mediated ribosomal entry mechanism for translation. Mutation analyses of the 5′ UTR of BCoV indicate that a scanning mechanism is used for entry of ribosomes onto ORF 1 (Senanayake and Brian 1999). Curiously in light of these results, a methylated cap on DI RNA transcripts is not required for initiation of replication of BCoV DI RNA, which contains a genomic 5′ UTR. This molecule has a *cis*-acting dependence on translation for replication (Chang et al. 1994; Chang and Brian 1996). It remains to be determined whether capping is required for translation and replication of the intact viral genome. It remains to be determined what enzyme functions to cap the viral RNAs (Ziebuhr et al. 2000).

In MHV it has been demonstrated that the viral nucleocapsid protein N binds tightly (K_d=14 nM) to the UCUAAAC intergenic region (also named transcription-regulating sequence, TRS) of the genomic leader and consequently may influence translation rate (Nelson et al. 2000; Tahara et al. 1998). Is this property of N common to all coronaviruses? If so, what role does it play in the regulation of genome replication?

Does the intra-5′ UTR short ORF play a role in translation (or in subsequent replication) of the genome? With reverse genetics, disruption of an analogous ORF in equine arterivirus had no apparent effect on virus replication in cell culture (Molenkamp et al. 2000), but the ORFs may not have homologous function in the two virus groups. Certainly, short upstream ORFs can have profound enhancing or suppressing effects on the translation of a downstream ORF (Morris and Geballe 2000), and their universal existence in coronavirus 5' UTRs, albeit with little or no conservation in size or amino acid sequence (Table 1), would suggest that they function in the regulation of replication or gene expression. One possibility is that the intra-5' UTR short ORF or some other 5' UTR element, such as the binding site for N described above, is responsible for the repression of translation from the ORF 1 start codon in virus-infected cells (Senanayake and Brian 1999).

Some observed phenomena in coronavirus genome and DI RNA replication hint that the 5' UTR might be bypassed altogether in order to meet the translation requirements for genome replication. One set of observations relates to a possible role for N in genome replication (Baric et al. 1988; Compton et al. 1987; Kim K and Makino 1995; Laude and Masters 1995; Nelson et al. 2000; Stohlman et al. 1988), a role that would set coronaviruses apart from arteriviruses in this regard because only gene 1 products have been shown to be sufficient for arterivirus genome replication (Molenkamp et al. 2000). N protein, for example, binds leader sequence with high affinity (Nelson et al. 2000), is present in a subpopulation of coronavirus RNA replication complexes (Sethna and Brian 1997; Sims et al. 2000), and is essential for infectivity of recombinant IBV full-length transcripts (Casais et al. 2001). If N is required, then some mechanism for the translation of N from the polycistronic genome, such as an internal entry of ribosomes onto genomic RNA or formation of an early subgenomic mRNA transcript, would be needed, at least when infection is initiated by the genome alone (as in transfection experiments). Some evidence for internal ribosomal entry has been demonstrated for IBV mRNA 3 (Liu and Inglis 1992), MHV mRNA 5 (Thiel and Siddell 1994; Jengrach et al. 1999), and TGEV mRNA 3 (O'Connor and Brian 2000), making it prudent to consider an internal entry at these or other sites on the genome for protein synthesis. Another set of observations relates to a requirement for translation in *cis* of the DI RNA molecule to be replicated. Although some DI RNAs with a single ORF do not appear to require translation *in cis* for replication (Liao and Lai 1995), others do (Chang and Brian 1996; De Groot et al. 1992; Van der Most et al. 1995). Might a *cis*-acting requirement for DI RNA translation reflect a similar *cis*-translation-dependent mechanism for genome replication as described for picornaviruses (Egger et al.

2000; Gamarnik and Andino 1998; Novak and Kirkegaard 1994) and flaviviruses (Khromykh et al. 1999)? If so, then perhaps an internal ribosomal entry for translation onto the 3' proximal region of the genome might be needed for coronavirus genome replication.

3.2
The Pseudoknot and Slippery Sequence Involved in the − 1 Ribosomal Frameshifting at the ORF 1a/1b Junction

Ribosomal frameshifting in coronaviruses was the first described non-retroviral example of ribosomal frameshifting in higher eukaryotes (Brierly 1987), and the earliest described higher-order RNA structure recognized as a *cis*-acting element in coronavirus genome replication was the pseudoknot located immediately downstream of the UUUAAAC slippery sequence in the IBV genome (Brierly et al. 1987, 1989; Brown and Brierly 1995) (Fig. 2). The pseudoknot in IBV was described as a hairpin-type and was shown by mutation analyses to be responsible for the highly efficient (25%–30%) frameshifting. Subsequently, a pseudoknot with similar properties was found in gene 1 of MHV (Bredenbeek et al. 1990) and BCoV (Yoo and Pei 2001). Interestingly, the pseudoknot found in gene 1 of HCoV-229E was found to be quite different in structure, possessing an extremely large loop 2 and a stem 3 (Fig. 2). This structure was termed an "elaborated" pseudoknot and was shown to function as such in in vitro measurements of frameshifting (Herold and Siddell 1993). The predicted pseudoknots in TGEV and PEDV gene 1 appear to be quite similar to that in HCoV-229E (Eleouet et al. 1995; Kocherhans et al. 2001). The pseudoknot-associated slippery sequence is UUUAAAC in all sequenced coronaviruses described to date.

3.3
Cis-Acting Elements Required for Membrane Association of the RNA with the Replication Complex

Once made, or possibly concurrent with synthesis, viral proteins and (possibly) associated cellular proteins function to form the membrane-associated RNA replication complexes. Membrane association is a hallmark of replication complexes of plus-strand RNA viruses, but the origin of the membrane and the anatomy of the replication complexes appear to differ among virus families. A preliminary understanding of the coronavirus replication complex has come primarily from studies with MHV and partly from studies with TGEV. The following features have been observed:

1. The membrane in the MHV replication complex has shown markers for the endoplasmic reticulum and Golgi (Shi et al. 1999; Gosert at al. 2002) and, alternatively, the late endosomes (van der Meer et al. 1999; Sims et al. 2000).
2. The replication complex is intimately associated with double membrane structures, and the anchored proteins are the hydrophobic sequence-containing intermediate cleavage products p290 and p150, and p210 and p44, of ORF 1a (Gossert et al. 2002).
3. There appear to be two populations of membrane-associated replication complexes separable by isopycnic sedimentation (Sethna and Brian 1997; Sims et al. 2000). In MHV the less dense fraction (1.05–1.09 g/ml) was found to contain p65 and p1a-22, products of ORF 1a, whereas the denser fraction (1.12–1.25 g/ml) contained p28 and helicase from ORF 1b, and N (Sims et al. 2000).

In TGEV two buoyant density populations (1.15–1.17 g/ml and 1.20–1.24 g/ml) were also found, and both had associated with them genome- and subgenome-length plus- and minus-strand RNAs (Sethna and Brian 1997). Some S, M, and N proteins were associated with the denser population. The TGEV membrane replication complexes, furthermore, appeared to have an unusual impermeability to micrococcal nuclease. It remains to be determined precisely what proteins, viral and cellular, function together to make up the coronavirus replication complexes and how they might be associated with the membranes and with one another. How might they differ between the processes of minus- and plus-strand synthesis? Between replication and transcription? Which proteins bind the RNA, both genomic and subgenomic, both plus and minus strands, within the complex? What is the stoichiometry of the components in the various complexes? What is the relationship between the RNA replication complex and the site of virus assembly at the Golgi and intermediate Golgi membranes? How is the genome selected and transported from the replication complex to the site of virus assembly? Does the evidence of resistance of coronaviral RNAs to ribonuclease suggest existence of a compartmentalized replication complex and have implications for resistance to RNA silencing (Ahlquist 2002) and long-term persistent coronaviral infections (Adami et al. 1995; Baric et al. 1999; Okumura et al. 1996; Stohlman et al. 1999)?

3.4
5′ and 3′-Proximal RNA *Cis*-Acting Elements for DI RNA (and Presumably Genome) Replication

Since the first description of their cloning and replication in helper virus-infected cells, coronavirus DI RNAs have been used in attempts to define the minimal *cis*-acting sequence requirements for their replication (Brian and Spaan 1997; Makino et al. 1985, 1988a, 1988b; van der Most et al. 1991). Through deletion analyses the regions harboring minimal *cis*-acting sequences have been mapped for DI RNAs from TGEV, MHV, BCoV, and IBV (noted as filled regions in the DI RNA maps in Fig. 3). For most of the DI RNAs it can be seen that these sequences reside at the termini of the viral genomes for distances of 467–1,348 at the 5′ end and 338–1,635 at the 3′ end. Further reduction in the sizes of these regions may result from further deletion analyses. Requirements for internal genome sequence elements appear to be DI RNA specific but may reflect requirements of the intact genome (see below). What is the nature of the terminal *cis*-acting RNA elements? Is a specific sequence alone sufficient, or are higher-order structures required? So far, these questions have focused primarily on the small (2.2–2.3 kb) DI RNAs of the group 2 coronaviruses MHV and BCoV.

With regard to the 3′ UTR of MHV-A59 and BCoV-Mebus, common replication signals exist between the two viruses. This was demonstrated by experiments in which the entire 3′ UTR of the MHV genome was replaced with the equivalent region of the BCoV genome without loss of virus viability (Hsue and Masters 1997) and in a BCoV DI RNA chimera in which the BCoV 3′ UTR was replaced with the MHV 3′ UTR with no detectable loss of replicating ability (Ku, Williams, and Brian, unpublished data). More recently, BCoV DI RNA has been shown to replicate in the presence of MHV as helper virus (Wu et al. 2003). To date, three higher-order *cis*-acting elements mapping within the 3′ UTR have been characterized in MHV and BCoV (Fig 4):

———————————————————————————▶

Fig. 3. Map positions of minimal *cis*-acting sequences for RNA replication (*solid boxes*) and signals for packaging (*stippled boxes*) as determined from studies on DI RNAs and their derivatives. The schematic diagrams of the four coronaviruses studied in this manner are shown. (a) Izeta et al. 1999; deletion analyses were done on derivatives of TGEV DI RNA C (9.7 kb) (Mendez et al. 1996); M21 contains minimal sequence elements for replication and inefficient packaging; M33 and M62 contain

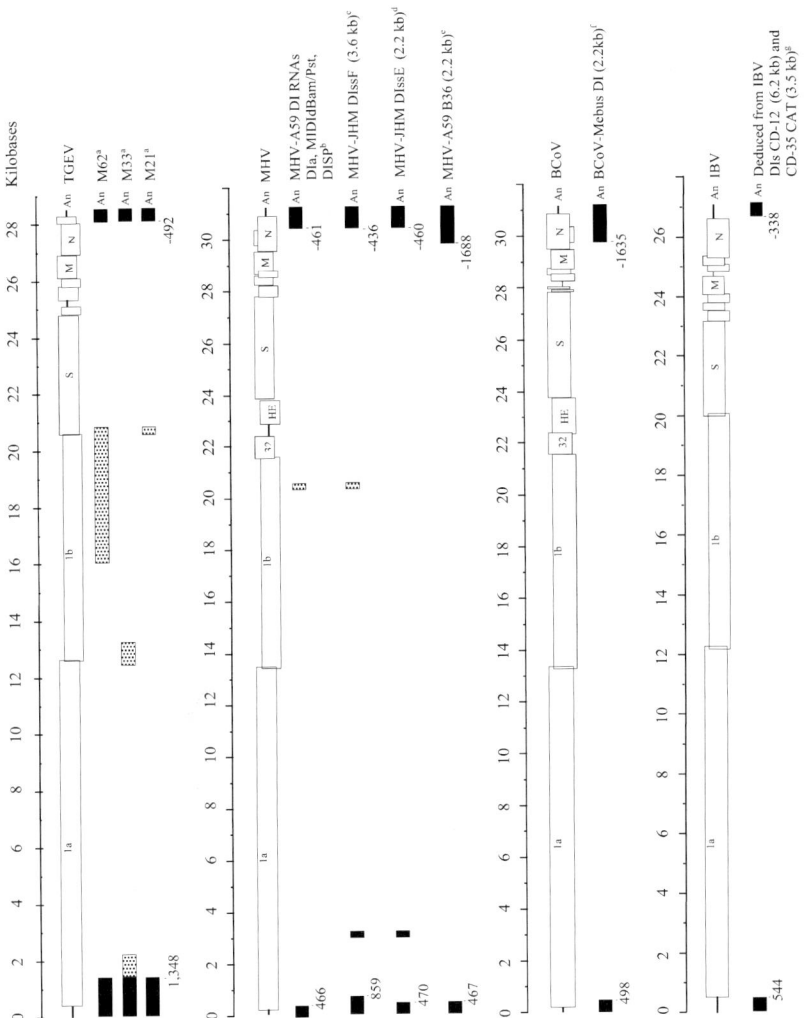

small nonoverlapping regions of ORFs 1a and 1b that contribute to packaging; (b) Luytjes et al. 1996; van der Most et al. 1991, 1995; deletion analyses were done on derivatives of MHV-A59 DIa RNA (5.5 kb); (c) Lin and Lai 1993; Makino et al. 1990; deletion analyses were done on DIssF; (d) Fosmire et al. 1992; Kim et al. 1993; Kim and Makino 1995; deletion analyses were done on DIssE; (e) Masters et al. 1994; DI B36 is synthetic and was designed after the BCoV-Mebus DI RNA; (f) Chang et al. 1994; deletion analyses were done on reporter-containing DI Drep1; (g) Dalton et al. 2001; deletion analyses were done on derivatives of 9.1-kb IBV DI RNA CD-91 (Penzes et al. 1994); unknown regions within the UTRs suffice for packaging of DI RNA, but packaging is inefficient

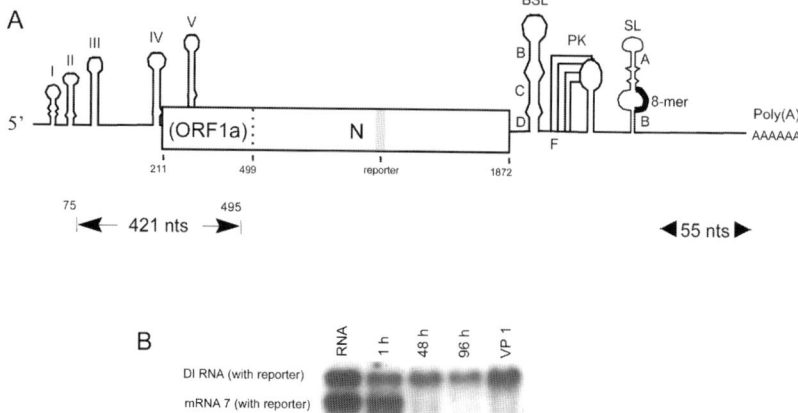

Fig. 4A, B. Terminal *cis*-acting replication sequences and higher-order structures identified to date in the smallest of the MHV and BCoV DI (group 2) RNAs. **A** The DI RNA illustrated is that for BCoV, but the structures drawn (with the exception of the 5'-proximal stem-loops I and II and the upper portion of the 3'-proximal octamer-associated stem loop) are phylogenetically conserved between MHV and BCoV. The *open rectangle* represents an open reading frame formed by the fusion of the first part of ORF1a and the entire N gene. The 3' higher-order structures are a 61-nt bulged stem-loop (Hsue et al. 2000), a hairpin-type pseudoknot (Williams et al. 1999), a helix formed at the base of a long stem-loop and adjacent to the phylogenetically conserved octameric sequence (Liu et al. 2001). The poly(A) tail is required for replication (Lin and Lai 1993;, Spagnolo and Hogue 2000), and the 5'-terminal 55 nt are the minimal sequence requirements for minus-strand RNA synthesis in MHV (Lin et al. 1994). The 5' higher-order structures are a stem-loop III and stem-loop IV within the 5' UTR (Raman et al. 2002) and stem V within the partial ORF 1a sequence (Brown et al. 2002). **B** Experimental evidence for replication (accumulation) of reporter-containing DI RNA but not mRNA7 containing the same reporter after transfection into helper virus-infected cells (Chang et al. 1994). The only difference between the two molecules is a sequence of 421 nt mapping between nt 74 and 497 in the BCoV DI RNA

1. A 68-nt bulged stem-loop beginning immediately downstream of the N stop codon consists in MHV of four stems (B, C, D, and F) and a 14-nt terminal loop (Hsue and Masters 1997; Hsue et al. 2000). Stems C, D, and F have been shown to be required for replication of both the DI RNA and virus genome.
2. A 54-nt hairpin-type pseudoknot beginning 60 nt downstream of the bulged stem-loop (Williams et al. 1999). Both stems of the pseudoknot have been shown to be required for replication. The pseudoknot sequence overlaps the downstream arm of stem F in the bulged stem-loop

such that the two structures cannot exist simultaneously. This led Hsue et al. (2000) to suggest a possible interaction between the two elements, with the alternative conformations acting as a possible "switching" mechanism. This switch has now been confirmed experimentally (Goebel et al. 2004).The pseudoknot appears phylogenetically conserved to some degree in all coronaviruses.

3. A 74-nt bulged stem-loop mapping from nt 68 to 142 from the 3' terminus in MHV contains two stems that demonstrated importance as cis-acting replication structures (referred to as stems A and B in Fig. 4) (Liu et al. 2001). Stem B, which shows greater importance in DI RNA replication, is phylogenetically conserved in structure between MHV and BCoV. Stem B is immediately adjacent downstream to the phylogenetically conserved 3' UTR octamer GGAAGAGC (Liu et al. 2001). Unidentified cellular proteins of 120, 55, 40, and 25 kDa molecular mass bind to nt 130–142 which is the upstream half of the internal loop in stem B (Liu et al. 1997; Yu and Liebowitz 1995).

Proteins identified to date that bind within the 3' region (or the minus-strand counterpart of this region) include the poly(A) binding protein (Spagnolo and Hogue 2000), mitochondrial aconitase, which binds within the 42-nt 3'-terminal region in MHV (Nanda and Leibowitz 2001), and the polypyrimidine tract-binding protein, which binds to minus-strand sequence complementary to nt 53–149 (strongly) and 270–307 (weakly) in MHV (Huang and Lai 1999). What roles the 3' UTR higher-order structures play in RNA replication are not known. Because the 3'-terminal 55 nt were shown to be a minimal sequence requirement for minus-strand synthesis in MHV (Lin et al. 1994), the higher-order structures mapping upstream of the 55-nt sequence possibly play no role in minus-strand synthesis. Do they play a role in initiating or regulating plus-strand synthesis? Precedents in picornaviruses (Barton et al. 2001; Herold and Andino 2001), alphaviruses (Frolov et al. 2001),and flaviruses (You et al. 2001) would suggest they might. Certainly the poly(A) tail through the poly(A)-binding protein is a candidate for such a process, perhaps through genome circularization (Spagnolo and Hogue 2000).

With regard to the 5' UTR it is known that the 5'-terminal sequence is required for DI RNA replication (Chang et al. 1994; Kim et al. 1993) and at least two stem-loops (stem-loops III and IV in Fig. 4) function as higher-order cis-acting signaling elements (Raman et al. 2003; Raman and Brian, unpublished data). A higher-order cis-acting structure mapping within the first 290 nt of ORF1 (stem-loop V in Fig. 4) has also been found (Brown, Nixon, Senanayake, and Brian, unpublished data). Proteins shown to bind within the 5' UTR include the viral N protein, which binds in and around the leader-adjacent intergenic sequence motif UCUAAAC (Nelson et al. 2000), the polypyrimidine tract binding protein, which also binds near the leader-adjacent UCUAAAC sequence motif (Li et al. 1999), and hnRNP A1, which binds the minus-strand complement of the leader-adjacent UCUAAAC sequence motif (Li et al. 1997). None of these has been reported to bind regions covered by stem-loops III, IV, or V depicted in Fig. 4. Might there be a process of leader priming of genome replication (Zhang and Lai 1996), as suggested by the phenomenon of high-frequency leader switching on DI RNAs during DI RNA replication (Chang et al. 1996; Makino and Lai 1989; Stirrups et al. 2000)?

The question of what *cis*-acting sequences act in coronavirus RNA replication has relevance not only for genome replication but also for poorly understood features of sgmRNA behavior. It has been suggested that coronavirus sgmRNAs amplify by a replication mechanism (Brian et al. 1994; Hofmann et al. 1990; Sethna et al. 1989). This hypothesis made use of the argument that the termini on the sgmRNAs and genome, identical at the 5' end for the length of the leader (65–98 nt, depending on the virus species) and at the 3' end for greater than the length of the 3' UTR (i.e., greater than 300 nt), are larger than the known promoters for a viral RdRp [replication promoters in influenza and Sindbis viruses are less than 20 nt in length (Levis et al. 1986; Li and Palese 1992)] and are therefore large enough to harbor promoters for replication. The hypothesis was also consistent with the observations that (1) the molar ratios of minus-strand to plus-strand RNA are equivalent for sgmRNA and genome (i.e., 1:100), (2) the rate of sgmRNA accumulation is inversely proportional to the length of the molecule, (3) the rate of sgmRNA minus strand disappearance parallels that of antigenome, and (4) sgmRNA minus strands possess 3'-terminal sequences complementary to the leader (Sethna et al. 1989). Furthermore, (5) double-stranded subgenomic mRNA-length RFs and RIs (Hofmann and Brian 1991; Hofmann et al. 1990; Sawicki and Sawicki 1990; Sethna et al. 1989) were shown to be active in subgenomic mRNA synthesis (Baric and Yont 2000; Sawicki and Sawicki 1995, 1998; Sawicki et al. 2001; Schaad and Baric 1994). If the 3'-terminal 55 nt are the only requirement for minus-strand RNA synthesis (Lin et al. 1994), the possibility is left open that the subgenomic mRNAs function as a templates for minus-strand synthesis. At no time, however, has it been directly demonstrated that sgmRNA transcripts, with or without a reporter, are replicated in the presence of a helper virus after transfection into helper virus-infected cells (Fig. 4B) (Chang et al. 1994; Makino et al. 1991). Therefore, what features enable the replication of the DI RNAs but not sgmRNAs on transfection into helper virus-infected cells? The answer could lie in the function of the 5'-proximal stem-loops III, IV, and V residing within the 421-nt region found in BCoV DI RNA but not found in sgmRNAs (Fig. 4A) (Chang et al. 1994). Do these higher-order structures bind viral or cellular proteins? Might they be signals working through long-distance RNA-RNA or RNA-protein interactions?

3.5
Internal *Cis*-Acting Signals
for DI RNA (and Possibly Also for Genome) Replication

Most DI RNAs described for coronaviruses are comprised of more than just the terminal genomic sequences. That is, they are mosaics of internal and terminal genome sequences. Replication of MHV-JHM DI RNAs has been found to be dependent on a 57-nt sequence mapping within ORF 1a (Kim and Makino 1995; Lin and Lai 1993). This sequence has been shown to form a secondary structure in the positive strand, and both the higher-order structure and its sequence are important for function as a replication signal (Repass and Makino 1998). Does this structure represent a *cis*-acting replication signal required for replication of the intact genome?

4
Packaging Signals

Perpetuation of coronavirus infection via cell-to-cell spread requires that the genome be packaged into virions via one or more *cis*-acting packaging signals. Inasmuch as several small DI RNAs containing only terminal sequences are packaged, some form of signal sufficient for incorporation into virions must reside in the termini. This idea is consistent with the observed packaging of subgenomic mRNAs in TGEV (Sethna et al. 1989), BCoV (Hofmann et al. 1990), and IBV (Zhao et al. 1993). However, these packaging signals may not be the ones used by the virus genome for packaging. A candidate 69-nt genome packaging signal has been identified in mosaic DI RNAs of MHV (Fosmire et al. 1992; Makino et al. 1990; van der Most et al. 1991) that maps to a region within ORF 1b, shows correlation of function with maintenance of secondary structure (Fosmire et al. 1992), and confers packaging on reporter RNA molecules (Bos et al. 1997; Woo et al. 1997). A homologous structure in BCoV ORF 1b also leads to packaging of nonviral RNAs (Cologna and Hogue 2000). Do these represent the bona fide packaging signals for the viral genome? Is there perhaps more than one packaging signal, as suggested by the ability of more than a single region of ORF 1b to contribute to packaging efficiency in large TGEV DI RNAs (Izeta et al. 1999)? Perhaps not since a recent study shows only a single packaging signal encoded within the 5'-terminal 649 nts of the TGEU genome is sufficient (Escors et al. 2003). In addition to the N protein (Laude and Masters 1995), might the packaging signals interact with other components of the virion? Perhaps so since in MHV the envelop (E) protein (Narayanan and Makino 2001) and M protein (Narayanan et al. 2003) have been shown to play roles in packaging.

5
Minimum Sequence Requirements for (Autonomous) Genome Replication

Although gene 1 products are the only ones required for arterivirus genome replication and sgmRNA synthesis (Molenkamp et al. 2000), the story might be different for coronaviruses. Gene 1 of HCoV-229E in the presence of the genomic 5' and 3' UTRs was shown to be sufficient for sgmRNA synthesis when the intergenic sequence for mRNA 7 (N mRNA) and an mRNA body (gene for the green fluorescence protein) were present just downstream of gene 1 (Thiel et al. 2001). The authors, however, were unable to conclude that these sequences alone were sufficient for RNA replication or to rule out a role for N as an enhancer for transcription. These results, therefore, leave open the possibility that another gene function is important for replication. Autonomous replicons of TGEV containing only genes 1, 2, part of 5, and all of 6 and 7 have been described (Curtis et al. 2002). Reverse genetics with these and other coronaviruses now make feasible the analysis of the minimal sequences required for genome replication and should lead to a definitive resolution of the question of the role of N protein in RNA replication.

6
Importance of Gene Order for Genome Replication

The gene order for coronaviruses, as for many positive- and negative-stranded RNA virus families, is highly conserved. In coronaviruses the essential genes pol, S, E, M, and N are invariably found in that order, 5' to 3', although they are sometimes interspersed with genes showing no essential function for virus growth in cell culture (discussed above). What is the significance of this gene order? If it is altered, what might the consequences be on virus growth? Might pathogenesis be altered such that the variants could be used as vaccines or vectors for other uses?

The presence of nonessential genes 3a and 3b in TGEV for cell culture growth has enabled development of TGEV as a heterologous expression vector (see the chapter by Enjuanes et al., this volume) and as a virus to study the effects of gene rearrangements. In initial studies on the effect of gene rearrangement, the N gene has been duplicated (producing the genotype SNEMN) and repositioned (producing the genotype SNEM) by making use of gene positions 3A and 3B (K. Curtis and R. Baric, unpublished data). The N gene was chosen for repositioning because it encodes the most abundantly expressed sgmRNA and is translated into the most abundant of the viral proteins. On the basis of general gene expression patterns relative to the 3' end of the genome in coronaviruses it was anticipated that expression of E and M would increase relative to N in the rearranged SNEM construct. When tested by transfection, the TGEV mutants SNEMN and SNEM were found to be viable but to replicate at about 10-fold and 1,000-fold less than wild-type virus levels, respectively. These results indicated that a specified gene order per se is not essential for coronavirus replication in cell culture, but that order contributes in some way to a more robust virus yield. When TGEV SNEM was serially passaged 15 times, the mutant gene order SNEM was maintained, but, surprisingly, virus growth was restored to near wild-type levels. Restoration of TGEV SNEM fitness as defined by virus yield was associated with changes within the N-(partial) Δ3B-E junction region. These included removal of most of the residual (partial) ΔORF3B sequence, deletion of the wt E intergenic sequence element, and activation of a new, highly transcriptionally active E intergenic sequence element just downstream of the newly inserted N gene (Fig. 5B). These results indicate that high-frequency RNA recombination does not function to restore a specific coronavirus gene order, at least over the short term, because the new N gene position in SNEM was stable for many passages. Rather, the

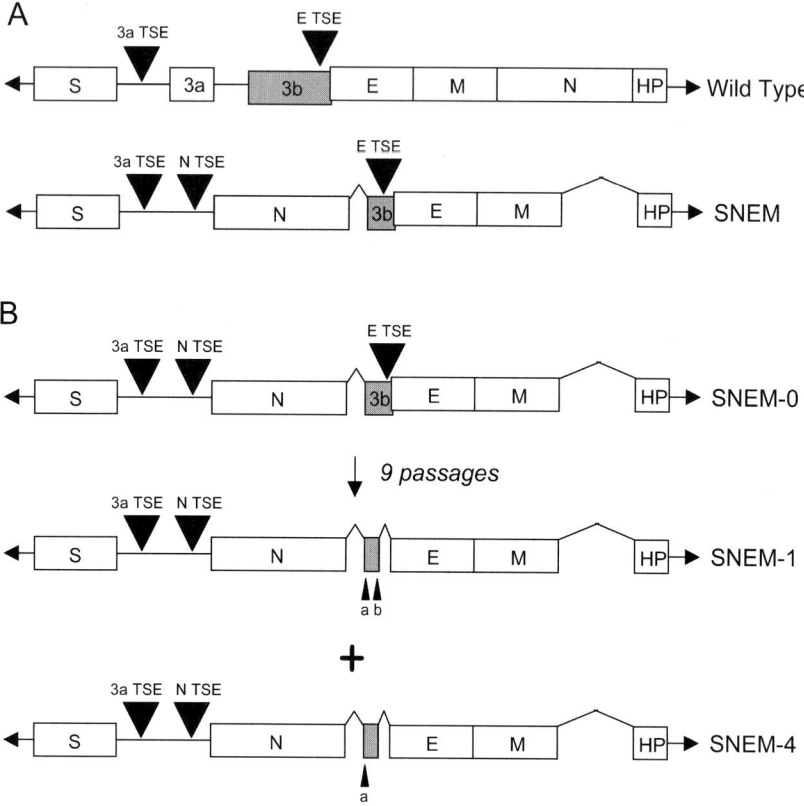

Fig. 5A, B. Effects of moving the N gene within the TGEV genome from its normal position to an upstream site. The N gene including its immediate transcription stimulating element (TSE)-containing upstream sequence of 24 nt was placed just downstream of the 3a TSE sequence in a TGEV genome from which the entire 3a and a portion of the 3b gene had been deleted (**A**). Transcripts of the recombinant TGEV genome, designated SNEM, were transfected into cells, and progeny viruses were studied (**B**). Immediately after transfection (passage 0) the titer of progeny was low ($<10^5$ PFU/ml) and the genome sequence was identical to the original construct. The progeny (SNEM-1 and SNEM-4) grew more efficiently ($\sim 5.0 \times 10^6$ PFU/ml) after 9 passages and reached wild-type levels ($\sim 1.0 \times 10^8$) after 24 passages. In all progeny the upstream 3a TSE sequence was used for leader fusion of the N transcript. For expression of the E gene, however, the story was different. At passage 0 (SNEM-0), transcripts of the E gene used the wt TSE as well as two additional sites, designated a and b within the ORF3b residual sequence, for leader fusion. In the SNEM-1 and SNEM-4 viruses the wt E TSE was deleted and transcripts of the E gene used the two new TSEs formed within the residual gene 3b sequence (a=4/5 clones, b=1/5 clones) in SNEM-1. In SNEM-4 only the a site was used for E gene expression. Thus the re-ordered TGEV genome was stable with regard to the new (upstream) position of N

coronavirus genome can rapidly develop compensatory changes to restore virus replication rate (fitness) while maintaining a new gene order. Mechanisms of fitness restoration appeared to include recombination events and point mutations (Baric et al., unpublished data). It is likely that gene order mutants will provide novel insights into the regulation of coronanvirus transcription and replication, identify protein-protein interactions that function cooperatively to maintain robust virus fitness and growth, and assist in the identification of core sequence elements that function in sgmRNA synthesis.

7
Future Directions

It is anticipated that reverse genetics, which now enables an alteration of any part of the coronavirus genome, will facilitate examination of the *cis*- and *trans*-acting elements in RNA replication and transcription within the context of the intact genome. These elements have until now been studied primarily in DI RNAs. In light of precedents established with many much smaller plus-strand RNA viruses of animals and plants, it would not be surprising to find novel long-distance RNA-RNA and protein-RNA interactions involving genome sequences not present in DI RNAs. Long-distance interactions are hinted at in comparative studies of DI RNAs (which replicate) and sgmRNAs (which do not replicate). What genes are important in regulation of replication and transcription, and how important is gene order in these processes? These questions can now be rigorously approached with reverse genetics. It is also anticipated that a greater understanding of the assembly, stoichiometry, and function of the RNA synthesizing complexes will be gained through similar rigorous analyses. It is anticipated that one practical outcome of reverse genetics will be the development of safe coronavirus-based replicon vectors, not necessarily only those that become packaged, for vaccine and other biomedical uses. Still in waiting is the development of an in vitro virus replication system such as that used for poliovirus (Molla

◀―――

for over 24 passages, but in SNEM-1 and SNEM-4 additional mutations were selected upstream of the 3aTSE and in the M gene that greatly enhanced virus fitness and N gene expression. In SNEM the sequences of the TSEs are AACTAAACT for 3a, and ACAAAAC for E, TAACTAAACT for N, AACTAAAG for a, and AACACAAAAC for b

et al. 1991), in which complete virus replication can be accomplished in cell lysates. This approach would enable still more detailed analyses of the requirements for genome replication beginning with the infectious genome transcript. All in all, it is likely that the next decade will bring significant breakthroughs regarding our understanding of the mechanisms involved in coronavirus genome replication and transcription, the function of the replication complexes, and the development and application of coronavirus recombinant vectors for the treatment of animal and human diseases.

Acknowledgements We thank Cary Brown, Kimberley Nixon, Sharmila Raman, Gwyn Williams, and Hung-Yi Wu in the Brian laboratory and Kristopher Curtis and Boyd Yount in the Baric laboratory for invaluable discussions and experimentation. Work in D. Brian's laboratory is supported by grant AI-14367 from the National Institutes of Health and work in R. Baric's laboratory by grants AI-23946 and GM-63228 from the National Institutes of Health.

References

Adami C, Pooley J, Glomb J, Stecker E, Fazal F, Fleming JO, Baker SC (1995) Evolution of mouse hepatitis virus (MHV) during chronic infection: quasispecies nature of the persisting MHV RNA. Virology 209:337–346

Ahlquist P (2002) RNA-dependent RNA polymerases, viruses, and RNA silencing. Science 296:1270–1273

Almazan F, Gonzalez JM, Penzes Z, Izeta a, Calvo E, Plana-Duran J, Enjuanes L (2000) Engineering the largest RNA virus genome as an infectious bacterial artificial chromosome. Proc Natl Acad Sci USA 97:5516–5521

Baric RS, Nelson GW, Fleming JO, Deans RJ, Keck JG, Casteel N, Stohlman SA (1988) Interactions between coronavirus nucleocapsid protein and viral RNAs: implications for viral transcription. J. Virol 62:4280–4287

Baric RS, Sullivan E, Hensley L, Yount B, Chen W (1999) Persistent infection promotes cross-species transmissibility of mouse hepatitis virus. J Virol 73:638–649

Baric RS, Yount B (2000) Subgenomic negative-strand RNAs function during mouse hepatitis virus infection. J Virol 74:4039–4046

Barton DJ, Donnell BJO, Flanegan JB (2001) 5′ Cloverleaf in poliovirus RNA is a *cis*-acting replication element required for negative-strand synthesis. EMBO J 20:1439–1448

Bos ECW, Dobbe JC, Luytjes W, Spaan WJM (1997) A subgenomic mRNA transcript of the coronavirus mouse hepatitis virus strain A59 defective interfering (DI) RNA is packaged when it contains the DI RNA packaging signal. J Virol 71:5684–5687

Boursnell MEG, Brown TDK, Foulds IJ, Green, PF, Tomley FM, Binns MM (1987) Completion of the sequence of the genome of the coronavirus avian infectious bronchitis virus. J Gen Virol 68:57–77

Bredenbeek PJ, Pachuk CJ, Noten AFH, Charite, J, Luytjes W, Weiss SR, Spaan WJM (1990) The primary structure and expression of the second open reading frame of the polymerase gene of the coronavirus MHV-A59: a highly conserved polymerase is expressed by an efficient ribosomal frameshifting mechanism. Nucleic Acids Res 18:1825–1832

Brian DA, Chang RY, Hofmann MA, Sethna PB (1994) Role of subgenomic minus-strand RNA in coronavirus replication. Arch Virol 9 (Suppl): 173–180

Brian DA, Dennis DE, Guy JS (1980) Genome of porcine transmissible gastroenteritis virus. J Virol 34:410–415

Brian DA, Spaan WJM (1997) Recombination and coronavirus defective interfering RNAs. Semin Virol 8:101–111

Brierly I, Boursnell MEG, Binns MM, Bilimoria B, Blok VC, Brown TDK, Inglis SC (1987) An efficient ribosomal frame-shifting signal in the polymerase-encoding region of the coronavirus IBV. EMBO J 6:3779–3785

Brierly I, Digard P, Inglis SC (1989) Characterization of an efficient coronavirus ribosomal frameshifting signal: requirement for an RNA pseudoknot. Cell 57:537–547

Brown TDK, Brierley I (1995) The coronavirus nonstructural proteins. In The Coronaviridae (S.G. Siddell, ed.), Plenum Press, New York and London, pp. 191–2171

Casais R, Thiel V, Siddell SG, Cavanagh D, Britton P (2001) Reverse genetics system for the avian coronavirus infectious bronchitis virus. J Virol 75:12359–12369

Cavanagh D, Brian DA, Brinton MA, Enjuanes L, Holmes KV, Horzinek MC, Lai MMC, Laude H, Plagemann PGW, Siddell, SG, Spaan W, Taguchi F, Talbot PJ (1997) Nidovirales: a new order comprising Coronaviridae and Arteriviridae. Arch Virol. 142:629–633

Chang RY, Brian DA (1996) *Cis* requirement for N-specific protein sequence in bovine coronavirus defective interfering RNA replication. J Virol 70:2201–2207

Chang RY, Hofmann MA, Sethna PB, Brian BA (1994) A *cis*-acting function for the coronavirus leader in defective interfering RNA replication. J Virol 68:8223–8231

Chang RY, Krishnan R, Brian DA (1996) The UCUAAAC promoter motif is not required for high-frequency leader recombination in bovine coronavirus defective interfering RNA. J Virol 70:2720–2729

Chouljenko VN, Lin XQ, Storz J, Kousoulas KG, Gorbalenya AE (2001) Comparison of genomic and predicted amino acid sequences of respiratory and enteric bovine coronaviruses isolated from the same animal with fatal shipping pneumonia. J Gen Virol 82:2927–2933

Cologna R, Hogue BG (2000) Identification of a bovine coronavirus packaging signal. J Virol 74:580–583

Compton SR, Rogers DB, Holmes KV, Fertsch D, Remenick J, McGowan JJ (1987) In vitro replication of mouse hepatitis virus strain A59. J Virol 61:1814–1820

Curtis K, Yount B, Baric RS (2002) Heterologous gene expression from transmissible gastroenteritis virus replicon particles. J Virol 76:1422–1434

Dalton K, Casais R, Shaw K, Stirrups K, Evans S, Britton P, Brown TDK, Cavanagh, D (2001) *cis*-Acting sequences required for coronavirus infectious bronchitis virus defective-RNA replication and packaging. J Virol 75:125–133

de Groot RJ, van der Most RG, Spaan WJM (1992) The fitness of defective interfering murine coronavirus DI-a and its derivatives is decreased by nonsense and frameshift mutations. J Virol 66:5898–5905

de Haan CAM, Masters PS, Shen X, Weiss S, Rottier PJM (2002) The group-specific murine coronavirus genes are not essential, but their deletion, by reverse genetics, is attenuating in the natural host. Virology 296:177–189

Denison MR, Spaan WJ, van der Meer Y, Gibson CA, Sims AC, Prentice E, Lu XT (1999) The putative helicase of the coronavirus mouse hepatitis virus is processed from the replicase gene poly protein and localizes in complexes that are active in viral RNA synthesis. J Virol 73:6862–6871

Dennis DE, Brian DA (1982) RNA-dependent RNA polymerase activity in coronavirus-infected cells. J Virol 42:153–164

de Vries AAF, Horzinek MC, Rottier PJM, de Groot RJ (1997) The genome organization of the Nidovirales: similarities and differences between arteri-, toro-, and coronaviruses. Semin Virol 8:33–47

Egger D, Teterina N, Ehrenfeld E, Bienz K (2000) Formation of the poliovirus replication complex requires coupled viral translation, vesicle production, and viral RNA synthesis. J Virol 74:6570–6580

Eleouet JF, Rasschaert D, Lambert P, Levy L, Vende P, Laude H (1995) Complete sequence (20 kilobases) of the polyprotein-encoding gene 1 of transmissible gastroenteritis virus. Virology 206:817–822

Enjuanes L, Brian D, Cavanagh D, Holmes K, Lai MMC, Laude H, Masters P, Rottier PJM, Siddell SG, Spaan WJM, Taguchi F, Talbot P (2000) Coronaviridae. In: Virus Taxonomy, Seventh Report of the International Committee on Taxonomy of Viruses (MHV van Regenmortel, CM Fauquet, DHL Bishop, EB Carstens, MK Estes, SM Lemon, J Maniloff, MA Mayo, DJ McGeoch, CR Pringle, RB Wickner, eds) Academic Press, San Diego. pp 835–849

Enjuanes L, Spaan WJM, Snijder E, Cavanagh D (2000) Nidovirales. In: Virus Taxonomy, Seventh Report of the International Committee on Taxonomy of Viruses (MHV van Regenmortel, CM Fauquet, DHL Bishop, EB Carstens, MK Estes, SM Lemon, J Maniloff, MA Mayo, DJ McGeoch, CR Pringle, RB Wickner, eds) Academic Press, San Diego. pp 827–834

Escors D, Izeta A, Capiscol C, Enjuanes L (2003) Transmissible gastroenteritis coronavirus packaging signal is located at the 5'end of the virus genome. J Virol 77:7890–7902

Fosmire JA, Hwang K, Makino S (1992) identification and characterization of a coronavirus packaging signal. J Virol 66:3522–3530

Frolov I, Hardy R, Rice CM (2001) *Cis*-acting RNA elements at the 5' end of Sindbis virus genome RNA regulate minus- and plus-strand RNA synthesis. RNA 7:1638–1651

Gamarnik AV, Andino R (1998) Switch from translation to RNA replication in a positive-stranded RNA virus. Genes Dev 12:2293–2304

Goebel SJ, Hsue B, Dombrowski TF, Masters PS (2004) Characterization of the RNA components of a putative molecular switch in the 3'untranstated region of the murine coronavirus. J Virol 78:669–682

Gosert R, Kanjanahaluethai A, Egger D, Bienz K, Baker SC (2002) RNA replication of mouse hepatitis virus takes place at double-membrane vesicles. J. Virol 76:3697–3708

Herold J, Andino R (2001) Poliovirus RNA replication requires genome circularization through a protein-protein bridge. Mol Cell 7:581–591

Herold J, Raabe T, Schelle-Prinz B, Siddell SG (1993) Nucleotide sequence of the human coronavirus 229E RNA polymerase locus. Virology 195:680–691

Herold, J, Siddell SG (1993) An "elaborated" pseudoknot is required for high frequency frameshifting during translation of HCV 229E polymerase mRNA. Nucleic Acids Res 21:5838–5842

Hofmann MA, Brian DA (1991) The 5' end of coronavirus minus-strand RNAs contains a short poly(U) tract. J Virol 65:6331–6333

Hofmann MA, Sethna PB, Brian DA (1990) Bovine coronavirus mRNA replication continues throughout persistent infection in cell culture. J Virol 64:4108–4114

Hsue B, Hartshorne T, Masters PS (2000) Characterization of an essential RNA secondary structure in the 3' untranslated region of murine coronavirus genome. J Virol 74:6911–6921

Hsue B, Masters PS (1997) A bulged stem-loop structure in the 3' untranslated region of the coronavirus mouse hepatitis virus genome is essential for replication. J Virol 71:7567–7578

Huang P, Lai MMC (1999) Polypyrimidine tract-binding protein binds to the complementary strand of the mouse hepatitis virus 3' untranslated region, thereby altering RNA conformation. J Virol 73:9110–9116

Izeta A, Smerdou C, Alonso S., Penzes Z, Mendez A, Plana-Duran J, Enjuanes, L (1999) Replication and packaging of transmissible gastroenteritis coronavirus-derived synthetic minigenomes. J Virol 73:1535–1545

Jengrach M, Thiel V, Siddell S (1999) Characterization of an internal ribosome entry site within mRNA 5 of murine hepatitis virus. Arch Virol 144:921–933

Khromykh AA, Sedlak PL, Westaway EG (1999) Trans-complementation analysis of the flavivirus Kunjin ns5 gene reveals an essential role for translation of its N-terminal half in RNA replication. J Virol 73:9247–9255

Kim KH, Makino S (1995) Two murine coronavirus genes suffice for viral RNA synthesis. J Virol 69:2313–2321

Kim Y, Makino S (1995) Characterization of a murine coronavirus defective interfering RNA internal cis-acting replication signal. J Virol 69:4963–4971

Kim YN, Jeong YS, Makino S (1993) Analysis of cis-acting sequences essential for coronavirus defective interfering RNA replication. Virology 197:53–63

Kim YN, Lai MMC, Makino S (1993) Generation and selection of coronavirus defective interfering RNA with large open reading frames by RNA recombination and possible editing. Virology 194:244–253

Kocherhans R. Bridgen A, Ackermann M, Tobler K (2001) Completion of the porcine epidemic diarrhoea coronavirus (PEDV) genome sequence. Virus Genes 23:137–144

Kozak M (1991) Structural features in eukaryotic mRNAs that modulate the initiation of translation. J Biol Chem 266:19867–19870

Lai MMC, Cavanagh D (1997) The molecular biology of coronaviruses. Adv Virus Res 48:1-100

Lai MMC, Holmes KC (2001) *Coronaviridae*: the viruses and their replication. In Fields Virology, Fourth Edition (Knipe DM, Howley PM, eds), Lippincott Williams and Wilkins, Philadelphia, pp. 1163–1185

Lai MMC, Patton CD, Stohlman SA (1982) Further characterization of mRNAs of mouse hepatitis virus: presence of common 5′-end nucleotides. J Virol 41:557–565

Laude H, Masters PS (1995) The coronavirus nucleocapsid protein. In The Coronaviridae (S.G. Siddell, ed.), Plenum Press, New York and London, pp. 141–163

Leparc-Goffart I, Hingley ST, Chua MM, Jiang X, Lavi E, Weiss SR (1997) Altered pathogenesis of a mutant of the murine coronavirus MHV-A59 is associated with a Q159L amino acid substitution in the spike protein. Virology 239:1-10

Levis R, Weiss BG, Tsiang M, Huang H, Schlesinger S (1986) Deletion mapping of Sindbis virus DI RNAs derived from cDNAs defines the sequences essential for replication and packaging. Cell 44:137–145

Li HP, Huang P, Park S, Lai MMC (1999) Polypyrimidine tract-binding protein binds to the leader RNA of mouse hepatitis virus and serves as a regulator of viral transcription. J Virol 73:772–777

Li HP, Zhang X, Duncan R, Comai L, Lai MMC (1997) Heterogeneous nuclear ribonucleoprotein A1 binds to the transcription-regulatory region of mouse hepatitis virus RNA. Proc Natl Acad Sci USA 94:9544–9549

Li X, Palese P (1992) Mutational analysis of the promoter required for influenza virus virion RNA synthesis. J Virol 66:4331–4338

Liao CL, Lai MMC (1995) A *cis*-acting viral protein is not required for the replication of a coronavirus defective interfering RNA. Virology 209:428–436

Lin Y, Lai MMC (1993) Deletion mapping of a mouse hepatitis virus defective interfering RNA reveals the requirement of an internal and discontiguous sequence for replication. J Virol 67:6110–6118

Lin YJ, Liao CL, Lai MMC (1994) Identification of the *cis*-acting signal for minus-strand RNA synthesis of a murine coronavirus: implications for the role of minus-strand RNA in RNA replication and transcription. J Virol 68:8131–8140

Liu DX, Inglis SC (1992) Internal entry of ribosomes on a tricistronic mRNA encoded by infectious bronchitis virus. J Virol 66:6143–6154

Liu Q, Johnson RF, Leibowitz JL (2001) Secondary structural elements within the 3′ untranslated region of mouse hepatitis virus strain JHM genomic RNA. J Virol 75:12105–12113

Liu Q, Yu W, Leibowitz JL (1997) A specific host cellular protein binding element near the 3′ end of mouse hepatitis genomic RNA. Virology 232:74–85

Luytjes W (1995) Coronavirus gene expression. In The Coronaviridae (S.G. Siddell, ed.), Plenum Press, New York and London, pp. 33–54

Luytjes W, Gerritsma H, Spaan WJ (1996) Replication of synthetic defective interfering RNAs derived from coronavirus mouse hepatitis virus-A59. Virology 216:174–183

Makino S, Fujioka N, Fujiwara K (1985) Structure of the intracellular defective viral RNAs of defective interfering particles of mouse hepatitis virus. J Virol 54:329–336

Makino S, Joo M, Makino JK (1991) A system for study of coronavirus mRNA synthesis: a regulated, expressed subgenomic defective interfering RNA results from intergenic site insertion. J Virol 65:6031–6041

Makino S, Lai MMC (1989) High-frequency leader sequence switching during coronavirus defective interfering RNA replication. J Virol 63:5285–5292

Makino S, Shieh CK, Keck JG, Lai MMC (1988) Defective interfering particles of murine coronavirus: mechanism of synthesis of defective viral RNAs. Virology 163:104–111

Makino S, Shieh CK, Soe LH, Baker SC, Lai MMC (1988) Primary structure and translation of a defective interfering RNA of murine coronavirus. Virology 166:1-11

Makino S, Yokomori K, Lai MMC (1990) Analysis of efficiently packaged defective interfering RNAs of murine coronavirus: localization of a possible RNA-packaging signal. J Virol 64:6045–6053

Masters PS (1999) Reverse genetics of the largest RNA viruses. Adv Virus Res 53:245–264

Masters PS, Koetzner CA, Kerr CA, Heo Y (1994) Optimization of targeted RNA recombination and mapping of a novel nucleocapsid gene mutation in the coronavirus mouse hepatitis virus. J Virol 68:328–337

Mendez A, Smerdou C, Izeta A, Gebauer F, Enjuanes L (1996) Molecular characterization of transmissible gastroenteritis coronavirus defective interfering genomes: packaging and heterogeneity. Virology 217:495–507

Molenkamp R, van Tol H, Rozier BCD, van der Meer Y, Spaan WJM, Snijder EJ (2000) The arterivirus replicase is the only viral protein required for genome replication and subgenomic mRNA transcription. J Gen Virol 81:2491–2496

Molla A, Paul AV, Wimmer E (1991) Cell-free, de novo synthesis of poliovirus. Science 254:1647–1651

Morris DR, Geballe AP (2000) Upstream open reading frames as regulators of mRNA translation. Mol Cell Biol 20:8635–8642

Nanda SK, Leibowitz JL (2001) Mitochondrial aconitase binds to the 3′ untranslated region of the mouse hepatitis virus genome. J Virol 75:3352–3362

Narayanan K, Chen C-J, Maeda J, Makino S (2003) Nucleocapsid-independent specific viral RNA packaging via viral envelope protein and viral RNA signal. J Virol 77:2922–2927

Narayanan K, Makino S (2001) Cooperation of an RNA packaging signal and a viral envelope protein in coronavirus RNA packaging. J Virol 75:9059–9067

Nelson GW, Stohlman SA, Tahara SM (2000) High affinity interaction between nucleocapsid protein and leader/intergenic sequence of mouse hepatitis virus RNA. J Gen Virol 81:181–188

Novak JE, Kirkegaard K (1991) Improved method for detecting poliovirus negative strands used to demonstrate specificity of positive-strand encapsidation and the ratio of positive to negative strands in infected cells. J Virol 65:3384–3387

Okumura A, Machii K, Azuma S, Toyoda Y, Kyuwa S (1996) Maintenance of pluripotency in mouse embryonic stem cells persistently infected with murine coronavirus. J Virol 70:4146–4149

Penzes Z, Gonzalez JM, Calvo E, Izeta A, Smerdou C, Mendez A, Sanchez CM, Sola I, Almazan F, Enjuanes L (2001) Complete genome sequence of transmissible gastroenteritis coronavirus PUR46-MAD clone and evolution of the Purdue virus cluster. Virus Genes 23:105–118

Penzes Z, Tibbles K, Shaw K, Britton P, Brown TDK, Cavanagh D (1994) Characterization of a replicating and packaged defective RNA of avian coronavirus infectious bronchitis virus. Virology 203:286–293

Penzes Z, Wroe C, Brown TDK, Britton P, Cavanagh D (1996) Replication and packaging of coronavirus infectious bronchitis virus defective RNAs lacking a long open reading frame. J Virol 70:8660–8668

Raman S, Bouma P, Williams GD, Brian DA (2003) Stem-loop III in the 5′ UTR is a cis-acting element in bovine coronavirus DI RNA replication. J Virol in press

Repass JF, Makino S (1998) Importance of the positive-strand RNA secondary structure of a murine coronavirus defective interfering RNA internal replication signal in positive-strand RNA synthesis. J Virol 72:7926–7933

Sarma JD, Hingley ST, Lai MMC, Weiss SR, Lavi E (1999) Direct submission to GenBank.

Sawicki SG, Sawicki DL (1990) Coronavirus transcription: subgenomic mouse hepatitis virus replicative intermediates function in mRNA synthesis. J Virol 64:1050–1056

Sawicki SG, Sawicki DL (1995) Coronaviruses use discontinuous extension for synthesis of subgenome-length negative strands. Adv Exp Med Biol 380:499–506

Sawicki SG, Sawicki DL (1998) A new model for coronavirus transcription. Adv Exp Med Biol 280:215–218.

Sawicki DL, Wang T, Sawicki SG (2001) The RNA structures engaged in replication and transcription of the A59 strain of mouse hepatitis virus. J Gen Virol 82:385–396

Schaad MC, Baric RS (1994) Genetics of mouse hepatitis virus transcription: evidence that subgenomic negative strands are functional templates. J Virol 68:8169–8179

Schochetman G, Stevens RH, Simpson RW (1977) Presence of infectious polyadenylated RNA in the coronavirus avian bronchitis virus. Virology 77:772–782

Senanayake SD, Brian DA (1999) Translation from the 5′ UTR of mRNA 1 is repressed, but that from the 5′ UTR of mRNA 7 is stimulated in coronavirus-infected cells. J Virol 73:8003–8009

Sethna PB, Brian DA (1997) Coronavirus subgenomic and genomic minus-strand RNAs copartition in membrane-protected replication complexes. J Virol 71:7744–7749

Sethna PB, Hofmann MA, Brian DA (1991) Minus-strand copies of replicating coronavirus mRNAs contain antileaders. J Virol 65:320–325

Sethna, PB, Hung SL, Brian DA (1989) Coronavirus subgenomic minus-strand RNA and the potential for mRNA replicons. Proc Natl Acad Sci USA 86:5626–5630

Shi ST, Schiller JJ, Kanjanahaluethai, A, Baker SC, Oh JW, Lai MMC (1999) Colocalization and membrane association of murine hepatitis virus gene 1 products and de novo-synthesized viral RNA in infected cells. J Virol 73:5957–5969

Siddell, SG (1995) The Coronaviridae. In The Coronaviridae (S.G. Siddell, ed.), Plenum Press, New York and London, pp. 1–10

Snijder EJ, Meulenberg JJM (1998) The molecular biology of arteriviruses. J Gen Virol 79:961–979

Spagnolo JF, Hogue BG (2000) Host protein interactions with the 3' end of bovine coronavirus RNA and the requirement of the poly(A) tail for coronavirus defective genome replication. J Virol 74:5053–5065

Stirrups K, Shaw K, Evans S, Dalton K, Cananagh D, Britton P (2000) Leader switching occurs during the rescue of defective RNAs by heterologous strains of the coronavirus infectious bronchitis virus. J Gen Virol 81:791–801

Stohlman SA, Baric RS, Nelson GN, Soe LH, Welter LM, Deans RJ (1988) Specific interactions between coronavirus leader RNA and nucleocapsid protein. J Virol 62:4288–4295

Stohlamn SA, Bergmann CC, Perlman S (1999) Selected animal models of viral persistence: mouse hepatitis virus. In Persistent Viral Infections (Ahmed R, Chen ISY, eds) John Wiley and Sons, New York, pp. 537–557

Tahara SM, Dietlin TA, Nelson GW, Stohlman SA, Manno DJ (1998) Translation effector properties of mouse hepatitis virus nucleocapsid protein. Adv Exp Med Biol 440:313–318

Thiel V, Herold J, Schelle B, Siddell SG (2001) Infectious RNA transcribed in vitro from a cDNA copy of the human coronavirus genome cloned in vaccinia virus. J Gen Virol 82:1273–1281

Thiel V, Herold J, Schelle B, Siddell SG (2001) Viral replicase gene products suffice for coronavirus discontinuous transcription. J Virol 75:6676–6681

Thiel V, Siddell SG (1994) Internal ribosomal entry in the coding region of murine hepatitis virus mRNA 5. J Gen Virol 75:3041–3046

van der Meer Y, Snijder EJ, Dobbe JC, Schleich S, Denison MR, Spaan WJ, Locker JK (1999) Localization of mouse hepatitis virus nonstructural proteins and RNA synthesis indicates a role for late endosomes in viral replication. J Virol 73:7641–7657.

van der Most RG, Bredenbeek PJ, Spaan WJM (1991) A domain at the 3' end of the polymerase gene is essential for encapsidation of coronavirus defective interfering RNAs. J Virol 65:3219–3226

van der Most RG, Luytjes W, Rutjes S, Spaan WJM (1995) Translation but not the encoded sequence is essential for the efficient propagation of the defective interfering RNAs of the coronavirus mouse hepatitis virus. J Virol 69:3744–3751

van der Most RG, Spaan WJM (1995) Coronavirus replication, transcription, and RNA recombination. In The Coronaviridae (S.G. Siddell, ed.), Plenum Press, New York and London, pp. 11–31

van Marle G, van Dinten LC, Spaan WJM, Luytjes W, Snijder EJ (1999) Characterization of an equine arteritis virus replicase mutant defective in subgenomic mRNA synthesis. J Virol 73:5274–5281

Williams AK, Wang L, Sneed LW, Collisson EW (1993) Analysis of a hypervariable region in the 3' non-coding end of the infectious bronchitis virus genome. Virus Res 28:19–27

Williams GD, Chang RY, Brian DA (1999) A phylogenetically conserved hairpin-type 3' untranslated region pseudoknot functions in coronavirus RNA replication. J Virol 73:8349–8355

Woo K, Joo M, Narayanan K, Kim KH, Makino S (1997) Murine coronavirus packaging signal confers packaging to nonviral RNA. J Virol 71:824–827

Wu HY, Guy JS, Yoo D, Vlasak R, Urbach E, Brian DA (2003) Common RNA replication signals exist among group 2 coronaviruses: evidence for in vivo recombination between animal and human coronavirus molecules. Virology 315:174–183

Yoo D, Pei Y (2001) Full-length genomic sequence of bovine coronavirus (31 kb). Adv Exp Med Biol 494:73–76

You S, Falgout B, Markoff L, Padmanabhan R (2001) In vitro RNA synthesis from exogenous dengue viral RNA templates requires long range interactions between 5′- and 3′-terminal regions that influence RNA structure. J Biol Chem 276:15581–15591

Yount B, Curtis KM, Baric RS (2000) Strategy for systematic assembly of large RNA and DNA genomes: the transmissible gastroenteritis virus model. J Virol 74:10600–19611

Yount B, Denison MR, Weiss SR, Baric RS (2002) Systematic assembly of a full-length infectious cDNA of mouse hepatitis virus strain A59. J Virol 76:11065–11078

Yu W, Leibowitz JL (1995) Specific binding of host cellular proteins to multiple sites within the 3′ end of mouse hepatitis virus genomic RNA. J Virol 69:2016–2023

Zhang X, Lai MMC (1996) A 5′-proximal RNA sequence of murine coronavirus as a potential initiation site for genomic-length mRNA transcription. J Virol 70:705–711

Zhao S, Shaw K, Cavanagh D (1993) Presence of subgenomic mRNAs in virions of coronavirus IBV. Virology 196:172–178

Ziebuhr J, Snijder EJ, Gorbalenya AE (2000) Virus-encoded proteinases and proteolytic processing in the Nidovirales. J Gen Virol 81:853–879

Coronavirus Transcription: A Perspective

S. G. Sawicki (✉) · D. L. Sawicki

Department of Microbiology, Medical College of Ohio, Toledo, OH 43614, USA
ssawicki@mco.edu

1	Introduction	32
2	Discontinuous Transcription by Coronaviruses	35
3	Kinetics of Plus- and Minus-Strand RNA Synthesis and Sensitivity to Translational Inhibition	41
4	The Kinetics of Synthesis of the Subgenomic Minus Strands	42
5	Subgenomic MHV RIs Exist in Infected Cells and Are Transcriptionally Active in mRNA 2–7 Synthesis	43
6	Characterization of Coronavirus Native RI/TIs and Native RF/TFs	44
7	Turnover of MHV Replicative/Transcriptive Intermediates	48
8	A Working Model	50
References		52

Abstract At the VIth International Symposium on Corona and Related Viruses held in Québec, Canada in 1994 we presented a new model for coronavirus transcription to explain how subgenome-length minus strands, which are used as templates for the synthesis of subgenomic mRNAs, might arise by a process involving discontinuous RNA synthesis. The old model explaining subgenomic mRNA synthesis, which was called leader-primed transcription, was based on erroneous evidence that only genome-length negative strands were present in replicative intermediates. To explain the discovery of subgenome-length minus strands, a related model, called the replicon model, was proposed: The subgenomic mRNAs would be produced initially by leader-primed transcription and then replicated into minus-strand templates that would in turn be transcribed into subgenomic mRNAs. We review the experimental evidence that led us to formulate a third model proposing that the discontinuous event in coronavirus RNA synthesis occurs during minus strand synthesis. With our model the genome is copied both continuously to produce minus-strand templates for genome RNA synthesis and discontinuously to produce minus-strand templates for subgenomic mRNA synthesis, and the subgenomic mRNAs do not function as templates for minus strand synthesis, only the genome does.

1
Introduction

Our studies focus on a Group 2 coronavirus, mouse hepatitis virus strain A59 (MHV-A59). MHV-A59 was isolated in 1961 from a colony of Balb/C mice that were being used to serially propagate leukemia that was caused by a virus (Manaker et al. 1961; David-Ferreira and Manaker 1965). Probably multiple serial passages of cell-free filtrates of leukemic cells in newborn mice selected from an endogenous mouse coronavirus a variant that replicated to high titer quickly and that produced, as a consequence, hepato-encephalopathy. In our laboratory MHV-A59 has the characteristic of replicating to high titer ($>10^9$ pfu/ml) when cells such as 17cl-1 cells that express mCEACAM, e.g., the receptor for MHV, are used. On 17cl-1 cells, which were derived from spontaneously transformed Balb/C 3T3 fibroblasts (Sturman and Takemoto 1972), MHV-A59 causes cell fusion, forms very efficiently clear plaques of ~5-mm diameter after 2 days, and produces about 10^4 virions per cell and 3×10^3 pfu/cell within 8 h (our unpublished observations). This makes MHV-A59 an ideal virus to study its replication at the molecular and genetic level. Essentially all of our studies used MHV-A59 and 17cl-1 cells.

Like all coronaviruses, MHV-A59 possesses a large (31.4 kb for MHV-A59), single-stranded, messenger- or plus-sense RNA genome that contains a poly(A) sequence at the 3' end and a methylated guanosine cap at the 5' end, although the length of the poly(A) sequence and the cap structure have not been extensively characterized. The genome is associated with a helical nucleocapsid composed of the N protein of 50–60 kDa. Surrounding the nucleocapsid is a membrane in which is embedded the envelope or spike (S) glycoprotein of 180 kDa, which may be cleaved in half by a host protease and which produces characteristic surface projections on the virions. Also embedded in the membrane are two other envelope proteins: a multi-spanning membrane (M) glycoprotein of 23 kDa essential for the assembly of virions and an E protein of 10 kDa, which although not absolutely essential for the formation of virions, is required for the release of large numbers of virions from infected cells (Kuo and Masters 2003). In addition, some strains of MHV produce virions containing an HE glycoprotein and an I protein, neither of which is essential for viral infectivity or growth. The virions form internally, accumulate in vesicles in the cytoplasm, and are released rapidly en masse by exocytosis. Recent publications (Siddell 1995; Lai and Cavanagh 1997; Lai and Holmes 2001) review and cite extensive literature on coronaviruses.

Presumably, a single molecule of genome RNA is capable of initiating a successful round of viral replication in the cytoplasm without the need of nuclear functions: MHV replicates in dactinomycin-pretreated cells. Viral structural proteins are not required for viral RNA synthesis because purified, deproteinized genomes obtained from virions are infectious. More recently it was shown that an infectious clone of the human coronavirus 229E with the genes encoded by the subgenomic mRNAs, including the structural genes for S, E, M, and N, deleted was able to replicate and produce a subgenomic mRNA encoding a nonviral fluorescent protein (Thiel et al. 2001a,b). Therefore, the proteins produced by the subgenomic mRNAs probably do not function in viral RNA synthesis. Further evidence of this is that the ratio of the synthesis of genome to subgenomic mRNA does not change from the earliest times postinfection (p.i.) to very late times p.i., even after virion production has ended and viral RNA synthesis is declining (our unpublished observations). Therefore, the accumulation of structural proteins such as N (nucleocapsid proteins) does not affect or change the transcriptional pattern, that is, the ratio of the synthesis of genome to subgenomic mRNA.

What makes coronaviruses particularly interesting is that they have the largest genome of the RNA viruses, which encodes a large number of gene products that are unique to viruses belonging to the *Nidovirales*, and they utilize a discontinuous RNA synthetic process to produce subgenomic messenger RNA molecules. During replication of MHV-A59, genomic (RNA-1) and six subgenomic mRNAs (RNA-2 of 9.6 kb, RNA-3 of 7.4 kb, RNA-4 of 3.4 kb, RNA-5 of 3.0 kb, RNA-6 of 2.4 kb, RNA-7 of 1.7 kb) are produced. These together comprise a 3′ coterminal nested set (Fig. 1). The mRNA species vary in size from one-third (mRNA 2) to about one-twentieth (mRNA 7) of the genome. For MHV-A59 plus strands (genomes and subgenomic mRNA) are produced in large amounts and, at about 1% of this number, minus strands of both genome- and subgenomic length that serve as the templates for genome and subgenomic mRNA synthesis (Sawicki and Sawicki 1986; Sawicki and Sawicki 1990).

On infection the first open reading frame (ORF-1) in the genome of MHV-A59, which is ~20,000 nt and divided into Rep1a and Rep1b by a ribosomal frameshift, is translated as two large polyproteins, pp1a and pp1ab (Fig. 2). Three functional proteinases (PRO), two papainlike (PL1 and PL2) and one poliovirus 3C-like (3CL), are located in pp1a and cleave the pp1a and pp1ab into 16 nonstructural proteins, nsp1–16 (for relevant references see Brockway et al, 2003). These proteins and/or the polyprotein precursors form the viral RNA-dependent RNA polymerase.

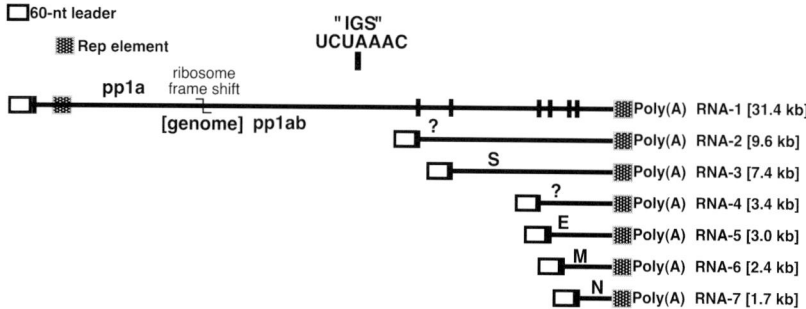

Fig. 1. The coronavirus genome RNA and six subgenomic mRNAs share identical 3′ sequences and form a 3′ nested set of RNAs. The genome is translated into two large polyproteins that are encoded in the large ORF 1 sequence. Only the ORF at the 5′ region of each of the subgenomic mRNAs is translated into a unique protein, making the genome and subgenomic mRNAs functionally monogenic

Very quickly after entry MHV-A59 produces first the minus-strand templates for genome and subgenomic mRNA synthesis and then the transcription complexes that transcribe the minus-strand templates into genomes and subgenomic mRNAs. Recently Alexander Gorbalenya and his colleagues (Snijder et al. 2003) described the actual and potential functions of the proteins encoded in ORF-1 of the SARS-CoV and related coronaviruses and members of the order *Nidovirales*. The question im-

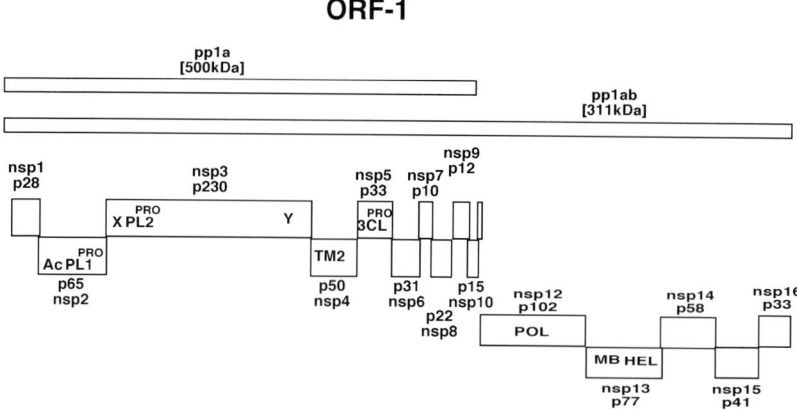

Fig. 2. A schematic showing the ORF1 of the coronavirus genome and its pp1a and pp1ab polyproteins and processed products. See Snijder et al. (2003) for detailed information

mediately arises as to why *Nidoviruses* possess so many individual proteins, especially where other plus-strand viruses make do with 1–4 proteins for viral RNA synthesis and one-half to one quarter the number of amino acids. These extra proteins could be needed to produce a unique replication and transcription machinery, to increase the fidelity of RNA-dependent RNA synthesis that is inherently error prone, to replace needed host factors that are required by other viruses, or to interact with and modulate the host cell or the immune system of the host animal. If mutations in all of the 15–16 genes of MHV give phenotypes that affect viral RNA synthesis, it would suggest that *Nidoviruses* utilize fundamentally different replication/transcription machinery than other plus-strand viruses. The viral gene products would also interact with those of the host cell because it is likely that the host cell provides functions not encoded by the viral genes.

2
Discontinuous Transcription by Coronaviruses

Whereas genome-sized plus and minus strands are made by the viral replicases continuously or processively copying their respective templates, discontinuous transcription is the mechanism responsible for generating the minus-strand templates for the subgenomic mRNA. The first evidence for discontinuous synthesis was the discovery that all of the viral plus strands possessed the 1.7-kb sequence of RNA-7, in addition to the poly(A) tract at their 3′ ends, and at their 5′ ends, the genome and all subgenomic mRNAs possessed an identical "leader" sequence of ~60 nt (Spaan et al. 1983). It was realized that because this leader sequence is restricted to the 5′ end of the genome, viral RNA synthesis must include a process by which the leader RNA is joined to the "body" of mRNAs 2–7 that are located at the 3′ end of the genome. At the end of the leader and before the body of each open reading frame of the subgenomic mRNAs is an "intergenic sequence," or "IGS or IG sequence" or "TRS," or "transcription regulating sequence" (van der Most et al. 1994; van Marle et al. 1995). Some also call it the "TAS," or "transcription activation sequence." For MHV-59 the IGS is $^{5'}$UCUAAAC$^{3'}$. The same replication strategy appears in the *Arteriviridae* that recently have been united with the *Coronaviridae* within the order *Nidovirales* (Cavanagh 1997).

Lai and his collaborators gave a stimulus to research on coronaviruses when they produced a model to explain the generation of subgenomic mRNAs of coronaviruses (Baric et al. 1983). They suggested that the viral polymerase (RdRp) initiated plus strand synthesis at the 3′ end of genome-length templates. In this model, after it produced the leader RNA and the IGS, the RdRp would jump from the 3′ end of the minus strand all the way to one of the internal complementary copies of the IGS and reinitiate RNA synthesis. The IGS served to redirect the RdRp to an internal site and then served as a primer for elongation. This "leader-primed transcription model" was based on three experimental findings. First, subgenomic mRNAs comprised a nested set of 3′ coterminal RNA molecules and each contained a 5′ leader sequence that was present only once at the 5′ end of the genome (Lai et al. 1983; Spaan et al. 1983; Lai et al. 1984). Second, formation of subgenomic RNAs by splicing was ruled out by finding that the UV light sensitivity for the synthesis of each species of viral plus-strand RNA was directly proportional to the RNA length (Jacobs et al. 1981; Stern and Sefton 1982). Third, cells infected with MHV were reported to contain only a full-length minus-strand copy of the genome; no subgenomic minus-strand RNA was detectable (Lai et al. 1982). They strengthened their argument by showing that only replication intermediates (RIs) with genome-length templates could be found in infected cells and that treatment of the RIs with ribonuclease (RNase) produced only RF RNA of genome length. The only model that could account for or explain these observations was their "leader-primed" model.

In the years following the proposal of the leader-primed transcription model it became clear that this model did not explain the newer experimental findings. Some of these experimental results directly invalidated the leader-primed transcription model or showed that experimental conditions used for detection of subgenomic minus-strand RNA were not optimal because of excessive digestion of the RNA complexes with RNase A or the ^{32}P-labeled RNA probes used (Lai et al. 1982: Baric et al. 1983) were likely of too low a specific activity to detect minus strands (Sawicki et al. 2001).

David Brian's lab (Sethna et al. 1989) showed that cells infected with a coronavirus (porcine transmissible gastroenteritis virus, TGEV) contained a 5′ nested set of minus strands that corresponded in size to the genome and the subgenomic mRNA. They did this by Northern blotting, using in vitro labeled probes to the minus and plus strands. After seeing their publication, we thought it was possible that the subgenome-length minus strands they observed were created during preparation of the infected cell lysates. They might have arisen from genome-length minus strands in RIs. Activation of endogenous RNase might have cleaved the native, genome-length RI molecules into subgenomic RFs by targeting certain nuclease-sensitive sites in the genome template. This occurs with alphavirus RIs, whose genome-length template can be cleaved specifically by RNase at a single, internal site that is near or at the site where the viral polymerase initiates 26S mRNA synthesis. The alphavirus RI population is cleaved into three RFs. The largest RF (20S in size) corresponds to a double-stranded form of the genome and represents intermediates active solely in genome synthesis. The other two RFs are derived from an RI active in subgenomic mRNA synthesis: The larger RF (18S in size) contains the double-stranded form of the sequence that is the 5' two-thirds of the genome; the smaller RF (15S in size) is the double-stranded form of the 3' third of the genome that is the 26S mRNA sequence (Simmons and Strauss 1972). Therefore, we set out to determine the origin of the MHV subgenomic minus strands. We found (Sawicki and Sawicki 1990) that cells infected with MHV-A59 contained both genomic- and subgenomic-sized RIs, the latter of which would contain templates corresponding in length to the viral subgenomic mRNA. We also showed that the genome-length RI was not converted to a smaller or subgenome-length RF but only to the genome-length RF on treatment with RNase. This allowed us to conclude that coronaviruses produce subgenomic mRNAs using templates that are of similar size to each mRNA.

David Brian's laboratory (Sethna et al. 1991) demonstrated that the TGEV subgenomic minus strands contained a sequence complementary to the 5' leader, and we confirmed this for the MHV-A59 subgenomic minus strands (Sawicki and Sawicki 1995). In light of these observations, two models could explain coronavirus transcription: the replicon model proposed originally by Sethna et al (1989) and our model (Sawicki and Sawicki 1995). With the replicon model, subgenomic mRNA would be made via leader-primed transcription using the full-length minus strands as a template. Once made, these subgenomic mRNA would be replicated into minus strands that would in turn serve as templates for more subgenomic mRNA synthesis. Also, subgenomic mRNA that were virion associated (i.e., packaged into virions) could serve directly as templates for minus strand synthesis and result in the replication of the subgenomic mRNA. In our model (Sawicki and Sawicki 1995), which we called by the awkward name of "discontinuous extension of minus-strand RNA", the minus-strand templates for subgenomic mRNAs would be made directly from the genome and would utilize a discontinuous transcription process. We proposed that, during minus strand synthesis, the viral polymerase would pause at the IG sequences and then relocate to the 5' end of the genome to finish transcription. This would result in the minus strand acquiring a sequence complementary to the 5' leader sequence and activating it as a template for subgenomic mRNA synthesis. Those polymerases that failed to pause, or resumed elongation after pausing, would create the full- or genome-length minus strand template. The crucial observation that makes our model more attractive than the replicon model is that sequences downstream of the leader are required for replication signals. Neither transfected, subgenomic mRNA (Brian et al. 1994) nor subgenomic DI-RNA (Makino et al. 1991) replicated in cells infected with a helper virus. It appeared that sequences downstream of the leader RNA, ~500 nt of the 5' end for MHV-A59 that are present only in genomes and DI RNAs, are required and sufficient for viral RNA replication (Masters et al. 1994). Therefore, our model could account for and explain the known experimental observations.

Although the mechanism by which coronaviruses accomplish the synthesis of minus-strand templates for subgenomic mRNA synthesis are not known presently, it is clear that a working and testable model for coronavirus transcription must be able to explain or account for all the known facts. As Agnostino Bassi remarked "When Fact speaks, Reason is silent, because Reason is the child of Fact, not Fact the child of Reason." The known facts for coronavirus transcription include:

1. The formation of the 3' coterminal nested set of subgenomic mRNA with a common leader sequence at their 5' ends (discontinuous transcription).
2. The presence of transcriptionally active RIs that contain either genome-length or subgenome-length templates.
3. Polarity in the synthesis of subgenomic mRNAs. With a few exceptions, the relative synthesis of one subgenomic mRNA to another is determined by its distance from the 5' end of the genomic minus strand, or the 3' end of the genome. The closer the region encoding a subgenomic mRNA or subgenomic DI RNA is relative to the 5' end of the genomic minus strand, the greater its synthesis level. Exceptions appear because of the IGS being too close to the 3' end of the genome to allow polymerase to always bind stably.
4. The ratio of genome to subgenomic mRNA synthesis remains constant from beginning to end of the infectious cycle.
5. Mutations in the IGS affect the synthesis of subgenomic mRNA but not in a uniform and predictable way; some abolish subgenomic mRNA and some have little or no effect.

6. A differential sensitivity to UV radiation proportional to the size of the synthesis of genomic and subgenomic mRNA, at least after subgenomic mRNA synthesis has started.
7. Synthesis of minus strands occurs simultaneously with the synthesis of plus strands. The only exception to date (An et al. 1998) may be due to the difficulty of the DI system chosen and very low numbers of viral molecules at early times p.i.
8. The capacity of subgenomic mRNA or a DI genome RNA or subgenomic DI RNA to acquire a leader RNA from a different genome (leader switching).
9. The high rate of recombination in coronaviruses.

The model of the discontinuous extension of the 3' end of the minus strands has many attractive features that fit the experimental data. We believe it is a unifying model to probe the mechanism underlying the generation of subgenomic mRNA. It provides a useful framework to design experiments and interpret results.

Eric Snijder's lab has used an arterivirus, equine arteritis virus (EAV), that is also a member of the *Nidovirales*, to provide direct evidence for minus, not plus, strand synthesis being discontinuous (van Marle et al. 1999a). They changed the sequence in the IGS (in the plus orientation) or the TRS (in the minus orientation). They substituted one or both "C"s in the IGS UCAAC sequence at the end of the leader, in a body IGS (adjacent to a downstream gene), or in both. Changes in the leader IGS led to loss of all subgenomic mRNAs. In contrast, changes to the body IGS-7 led to loss of only mRNA-7. The very low amounts of mRNA-7 made under the latter conditions contained the mutated body IGS-7 sequence in their 5' leader and not the original leader IGS sequence. Significantly, when both gene-7 and the 5' leader contained the same mutated sequence, full subgenomic mRNA-7 synthesis was restored. Thus, (1) complementary base pairing between TRS at the 3' end of nascent minus strand "body" sequences and IGS on the end of the leader at the 5' end of genome template appears necessary to obtain maximal subgenomic mRNA synthesis (i.e., formation of the templates for subgenomic mRNA). (2) It is unlikely that the recombination requires recognition of a specific sequence of IGS or TRS. Finally, (3) the discontinuous direction of synthesis was from body to leader and, thus, the discontinuous transcription step must happen during minus strand synthesis, that is, during the formation of subgenomic minus strands.

Presently, there is no experimental information on how this process occurs with coronaviruses. Only with arteriviruses has Eric Snijder's laboratory obtained direct information on a mechanism (van Marle et al. 1999a; Pasternak et al. 2003) and on the viral genes involved in subgenomic mRNA synthesis (van Marle et al. 1999b; van Dinten et al. 2000; Tijms and Snijder 2003). According to our model of 3' discontinuous extension (Sawicki and Sawicki 1995), the viral polymerase would begin transcription at the 3' end of the genome. It would pause after transcribing each of the IGS UCUAAAC elements. Each polymerase can then either elongate through the pause site, that is, continue transcription, or move to the 5' end of the genome without copying the intervening sequences, that is, discontinuous transcription. The coronavirus polymerase might copy the template RNA in a fashion analogous to DNA-dependent RNA polymerases that retract at pause sites (Komissarova and Kashlev 1997a,b) and where the polymerase remains associated with the growing nascent strand rather than with the template. Of interest, a similar mechanism was recently proposed for proofreading by RNA polymerases, where the polymerase pauses, retracts, removes several nucleotides from the nascent 3' end, and then resynthesizes this region (Shaevitz et al. 2003). Several gene products of ORF1b of coronaviruses and SARS CoV were identified that might function in such a proofreading mechanism, some of which are predicted to have nuclease activity (Snijder et al. 2003).

If retraction occurred when the coronavirus polymerase was transcribing the IGS, the exposed 3' end of the nascent minus strand would relocate and align precisely to complementary sequences at the 3' side of the leader RNA at the 5' end of the genome and complete transcription of the subgenomic minus-strand RNA. The interaction of the nascent minus strand with the 5' end of the genome might be mediated by protein:protein interactions between the polymerase attached to the growing minus strand and a protein associated with the 5' end of the genome. This also could occur by the polymerase binding directly to the sequences downstream of the leader. The minus-sense or anti-IGS would align the polymerase at the end of the leader sequence in the genome template. If the 3'-terminal nucleotide of the anti-IGS matched, the polymerase would elongate and copy the leader sequence to put an antileader (complementary to the 5' plus-strand leader sequence) on the 3' end of the nascent, subgenomic minus strand. Thus the anti-IGS acts as a primer to complete transcription and copy the leader RNA. The IGS in the context of the sequences surrounding it would function as an attenuator element. The result of 3' discontinuous extension during minus strand synthesis is that the subgenomic minus strands acquire the antileader sequence at their 3' end. The replication complex that synthesized a subgenomic minus strand may retain it as a template and go on to synthesize subgenomic mRNA. The promoter for plus strand synthesis would be within the antileader sequence. As stated above, because sequences downstream of the leader are required for minus strand synthesis (Masters et al. 1994), only genomes can serve as templates for genomic and subgenomic minus strands; therefore, a great advantage of this strategy is that subgenomic mRNAs cannot be replicated and cannot behave as DI RNAs.

3
Kinetics of Plus- and Minus-Strand RNA Synthesis and Sensitivity to Translational Inhibition

Our interest in coronavirus RNA synthesis initially came from a report in 1982 by Brayton et al (Brayton et al. 1982). They claimed that minus strand synthesis was temporally distinct from plus strand synthesis when MHV-infected cell extracts were assayed in vitro. They reported that extracts of A59-infected cells at 1 h p.i. made only minus strands and extracts of cells harvested at 6 h p.i. made only plus strands. Other plus-strand RNA viruses that replicate via RNA-dependent RNA polymerases show concurrent synthesis of minus and plus strands, unless minus strand synthesis ceases at some later time in infection [as is the case for alphaviruses (Sawicki and Sawicki 1998)]. Thus minus strand synthesis in the absence of plus strand synthesis would be a very unusual transcription pattern. We began a comparative analysis of minus and plus strand syntheses by MHV-A59 and the alphavirus Sindbis to probe the veracity of this observation and its implications for the regulation of transcription.

We first established optimal conditions for infection of 17cl-1 cells. If infection was done in low-pH medium (pH 6.8), infected cultures delayed the expression of the typical cytopathic effect and did not undergo extensive syncytium formation until after 10 h p.i., thus allowing us to examine a broader range of times after infection. Fusion of the monolayer at pH 7–7.5 normally resulted in the shutdown of viral replication much earlier and yielded only low titers of progeny. We found that MHV minus-strand RNA synthesis was detectable as early as 3 h p.i. in pulse-labeling experiments using ^{3}H-uridine in the presence of dactinomycin and that the minus strands were exclusively in the partially double-stranded RI population, as expected for template molecules. The rate of MHV minus strand synthesis peaked at 5–6 h p.i. at 37°C (or at about

10–12 h p.i. if infection was at 28°C–30°C) and then declined to about 20% of its maximum rate but interestingly did not cease. Adding cycloheximide before 3 h p.i. prevented viral RNA synthesis; addition of the drug after RNA synthesis was ongoing led in its inhibition. Unlike alphavirus but similar to picornavirus replication, both MHV plus- and minus-strand polymerase activities were unstable but in MHV they differed in their turnover rates: Minus strand synthesis was almost immediately inhibited after translation inhibition, that is, addition of cycloheximide or puromycin, whereas both genome and subgenomic mRNA syntheses continued for 30–60 min before declining slowly. The short half-life of the coronavirus minus-strand polymerase activity may reflect a possible role in minus strand synthesis of intermediate or partially processed polyproteins, whose further processing leads to loss of this enzymatic activity. Whether loss of minus-strand polymerase activity is correlated with its conversion to another type of polymerase activity (plus strand synthesis) is not known. The latter has been demonstrated for alphaviruses (Lemm et al. 1994; Shirako and Strauss 1994; Wang et al. 1994).

4
The Kinetics of Synthesis of the Subgenomic Minus Strands

This was crucial to supporting or not supporting different models for coronavirus transcription. If, at all times, it was found that minus strand populations were composed of genome- and subgenomic-sized molecules in fixed but nonequal ratios to each other, then the 3′ discontinuous minus strand transcription model would be supported. Therefore, in an attempt to disprove our model, we used several approaches to ask which size minus-strand molecules first appeared and what the plus strand products were at that time. PCR technology allows the detection of very small numbers of target sequences. We designed specific primer sets to allow the amplification of genome (RNA 1) plus- or minus-strand sequences separately from the amplification of subgenome RNA 7 plus- and minus-strand sequences. One of the primers in each set was targeted to the 5′ leader (or 3′ antileader) sequences at the ends of the RNAs and the other of each pair to the "body" sequence that would be a known distance downstream or upstream from the end. Our first studies used infected cell RNAs isolated at 7 h p.i., when viral transcription is readily detected by pulse-labeling in vivo (Sawicki and Sawicki 1995). The results indicated that at 7 h p.i. cells contained minus strands corresponding to both minus-sense RNA1 and RNA7 and that both of these minus-strand molecules contained an antileader on the 3' ends. Uninfected cells did not amplify a PCR product with these primers, but cells infected with MHV in the cold (4°C) and harvested at 4°C after removing unadsorbed virus contained only genome RNA and not the minus-strand copy of the genome. Nor did they contain plus- or minus-strand copies of RNA 7. Our twice isopycnic gradient-purified virion preparations contained small amounts of mRNA 7, but these were in different "particles" than the virions because they were not adsorbed to cells during the 1-h period at 4°C. The results demonstrated directly that the subgenomic mRNAs and subgenomic minus-strand RNAs arise from genome plus strands because that was the only RNA in virions that initiated the infection (Sawicki and Sawicki 1995).

Our experiments consistently found that radiolabel was detectable in the subgenomic RI population as early as 1.5 h p.i. and at the same time that radiolabel was detectable in the genomic RI population. Because the genome RNA is 15–20 times larger than RNA 7, each molecule of 60S RNA contains on average of 15–20 times more uridine residues and is thus "hotter" than a molecule of RNA 7 and easier to detect in pulse labels. Increased exposure of the gels to compensate for this imbalance verified the presence at each time period of the subgenomic RI population. Moreover, the relative proportion of each of the seven size classes of RI was maintained at all times studied, both early and late. We interpreted our results to indicate that both genome and subgenomic minus strands are produced at the earliest times in infection and thus must be derived from input genome plus strands. We have failed so far to detect a time early in infection when only genomic minus strands are made. Thus the synthesis of subgenomic mRNAs observed at 1.5–2 h p.i., or other early times, must arise from copying of complementary and subgenomic minus-strand templates.

5
Subgenomic MHV RIs Exist in Infected Cells and Are Transcriptionally Active in mRNA 2–7 Synthesis

Up until 1989, the prevailing model for coronavirus transcription and the formation of the 3' nested set of mRNAs was "leader-primed transcription" (Baric et al. 1983). The first report that coronaviruses might use subgenomic templates came from the lab of David Brian (Sethna et al. 1989), who reported finding a 5' nested set of minus strands in

TGEV-infected cells by Northern blotting. Thus the number and kind of minus-strand RNAs matched those found for the virus's plus-strand RNAs. This exciting report led us to probe whether MHV A59-infected cells contained more than one size class of RF RNA and, if so, whether these were derived from genome-length RI molecules. Our results showed seven double-stranded RF cores in infected cell extracts digested with limited amounts of RNase and demonstrated that they were proportional in size and amounts to template the 3′ nested set of viral mRNAs and account for their nonequal ratios (Sawicki and Sawicki 1990). Thus the abundance of individual MHV mRNAs is a direct result of the number of their minus-strand templates. We also showed that this family of RF molecules did not come from an originally genome-length RI by demonstrating the existence of a family of RI molecules in the absence of RNase treatment. Finally, nascent labeling experiments using very short pulses with 900 μCi ^{3}H-uridine/ml demonstrated that the subgenomic RI RNA accumulated label first and only later was radiolabel chased into single-stranded mRNA (Sawicki and Sawicki 1990). This precursor-product relationship is what is expected for transcription templates, leading us to conclude that discontinuous transcription must occur during minus strand synthesis and its result was the 5′ nested set of minus-strand templates. If these contained an antileader sequence at their 3′ end and a poly U at their 5′ end, they could be directly transcribed into the subgenomic mRNAs. Thus the presence of subgenomic minus strands suggested a replication strategy that would be common to coronaviruses (and arteriviruses) and unique among the single-stranded RNA animal viruses. We suggested and favor the hypothesis that subgenomic minus strands are produced by discontinuous transcription directly from genome plus strands and are the only templates for subgenomic mRNAs.

6
Characterization of Coronavirus Native RI/TIs and Native RF/TFs

The replicative structures used by MHV to produce its genome and subgenomic mRNAs have the classic properties exhibited by viral RIs and RFs but are of genome or subgenomic length. Replicative intermediates are of genome length and have multiple replicases and nascent RNA chains. We proposed the term transcriptive intermediates (TI or TF) to refer to the same type of transcriptionally active coronavirus template structure but whose templates are of subgenome-length and whose

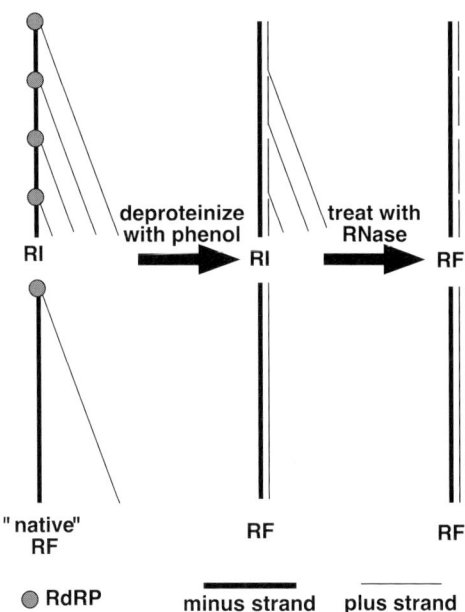

Fig. 3. Models for the structures of the coronavirus RI (TI for subgenomic templates) and RF (TF) structures. In infected cells, these molecules are predicted to have the "native" structures and to be essentially single stranded, with one or more RNA-dependent RNA polymerases elongating along the length of the template. When isolated from cells and deproteinized with phenol and chloroform (or proteases), the polymerase and other proteins are removed, allowing the nascent strands to collapse and bind to complementary sequences of the template. The RI/TI have multiple nascent strands and would be only partially double stranded and mostly single stranded; the RF/TF have only one or two nascent strands and would be almost or fully double stranded even without nuclease digestion. Finally, limited digestion of the deproteinized RNAs with RNase cleaves off all single-stranded RNA chains and converts these structures to fully double-stranded RF or TF molecules

product is one of the six size classes of subgenomic mRNA. Such transcriptionally active structures are likely essentially single stranded in vivo, with only small regions near the polymerase in double-stranded form. When proteins are removed from RI and TI structures, and the resulting RNA is digested with low concentrations of RNase, a double-stranded core or replicative form is recovered from each (also called a RF or TF). A schematic of these structures is shown in Fig. 3.

The RI/TI and RF/TF differ in their relative proportion of single-stranded and double-stranded character. Viral infections also form native RF and TF under conditions when replicase or transcriptases are limited or initiation is reduced. Deproteinized native RF/TF would be completely or nearly completely double stranded as a result of having its one or few polymerases engaged in transcription and near the end of each template at the time of cell lysis. Thus plus and minus strands in native RF/TF would be mostly equal in length. Native RF are soluble in high-salt solutions (2 M LiCl or 1 M NaCl) and they function as transcription intermediates based on their rapid incorporation of radiolabeled nucleotide precursors. On the other hand, RI is a multistranded intermediate, whose multiple polymerases are associated with an equal number of nascent RNA chains of varying lengths. Its large percentage of single-strandedness makes the RI insoluble in high-salt solutions, allowing its physical separation from RF/TF, tRNA, and DNA. The MHV structures characterized so far are those active in plus strand synthesis (Sawicki et al. 2001). It is assumed that intermediates active in minus strand synthesis share some or all of these classic features, but these intermediates have not yet been isolated free of RI/TI active in plus strand synthesis.

Our analysis of RNA structures formed in MHV-infected cells found that, in addition to the RI and native RF, MHV-infected cells contained six species of RNA intermediates active in transcribing subgenomic mRNA (Sawicki et al. 2001). When radiolabeled between 5 and 6 h p.i., 70% of MHV replicating and transcribing structures in 17cl-1 cells had the physical properties of RI/TI and 30% were native RF/TF. The seven size classes of intermediates or native forms can be separated partially from each other by velocity sedimentation on sucrose gradients or by gel filtration chromatography on Sepharose 2B and Sephacryl S-1000. The MHV RI/TI and native RF/TF double-stranded core structures are resistant to a range of RNase T1 concentrations (1-10 units/mg RNA) but are degraded during exposure to intermediate to high concentrations of pancreatic RNase (RNase A). There is an exposed poly(A) sequence on each of the size classes of TI and TF long enough to allow their adsorption to magnetic beads containing oligo(dT)25. It is likely that the genome-sized RF and RI also possess poly(A) sequences, but the large overall size of these molecules prevented their stable attachment to the beads. It was demonstrated directly that RI and TI are transcriptionally active structures in vivo, and thus biologically significant, from the finding that each size class of replicative molecule was able to rapidly incorporate radiolabeled precursors into nascent RNA chains (Fig. 4).

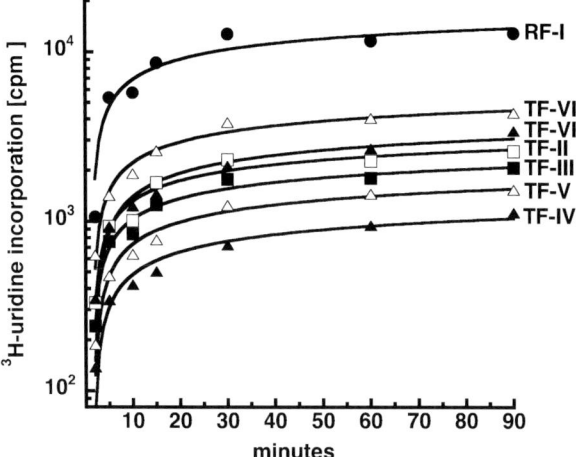

Fig. 4. The subgenome-length TI and native TF are transcriptionally active and incorporate radioactive nucleotide precursors at the same rate (kinetics) as the genome-length RI and native RF. See also Sawicki et al (2001). This evidence, which was also shown in our 1990 publication (Sawicki and Sawicki, 1990), dispelled any notion that the incorporation of ^{3}H-uridine into RIs with subgenome-length templates is into dead-end products

In summary, the discontinuous transcription process occurs during minus strand synthesis and produces a 5' nested set of minus strands each of which contains a copy of the antileader sequence at its 3' end. This sequence likely functions or aids in polymerase initiation for synthesis of the subgenomic mRNA. This strategy has several key advantages: It enables the coronavirus (nidovirus) transcriptase to skip over sequences in the template that are not needed in a particular subgenomic mRNA, thus conserving cellular substrates. Perhaps more importantly, it allowed the evolution of the "nested set" strategy, whereby each open reading frame is expressed at the 5' end of a specific mRNA. This in turn allows for regulation at the transcription level. The number of minus-strand templates controls the number of each mRNA species, which controls the amount of its specific protein product(s). It is not surprising to find that mRNA 7 is produced in highest abundance and it encodes the viral capsid protein. This strategy also would enable the genome to encode essential regulatory sequences once, conserving linear genome space, and at the same time to provide for these sequences to be added to each of the resulting RNA species that require them to function efficiently (ribosome binding; polymerase initiation).

7
Turnover of MHV Replicative/Transcriptive Intermediates

Coronavirus RI/TI and RF/TF are short-lived and turn over during infection and after exposure of infected cells to temperatures of 0°C–4°C. We were the first to show that inhibition of protein synthesis in cells infected with MHV-A59 caused minus strand synthesis to stop almost immediately (Sawicki and Sawicki, 1986). Plus strand synthesis was also inhibited when translation was prevented, but its loss occurred more slowly over time. This, as we pointed out, was different from alphaviruses that make a very stable transcription complex that functions for many hours in the absence of new protein synthesis. The unstable nature of the coronavirus transcription complex has now also been shown for the arteriviruses (Den Boon et al. 1995; Snijder and Meulenberg, 1998). Recently we discovered (Wang and Sawicki 2001) that the minus-strand templates but not the plus strands made in MHV-infected cells are unstable and turn over. In most plus-strand viruses the minus strands are stable and are found not as free, single-stranded RNA but as components of RI or RF structures. Therefore, we were surprised to find that the RI/TI and RF/TF that had already accumulated in infected cells between 1–6 h p.i. disappeared over time after 6 h p.i. Minus strands made after 6 h p.i. accumulated temporarily and replaced some of the minus strands synthesized earlier. Treatment with cycloheximide for extended periods at early or late times p.i. led to the loss of 90% or more of the pre-existing native RF/TFs and RI/TIs and prevented the synthesis of new ones. The amounts remaining at each time subsequent to drug addition were proportional to the reduced rates of plus strand synthesis observed. Removal of the drug led to a burst of minus strand synthesis and full recovery of transcription. The turnover of RI/TI and RF/TF was specific to MHV and did not affect the RI/RFs of alphaviruses (Sindbis virus) in cells coinfected with both viruses, as shown in Fig. 5.

Thus MHV RI/TIs function for a limited time, whereas the Sindbis RI is a stable replication/transcription complex. The MHV turnover process involved release of minus strands, possibly with dissociation of the entire replication/transcription complex, after which these minus strands were degraded. Minus strands not associated with plus strands were present in infected cells late in infection and after cycloheximide treatment. These "free" minus strands could be captured as RNase-resistant, double-stranded RNA if they were allowed to hybridize with the excess plus strands in the cell lysates before RNase treatment. The results suggest that the continuation of minus strand synthesis is essential for viral

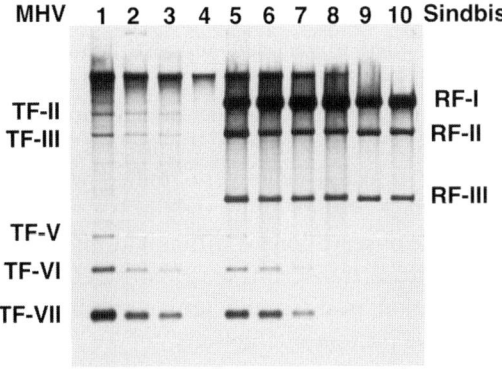

Fig. 5. MHV replicative and transcriptive intermediates are unstable and turn over throughout infection, in contrast to alphavirus (Sindbis, strain HR) intermediates that are stable once formed. Cultures of 17cl-1 cells were infected (at an MOI of 100) with MHV A59 (*lanes 1, 2, 3,* and *4*), with MHV + Sindbis (*lanes 5, 6, 7,* and *8*), or with Sindbis alone (*lanes 9* and *10*). The infected cells were labeled with ³H-uridine from 3 to 7 h p.i. (*lanes 1* and *5*), from 3 to 9 h p.i. (*lanes 2* and *6*), from 3 to 11 h p.i. (*lanes 3, 7,* and *9*), or from 3 to 13 h p.i. (*lanes 4, 8,* and *10*). The deproteinized and RNase-treated RF/TF cores of the intermediates and native forms were obtained and electrophoresed on a 1% agarose-TBE gel. The gel was prepared for fluorography

transcription seen late in infection and that the coronavirus minus-strand templates need to be continuously replaced. Failure to continue minus strand synthesis would cure the infection. Thus we would argue that the persistence of MHV infections would be dependent on host cell environments that either allow the virus to continue to form new polymerases for minus strand synthesis or block the turnover of previously made template complexes. Although this turnover resembles that seen for poliovirus RI being converted to an inactive RF, release of minus strands from RI/RF and TI/TF RNA is unique to coronaviruses (and probably arteriviruses) among animal RNA viruses. Somewhat surprisingly, it resembles what is known for phage Qβ transcription, where the template and product strands are released as single strands after the polymerase has copied the template (Dobkin et al. 1979).

Even more unexpected was our finding (Wang T. and Sawicki, unpublished results) that exposure to cold (0°C–4°C for as little as 20 min) led to the disappearance of RI/RF and TI/TF RNA. Viral double-stranded RNA that had been present in cells labeled with ³H-uridine for several hours or even 30 min before exposure to cold was no longer detectable. We could recover RI/RF and TI/TF RNA by returning the cells to 37°C.

However, if we added cycloheximide when we returned the cells to 37°C, viral RNA synthesis failed to recover, suggesting that viral polymerase proteins were needed. Recovery of viral RNA synthesis was preceded by a burst of minus strand synthesis, and these new products were utilized as templates for plus strand synthesis. Rapid loss of RI/RF and TI/TF was also found after infected cells were treated with trypsin or were scraped from the surface of the culture dish at 0°C, 25°C, and 37°C. The interpretation of these observations is difficult at this time, but they may indicate that the replicase-transcriptase is associated with the cytoskeleton and this association is disrupted under conditions that alter cell shape or cold-sensitive protein interactions.

8
A Working Model

Figure 6 presents our working model for coronavirus replication and transcription. The infecting genome is translated into pp1a and pp1ab. Within 1 h of infecting the cells we can detect genome and subgenomic mRNA being made. As far as we can determine, both genomic and subgenomic minus strands are made very early after the genome enters the cell. We believe that parts of the pp1ab function, at least initially, as uncleaved or partially cleaved polyprotein forms to make minus strands. Once the minus strands are made they would be rapidly converted to templates for use in plus strand synthesis. There does not appear to be any difference early or late in the ratio of genome and subgenomic mRNA; it remains fixed from the earliest moments at which the synthesis of viral RNA can be detected. As we published many years ago (Sawicki and Sawicki 1986), minus strands are made throughout infection although their synthesis declines after 6–7 h p.i., but so does the amount of plus strand synthesis. Only the genomes are capable of serving as templates for minus strand synthesis because only they have a replication signal at their 5' and 3' ends. The subgenomic mRNA cannot serve as templates for minus strand synthesis because they lack the 5' replication signal. After the minus strands serve as templates for several rounds of plus strand synthesis, they are released from the replication/transcription complexes, which do not appear to be different for those making genomes compared with those making subgenomic mRNA. After their release they are subject to degradation by a RNase activity of unknown origin that seems to be specific for them because the genome and subgenomic mRNA are stable, perhaps because they are engaged with the ribo-

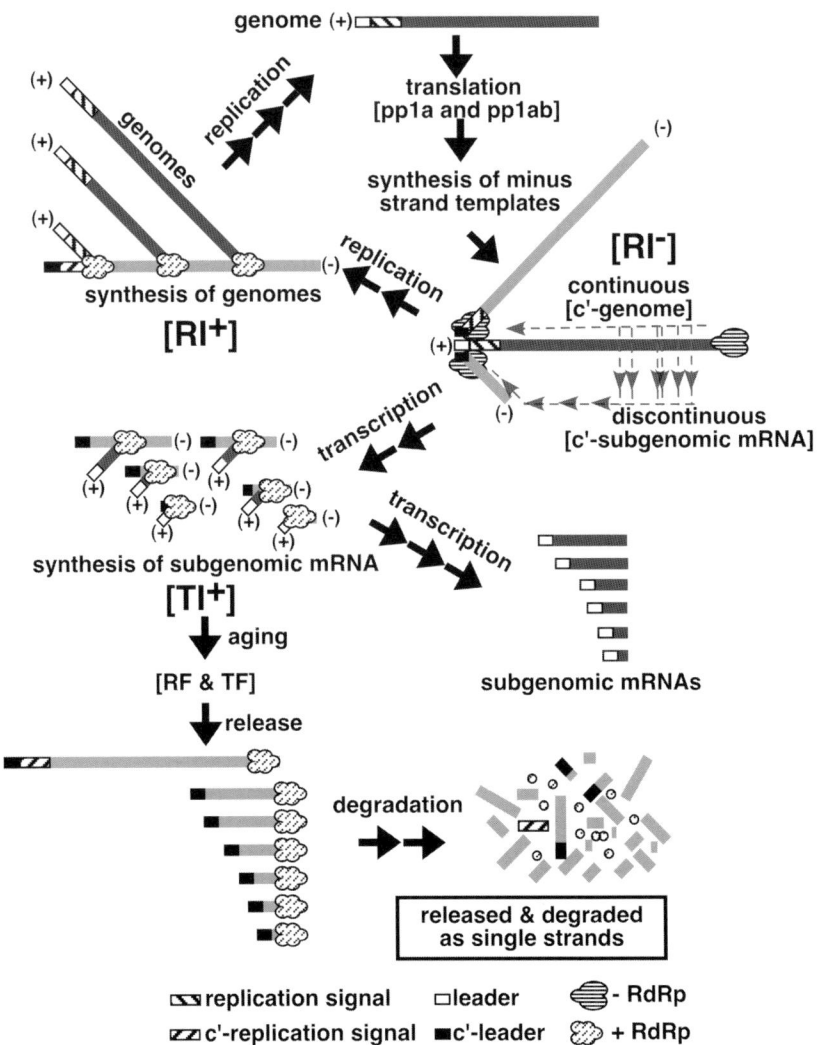

Fig. 6. Depiction of our model for the replication and transcription strategy of coronaviruses

some or in the process of being encapsidated by N proteins that act to protect them. With MHV-A59 cell fusion starts about the time viral RNA synthesis reaches a maximum rate and cell fusion results in early killing of the cells. With mutants we have isolated that do not cause cell fusion, the cells survive until 24 h after infection even though the virus was released by 8–10 h p.i. We can delay the cell fusion and cell death with MHV-A59 if we use medium that has a pH below 6.5 and if we lower the temperature to 33°C. If we keep the cells alive, the loss of RI/TI is easily observable. We speculate that coronaviruses encode in pp1a or pp1ab an RNase, for example, the XendoU in nsp15 perhaps (Snijder, Bredenbeek et al. 2003), that is responsible for the degradation of the minus strands. Our hypothesis, or working model, is that the replication/transcription complex ages after its synthesis: First it has minus-strand activity, then it has plus-strand activity, and then it further ages and degrades the template it is using to make plus strands. Aging may be through a proteolytic cleavage pathway by the 3CL^{PRO} in nsp5, and the final cleavages release or activate the RNase activity of the complex. This working model provides a framework for our future studies of coronavirus transcription and replication. Important for this effort will be the examination of *ts* RNA-negative mutants and designing mutants with reverse genetics.

Acknowledgements We wish to acknowledge support from the National Institutes of Health (SGS, AI-28506). We especially wish to acknowledge the contributions to this work of colleagues and friends, whose intellectual input and challenges over the years have assisted our efforts. In particular, we wish to thank Larry Sturman for starting us on this exciting research path.

References

An S, Maeda A, Makino S (1998) Coronavirus transcription early in infection. J Virol 72:8517–8524

Baric RS, Stohlman SA, Lai MMC (1983) Characterization of replicative intermediate RNA of mouse hepatitis virus: presence of leader RNA sequences on nascent chains. J Virol 48:633–640

Brayton PR, Lai MMC, Patton CD, Stohlman S (1982) Characterization of two RNA polymerase activities induced by mouse hepatitis virus. J. Virol 42:847–853

Brian DA, Chang RY, Hofmann MA, Sethna PB (1994) Role of subgenomic minus-strand RNA in coronavirus replication. Arch Virol Suppl 9:173–180

Brockway SM, Clay CT, Denison MR (2003) Characterization of the expression, intracellular localization and replication complex association of the putative mouse hepatitis virus RNA-dependent RNA polymerase. J Virol 77:10515–10527

Cavanagh D (1997) Nidovirales: a new order comprising Coronaviridae and Arteriviridae. Arch Virol 142:629–633

David-Ferreira JF, Manaker RA (1965) An electron microscope study of the development of a mouse hepatitis virus in tissue culture cells. J Cell Biol 24:57–64

Den Boon JA, Spaan WJ, Snijder EJ (1995) Equine arteritis virus subgenomic RNA transcription: UV inactivation and translation inhibition studies. Virology 213:364–372

Dobkin C, Mills DR, Kramer FR, Spiegelman S (1979). RNA replication: required intermediates and the dissociation of template, product and the Qbeta replicase. Biochemistry 18:2038–2044

Jacobs L, Spaan WJ, Horzinek MC, van der Zeijst BA (1981) Synthesis of subgenomic mRNAs of mouse hepatitis virus is initiated independently: evidence from UV transcription mapping. J Virol 39:401–406

Komissaarova N, Kashlev M (1997a) RNA polymerase switches between inactivated and activated states by translocating back and forth along the DNA and the RNA. J Biol Chem 272:15329–15338

Komissaarova N, Kashlev M (1997b) Transcriptional arrest: *Escherichia coli* RNA polymerase translocates backward, leaving the 3' end of the RNA intact and extruded. Proc Natl Acad Sci USA 94:1755–1760

Kuo L, Master P (2003) The small envelope protein E is not essential for mouse coronavirus replication. J Virol 77:4597–4608

Lai MMC, Patton CD, Baric RS, Stohlman SA (1983) Presence of leader sequences in the mRNA of mouse hepatitis virus. J Virol 46:1027–1033

Lai MMC, Patton CD, Stohlman SA (1982) Replication of mouse hepatitis virus: negative strand RNA and replicative form RNA are of genome length. J Virol 44:487–492

Lai MMC, Baric RS, Brayton PR, Stohlman SA (1984) Characterization of leader RNA sequences on the virion and mRNAs of mouse hepatitis virus, a cytoplasmic RNA virus. Proc Natl Acad Sci USA 81:3626–3630

Lai MMC, Cavanagh D (1997) The molecular biology of coronaviruses. Adv Virus Res 48:1-100

Lai MMC, Holmes KV (2001) Coronaviridae: The Viruses and Their Replication. In Fields BN, Knipe DM, Howley PM and Griffin DE (eds). Field's Virology, 4th Edition, volume 1, Lippincott, Williams and Wilkins, Philadelphia.

Lemm JA, Rumenapf T, Strauss EG, Strauss JH, Rice CM (1994) Polypeptide requirements for assembly of functional Sindbis virus replication complexes: a model for temporal regulation of minus strand and plus strand RNA synthesis. EMBO J 13:2925–2934

Makino S, Joo M and Makino JK (1991) A system for study of coronavirus mRNA synthesis: a regulated, expressed subgenomic defective interfering RNA results from intergenic site insertion. J Virol 65:6031–6041

Manaker RA, Piczak CV, Miller AA, Stanton MF (1961) A hepatitis virus complicating studies with mouse leukemia. Natl Cancer Inst 27:29

Masters PS, Koetzner CA, Kerr CA, Heo Y (1994) Optimization of targeted RNA recombination and mapping of novel nucleocapsid gene mutations in the coronavirus mouse hepatitis virus. J Virol 68:328–337

Pasternak AO, van den Born E, Spaan WJ, Snijder EJ (2003) The stability of the duplex between sense and antisense transcription-regulating sequences is a crucial factor in arterivirus subgenomic mRNA synthesis. J Virol 77:1175–1183

Sawicki DL, Sawicki SG (1998) Role of the nonstructural polyproteins in alphavirus RNA synthesis. Adv Exp Med Biol 440:187–198

Sawicki SG, Sawicki DL (1986) Coronavirus minus strand synthesis and effect of cycloheximide on coronavirus RNA synthesis. J Virol 57:328–334

Sawicki SG, Sawicki DL (1990) Coronavirus transcription: subgenomic mouse hepatitis virus replicative intermediates function in RNA synthesis. J Virol 64:1050–1056

Sawicki SG, Sawicki DL (1995) Coronaviruses use discontinuous extension for synthesis of subgenome-length negative strands. Adv Exp Med Biol 380:499–506

Sawicki DL, Wang T, Sawicki SG (2001) The RNA structures engaged in replication and transcription of the A59 strain of mouse hepatitis virus. J Gen Virol 82:385–396

Sethna PB, Hofmann MA, Brian DA (1991) Minus-strand copies of replicating coronavirus mRNAs contain antileaders. J. Virol 65:320–325

Sethna PB, Hung SL, Brian DA (1989) Coronavirus subgenomic minus strand RNAs and the potential for mRNA replicons. Proc Natl Acad Sci USA 86:5626–5630

Shaevitz JW, Abbondanzieri EA, Landick R, Block SM (2003) Backtracking by single RNA polymerase molecules observed at near-base-pair resolution. Nature 426:684–687

Shirako Y, Strauss JH (1994) Regulation of Sindbis virus RNA replication: uncleaved P123 and nsP4 function in minus strand RNA synthesis, whereas cleaved products from P123 are required for efficient plus strand synthesis. J Virol 68:1874–1885

Siddell SG (1995) The Coronaviridae. Plenum Press, New York.

Simmons DT, Strauss JH (1972) Replication of Sindbis virus II. Multiple forms of double-stranded RNA isolated from infected cells. J Mol Biol 71:615–631

Snijder EJ, Bredenbeek PJ, Dobbe JC, Thiel V, Ziebuhr J, Poon LL, Guan Y, Rozanov M, Spaan WJ, Gorbalenya AE (2003) Unique and conserved features of genome and proteome of SARS-coronavirus, an early split-off from the coronavirus group 2 lineage. J Mol Biol 331:991–1004

Snijder EJ, Meulenberg JJ (1998) The molecular biology of arteriviruses. J Gen Virol 79:961–979

Spaan WJ, Delius H, Skinner M, Armstrong J, Rottier P, Smeekens S, van der Zeijst BA, Siddell SG (1983) Coronavirus mRNA synthesis involves fusion of non-contiguous sequences. EMBO J 2:1839–1844

Stern DF, Sefton BM (1982) Synthesis of coronavirus mRNAs: kinetics of inactivation of infectious bronchitis virus RNA synthesis by UV light. J Virol 42:755–759

Sturman, LS, Takemoto KK (1972) Enhanced growth of a murine coronavirus in transformed mouse cells. Infect Immun 6:501–507

Thiel V, Herold J, Schelle B, Siddell S (2001a) Infectious RNA transcribed in vitro from a cDNA copy of the human coronavirus genome cloned in vaccinia virus. J Gen Virol 82:1273–1281

Thiel V, Herold J, Schelle B, Siddell S. (2001b) Virus replicase gene products suffice for coronavirus discontinuous transcription. J Virol 75:6676–6681

Tijms MA, Snijder EJ (2003) Equine arteritis virus non-structural protein 1, an essential factor for viral subgenomic mRNA synthesis, interacts with the cellular transcription co-factor p100. J Gen Virol 84:2317–2322

van der Most RG, deGroot RJ, Spaan WJ (1994) Subgenomic RNA synthesis directed by a synthetic defective interfering RNA of mouse hepatitis virus: a study of coronavirus transcription initiation. J Virol 68:3656–3666

van Dinten LC, van Tol H, Gorbalenya AE, Snijder EJ (2000) The predicted metal binding region of the arterivirus helicase protein is involved in subgenomic mRNA synthesis, genome replication and virion biogenesis. J Virol 74:5213–5223

van Marle G, Dobbe JC, Gultyaev AP, Luytjes W, Spaan WJ, Snijder EJ (1999a) Arterivirus discontinuous mRNA transcription is guided by base pairing between sense and antisense transcription-regulating sequences. Proc Natl Acad Sci USA 96:12056–12061

van Marle G, Luytjes W, van der Most RG, van der Straaten T, Spaan, WJ (1995) Regulation of coronavirus mRNA transcription. J Virol 69:7851–7856

van Marle G, van Dinten LC, Spaan WJ, Lyuytjes W, Snijder EJ (1999b) Characterization of an equine arteritis virus replicase mutant defective in subgenomic mRNA synthesis. J Virol 73:5274–5281

Wang T, Sawicki SG (2001) Mouse hepatitis virus minus strand templates are unstable and turnover during viral replication. Adv Exp Med Biol 494:491–497

Wang YF, Sawicki SG, Sawicki DL (1994) Alphavirus nsP3 functions to form replication complexes transcribing negative strand RNA. J Virol 68:6466–6475

The Coronavirus Replicase

J. Ziebuhr

Institute of Virology and Immunology, University of Würzburg, Versbacher Str. 7, 97078 Würzburg, Germany
j.ziebuhr@mail.uni-wuerzburg.de

1	Introduction	58
2	Organization and Expression of the Replicase Gene	59
3	Replicase Polyproteins	61
3.1	Functional Domains	61
3.2	Proteolytic Processing by Viral Cysteine Proteinases	64
3.2.1	Accessory Proteinases	66
3.2.2	Main Proteinase	69
3.3	Helicase	76
3.4	RNA-Dependent RNA Polymerase	78
4	Subcellular Localization of the Coronavirus Replicase	79
5	Concluding Remarks	83
References		83

Abstract Coronavirus genome replication and transcription take place at cytoplasmic membranes and involve coordinated processes of both continuous and discontinuous RNA synthesis that are mediated by the viral replicase, a huge protein complex encoded by the 20-kb replicase gene. The replicase complex is believed to be comprised of up to 16 viral subunits and a number of cellular proteins. Besides RNA-dependent RNA polymerase, RNA helicase, and protease activities, which are common to RNA viruses, the coronavirus replicase was recently predicted to employ a variety of RNA processing enzymes that are not (or extremely rarely) found in other RNA viruses and include putative sequence-specific endoribonuclease, 3'-to-5' exoribonuclease, 2'-O-ribose methyltransferase, ADP ribose 1"-phosphatase and, in a subset of group 2 coronaviruses, cyclic phosphodiesterase activities. This chapter reviews (1) the organization of the coronavirus replicase gene, (2) the proteolytic processing of the replicase by viral proteases, (3) the available functional and structural information on individual subunits of the replicase, such as proteases, RNA helicase, and RNA-dependent RNA polymerase, and (4) the subcellular localization of coronavirus proteins involved in RNA synthesis. Although many molecular details of the coronavirus life cycle remain to be investigated, the available information suggests that these viruses and their distant nidovirus relatives employ a unique collection of enzymatic activities and other protein functions to synthesize a set of 5'-leader-containing subgenomic mRNAs and to replicate the largest RNA virus genomes currently known.

1
Introduction

Plus-strand (+) RNA viruses exhibit an enormous genetic diversity that also applies to their RNA synthesis machinery. The RNA-dependent RNA polymerase (RdRp) is the only enzyme to be absolutely conserved, whereas other replicative and accessory protein domains vary considerably, in terms of both number and arrangement in the polyprotein (Koonin and Dolja 1993). Despite this diversity, phylogenetic relationships have been identified and used to group +RNA viruses into large superfamilies (or classes) (Goldbach 1987; Strauss and Strauss 1988; Koonin and Dolja 1993). As few as three superfamilies, the picornavirus-like, flavivirus-like and alphavirus-like viruses, were proposed to accommodate the vast majority of +RNA viruses infecting animals, plants, and microorganisms (Koonin and Dolja 1993). Interestingly, coronaviruses were among the few exceptions that did not easily fit into one of the established superfamilies; and the sequence analysis and characterization of arteri-, toro-, and roniviruses suggested that coronaviruses and their relatives may indeed exemplify a viral life form that, in several fundamental aspects, differs from that of other +RNA viruses (Gorbalenya et al. 1989c; Snijder et al. 1990a; den Boon et al. 1991; Snijder and Horzinek 1993; de Vries et al. 1997; Lai and Cavanagh 1997; Snijder and Meulenberg 1998; Cowley et al. 2000). Thus coronaviruses (and all their relatives) (1) produce a nested set of 3′-coterminal mRNAs (Lai et al. 1983; Spaan et al. 1983), (2) use ribosomal frameshifting into the −1 frame to express their key replicative functions (Brierley et al. 1987, 1989), (3) have a unique set of conserved functional domains that are arranged in the viral polyproteins in the following order: chymotrypsin-like proteinase, RdRp, helicase, and endoribonuclease (from N- to C-terminus) (Gorbalenya et al. 1989c; Gorbalenya 2001; Snijder et al. 2003), and (4) use RdRp and helicase activities that, based on the conservation of signature motifs, have been classified as belonging to the RdRp and helicase superfamilies 1, respectively (Koonin and Dolja 1993). Both the combination of two superfamily 1 domains and their sequential order in the polyprotein, with RdRp preceding the helicase, is extremely unusual (if not unique) among +RNA viruses. On the basis of these and other common properties, a new virus order, the *Nidovirales*, was introduced several years ago (Cavanagh 1997). At present, there is only little information on the toro- and ronivirus replicases, whereas information on the replicases of corona- and arteriviruses is accumulating rapidly. On the basis of both serological relationships and sequence similarity, coronaviruses have been classified into three groups (Siddell 1995), with human coronavirus 229E (HCoV-229E, group 1), porcine transmissible gastroenteritis virus (TGEV, group 1), mouse hepatitis virus (MHV, group 2), and avian infectious bronchitis virus (IBV, group 3) being the best-studied coronaviruses to date. Because of its medical importance, SARS coronavirus (SARS-CoV) (tentatively classified as belonging to group 2) (Snijder et al. 2003) is currently becoming a major topic of coronavirus research.

2
Organization and Expression of the Replicase Gene

Complete genome sequences are currently available for seven species of coronaviruses, IBV (Boursnell et al. 1987), MHV (Bredenbeek et al. 1990; Lee et al. 1991; Bonilla et al. 1994), HCoV-229E (Herold et al. 1993), TGEV (Eleouet et al. 1995; Penzes et al. 2001), porcine epidemic diarrhea virus (PEDV) (Kocherhans et al. 2001), bovine coronavirus (Chouljenko et al. 2001), and SARS-CoV (Marra et al. 2003; Rota et al. 2003). In some cases (for example, SARS-CoV) complete genome sequences are available for several or even multiple isolates (Ruan et al. 2003). The genome sizes of coronaviruses range between 27.3 (HCoV-229E) and 31.3 (MHV) kb, making coronaviruses the largest RNA viruses currently known. About two-thirds of the coronavirus genome (~20,000 bases) are devoted to encoding the viral replicase that mediates viral RNA synthesis (Thiel et al. 2001b) and, possibly, other functions. The replicase gene is comprised of two large open reading frames, designated ORF1a and ORF1b, that are located at the 5' end of the genome. The upstream ORF1a encodes a polyprotein of 450–500 kDa, termed polyprotein (pp)1a, whereas ORF1a and ORF1b together encode pp1ab (750–800 kDa) (Fig. 1). Expression of the C-terminal, ORF1b-encoded half of pp1ab requires a (–1) ribosomal frameshift during translation. It is generally accepted that frameshifting depends on two critical elements, the "slippery" sequence, UUUAAAC, at which the ribosome shifts into the (–1) reading frame and a tripartite RNA pseudoknot structure located more downstream, near the ORF1a/1b junction (Brierley et al. 1987, 1989; Herold and Siddell 1993). In vitro experiments using reticulocyte lysates indicate that frameshifting occurs in about 20%–30% of the translation events, but it is not known whether this reflects the situation in vivo. The fact that the core replicative functions, RdRp and helicase, are encoded by ORF1b implies that their expression critically depends

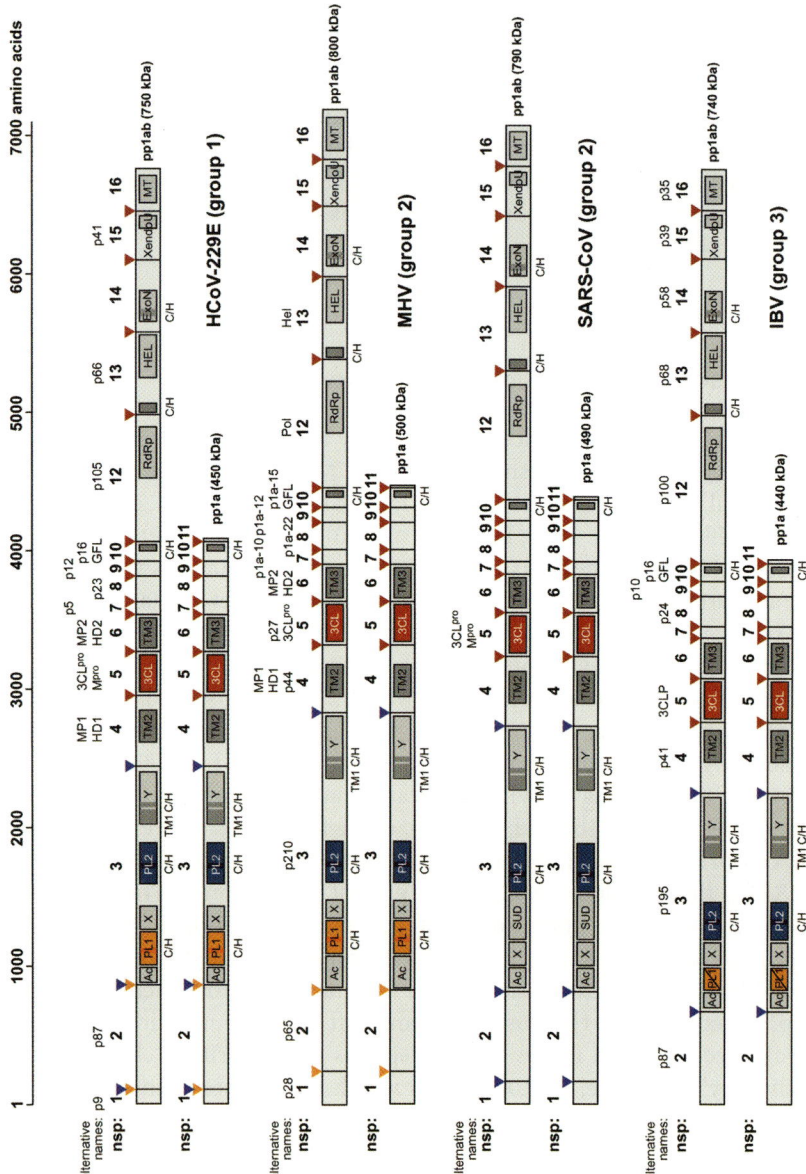

Fig. 1. Overview of the domain organization and proteolytic processing of coronavirus replicase polyproteins. Shown are the replicase polyproteins pp1a and pp1ab of human coronavirus 229E (*HCoV-229E*), mouse hepatitis virus (*MHV*), SARS coronavirus (*SARS-CoV*), and avian infectious bronchitis virus (*IBV*). The processing end-products of pp1a are designated nonstructural proteins (nsp) 1 to nsp11, and those

on ribosomal frameshifting, suggesting a requirement for a specific molar ratio between ORF1a- and ORF1b-encoded protein functions.

3 Replicase Polyproteins

3.1 Functional Domains

Initial sequence analyses in the late 1980s suggested a large divergence of the coronavirus replicase from the replicative machinery of other +RNA viruses. Accordingly, at this time, only very few functional predictions could be made for the ~800-kDa replicative polyproteins of coronaviruses (Boursnell et al. 1987). In 1989, a detailed comparative sequence analysis of the IBV replicase gene (Gorbalenya et al. 1989c) was pub-

◄───

of pp1ab are designated nsp1 to nsp10 and nsp12 to nsp16. Note that nsp1 to nsp10 may be released by proteolytic processing of either pp1a or pp1ab, whereas nsp11 is processed from pp1a and nsp12 to nsp16 are processed from pp1ab. nsp11 and nsp12 share a number of residues at the N-terminus. Alternative names that have been used in the past to designate specific processing products are given. Cleavage sites that are processed by the viral main proteinase are indicated by *red arrowheads*, and sites that are processed by the accessory papainlike proteinases 1 and 2 are indicated by *orange* and *blue arrowheads*, respectively. *Ac*, acidic domain (Ziebuhr et al. 2001); *PL1*, accessory papainlike cysteine proteinase 1 (Baker et al. 1989, 1993; Gorbalenya et al. 1991; Herold et al. 1998); *X*, X domain (Gorbalenya et al. 1991), which is predicted to have adenosine diphosphate-ribose $1''$-phosphatase activity (Snijder et al. 2003); *SUD*, SARS-CoV unique domain (Snijder et al. 2003); *PL2*, accessory papainlike cysteine proteinase 2 (Gorbalenya et al. 1991; Liu et al. 1995; Kanjanahaluethai and Baker 2000; Ziebuhr et al. 2001); *Y*, Y domain containing a transmembrane domain and a putative Cys/His-rich metal-binding domain; *TM1, TM2,* and *TM3,* putative transmembrane domains 1 to 3; *3CL,* 3C-like main proteinase (Gorbalenya et al. 1989c; Liu and Brown 1995; Ziebuhr et al. 1995; Lu et al. 1995); *RdRp*, putative RNA-dependent RNA polymerase domain (Gorbalenya et al. 1989c); *HEL*, helicase domain (Seybert et al. 2000a); *ExoN*, putative $3'$-to-$5'$ exonuclease (Snijder et al. 2003); *XendoU*, putative poly(U)-specific endoribonuclease (Snijder et al. 2003); *MT*, putative S-adenosylmethionine-dependent ribose $2'$-O-methyltransferase (Snijder et al. 2003); *C/H*, Cys/His-rich domains predicted to bind metal ions. Note that IBV pp1a and pp1ab do not have a counterpart of nsp1 of other coronaviruses. The papainlike cysteine proteinase 1 of IBV is *crossed out* to indicate that the domain is proteolytically inactive

lished in which the RdRp and NTPase/helicase domains were predicted to be encoded by the 5′ region of ORF1b. Furthermore, a putative chymotrypsin-like (picornavirus 3C-like) cysteine proteinase domain (3CL^{pro}) was identified in ORF1a and predictions on putative cleavage sites in the C-terminal regions of pp1a and pp1ab were made. The proteinase was found to be flanked by membrane domains on both sides. The coronavirus replicative proteins were proposed to be only extremely distantly related to the corresponding homologs of other +RNA viruses, and many of the pp1a/pp1ab-encoded enzymes appeared to have unique structural properties. Thus, for example, the helicase was proposed to be linked at its N-terminus to a complex zinc-binding domain (ZBD) consisting of 12 Cys/His residues (see below). In several cases, mutations in otherwise strictly conserved signature sequences were found. Thus the typical G–D–D signature of the conserved RdRp motif VI (Koonin 1991) was found to be replaced by S–D–D in the coronavirus homolog and the G(A)–X–H motif conserved in the S1 subsite of the substrate-binding pocket of picornavirus 3C proteinases (Gorbalenya et al. 1989a, 1989c) was substituted with Y–M–H. The predictions on functional domains, putative active-site residues, and proteinase cleavage sites were continuously elaborated and extended when more coronavirus replicase sequences became available (Gorbalenya et al. 1991; Lee et al. 1991; Herold et al. 1993; Eleouet et al. 1995; Chouljenko et al. 2001; Kocherhans et al. 2001; Penzes et al. 2001; Ziebuhr et al. 2001; Snijder et al. 2003). In these studies, papainlike cysteine proteinase (PL^{pro}) domains (Gorbalenya et al. 1991), a conserved domain of corona-, alpha-, and rubiviruses, termed X[1] (Gorbalenya et al. 1991), an acidic domain (Ac) of unknown function, and a domain (termed Y) with putative metal-binding and membrane-targeting functions (Ziebuhr et al. 2001) were identified in the coronavirus ORF1a sequence (Fig. 1). Overall, the sequence similarities between the replicase genes of prototypic viruses from the three coronavirus groups corresponded well to those of the structural protein regions, providing support for the traditional classification of coronaviruses into three groups, which previously was based on structural protein sequence relationships and serological cross-reactivities (Siddell 1995).

Recently, the list of putative enzymes involved in coronavirus RNA synthesis was extended considerably. Thus, in the context of a bioinformatics study of the SARS-CoV genome, as many as five (putative) coro-

---

[1] The X domain has recently been predicted to be an adenosine diphosphate-ribose 1″-phosphatase (ADRP).

naviral RNA processing activities were identified (Snijder et al. 2003) (Fig. 1). These include (1) a 3′-to-5′ exonuclease (ExoN) of the DEDD superfamily (Zuo and Deutscher 2001), (2) a poly(U)-specific endoribonuclease (XendoU) (Laneve et al. 2003), (3) an S-adenosylmethionine-dependent ribose 2′-O-methyltransferase (2′-O-MT) of the RrmJ family (Bügl et al. 2000), (4) an ADRP (Martzen et al. 1999), and (5) a cyclic phosphodiesterase (CPD) (Martzen et al. 1999; Nasr and Filipowicz 2000). Four of the activities are conserved in all coronaviruses, indicating their essential role in the coronaviral life cycle. In fact, the number of enzymes predicted to be involved in coronavirus RNA synthesis and modification is unique in RNA viruses and indicates a remarkable functional complexity, which approaches that of DNA replication. Three of the newly identified activities, ExoN (nsp14), XendoU (nsp15), and 2′-O-MT (nsp16), are arranged in pp1ab as a single protein block downstream of the RdRp (nsp12) and helicase (nsp13) domains (Fig. 1), suggesting that their activities cooperate in the same metabolic pathway(s). This conclusion is supported by the identification of a stable processing intermediate in IBV-infected cells that exactly comprises these three domains (Xu et al. 2001). It is also supported by the fact that nsp14–16 expression involves common regulatory mechanisms, (1) ribosomal frameshifting and (2) 3CL^{pro}-mediated proteolysis. As a first clue to possible functions encoded by this gene block in ORF1b, an exciting parallel to cellular RNA processing pathways was found by Snijder et al. (2003). Thus homologs of the coronavirus nsp14–16 processing products cleave and process mRNAs to produce small nucleolar (sno) RNAs that, in turn, guide specific 2′-O-ribose methylations of rRNA (Kiss 2001; Filipowicz and Pogacic 2002).

Two other coronavirus domains, CPD and ADRP, both of which do not require ribosomal frameshifting for expression, were speculated to cooperate in a pathway that again has parallels in the cell. Thus two cellular homologs are known to mediate two consecutive steps in the downstream processing of tRNA splicing products. In this pathway, CPD converts adenosine diphosphate ribose 1″-2″ cyclic phosphate (Appr>p) to adenosine diphosphate ribose 1″-phosphate (Appr-1″-p) (Culver et al. 1994) that, in a second reaction, is further processed (probably dephosphorylated) by an ADRP homolog (Martzen et al. 1999).

Obviously, the characterization of the substrate specificities of the newly identified enzymes will now be of major interest and may allow predictions or even conclusions on the functions of these proteins. Both (reverse) genetic and biochemical data will be required to answer the question of whether the RNA processing enzymes are directly involved in the synthesis and/or processing of viral RNA or rather interfere with (and thereby reprogram) cellular pathways for the benefit of viral replication (or even have other functions).

The observed pattern of conservation in different nidovirus families suggests a functional hierarchy for the five RNA processing activities, with XendoU playing a central role. This enzyme is universally conserved in nidoviruses and was previously referred to as "nidovirus-specific conserved domain" (Snijder et al. 1990b; den Boon et al. 1991; de Vries et al. 1997). In contrast, CPD is only encoded by toroviruses and a subset of group 2 coronaviruses (excluding SARS-CoV) (Snijder et al. 2003). Given that coronaviruses and arteriviruses are generally believed to use very similar replication and transcription strategies, it is intriguing that, out of the four activities conserved in all coronaviruses (ExoN, XendoU, 2′-O-MT, and ADRP), only one activity (XendoU) is conserved in arteriviruses. One may therefore speculate that (1) arterivirus and coronavirus RNA synthesis mechanisms differ in several molecular details or (2) the viruses interact differentially with RNA processing pathways of the host cell. Alternatively, the extra functions encoded by corona- and toroviruses (and, to a lesser extent, roniviruses) may be required to synthesize and maintain the extremely large (~30 kb) RNA genomes of these viruses. Thus, on the basis of its sequence similarity with cellular 3′-to-5′ exonucleases involved in proofreading, repair, and/or recombination, ExoN has been speculated to be involved in related mechanisms that may be required for the life cycle of corona-, toro-, and roniviruses but may be dispensable for the much smaller arteriviruses (Snijder et al. 2003). The significance of the observation that overexpression of nsp14 induces apoptotic changes in the host cell (Liu et al. 2001) remains to be further investigated.

3.2
Proteolytic Processing by Viral Cysteine Proteinases

In common with many other +RNA viruses (Kräusslich and Wimmer 1988; Dougherty and Semler 1993), coronaviruses employ proteolytic processing as a key regulatory mechanism in the expression of their replicative protein functions (Ziebuhr et al. 2000). Proteinase inhibitors that block proteolytic processing also obviate coronavirus replication, illustrating the essential role of pp1a/pp1ab processing for viral RNA synthesis (Kim et al. 1995). On the basis of their physiological role, coronavirus proteinases can be classified into *accessory* proteinases, which are responsible for cleaving the more divergent N-proximal pp1a/pp1ab regions at two or three sites, and *main* proteinases, which cleave the major part of the polyproteins at 11 conserved sites and also release the conserved key replicative functions, such as RdRp, helicase, and three of the RNA processing domains (Ziebuhr et al. 2000; Snijder et al. 2003). All coronaviruses encode one main proteinase and, depending on the virus (see below and Fig. 1), one or two accessory proteinases. The accessory proteinases are papainlike cysteine proteinases that are designated PL^{pro} ($PL1^{pro}$ and $PL2^{pro}$). The main proteinase is a cysteine proteinase with a serine proteinase-like structure (Anand et al. 2002). In previous publications, two alternative designations have been used for this protein. The name *main proteinase*, M^{pro}, is generally used to stress the dominant physiological role of this proteinase in coronavirus gene expression, whereas the name *3C-like proteinase* is used to stress the (distant) relationship with picornavirus 3C proteinases, which is based on a common chymotrypsin-like two-β-barrel structure and similar substrate specificities (Gorbalenya et al. 1989a,c; Ziebuhr et al. 2000). Despite this relationship, there are also important structural differences between picornavirus and coronavirus chymotrypsin-like proteinases (see below).

Peptide cleavage data obtained for several coronavirus main proteinases revealed differential processing kinetics for specific sites. The order of cleavages was found to be conserved among coronaviruses and appears to depend on the accessibility of specific sites in the context of the polyprotein (Piñon et al. 1999) as well as the primary and secondary structures of a given cleavage site. Thus deviation from the $3CL^{pro}$ cleavage site consensus sequence, L-Q|(A,S,G), resulted in most cases in significantly reduced cleavage efficiencies (Ziebuhr and Siddell 1999; Hegyi and Ziebuhr 2002; Fan et al. 2003). Furthermore, substrate peptides adopting extended β-strand structures appear to be favored by $3CL^{pro}$ over α-helical or disordered structures (Fan et al. 2003). On the basis of these data, it is reasonable to postulate that coronavirus polyprotein processing occurs in a temporally coordinated manner, which might lead to activation and inactivation of specific functions in the course of the viral life cycle, as has been demonstrated for other +RNA viruses (Lemm et al. 1994; Vasiljeva et al. 2003).

The combined data of numerous studies published in the past 15 years provide a (nearly) complete picture of the pp1a/pp1ab processing pathways of prototypic viruses from all three coronavirus groups (Fig. 1). Throughout this chapter, the replicase processing end products will be continuously numbered from nonstructural protein (nsp) 1 to nsp16 (from N- to C-terminus^2) to facilitate their comparison with homologs from other coronaviruses.

3.2.1
Accessory Proteinases

The N-proximal regions of the MHV and HCoV-229E replicase polyproteins are processed by two PL^{pro}s at three sites to produce nsp1–4, with the C-terminus of nsp4 being cleaved by the main proteinase (Fig. 1). The proteolytic activities of the MHV and HCoV-229E PL1^{pro} and PL2^{pro} domains and the IBV PL2^{pro}, which all reside in nsp3, have been characterized in detail (Ziebuhr et al. 2000). Briefly, the MHV PL1^{pro} cleaves the nsp1|nsp2 and nsp2|3 sites, while PL2^{pro} processes the third site, nsp3|nsp4 (Baker et al. 1989, 1993; Dong and Baker 1994; Denison et al. 1995; Hughes et al. 1995; Bonilla et al. 1997; Teng et al. 1999; Kanjanahaluethai and Baker 2000; Kanjanahaluethai et al. 2003). Also in HCoV-229E, PL1^{pro} was shown to cleave the nsp1|nsp2 and nsp2|nsp3 sites (Herold et al. 1998; Ziebuhr et al. 2001). However, in the case of HCoV-229E, the regulation of proteolytic processing was shown to be more complex than previously thought. Thus PL2^{pro} (originally believed to process only the nsp3|nsp4 site) was demonstrated also to process the nsp2|nsp3 site. The nsp2|nsp3 cleavages mediated by PL1^{pro} and PL2^{pro}, respectively, were shown to occur at exactly the same scissile bond (Herold et al. 1998; Ziebuhr et al. 2001). Whereas the PL1^{pro}-mediated cleavage proved to be slow and incomplete in vitro, PL2^{pro} cleaved this site efficiently under the same experimental conditions. Furthermore, evidence was obtained to suggest that the proteolytic activity of PL1^{pro} at the nsp2|nsp3 site is downregulated by PL2^{pro} by a noncompetitive mechanism (Ziebuhr et al. 2001). It was concluded that the activities of the two proteinase domains present in nsp3 are tightly regulated in HCoV-229E and, probably, also other coronaviruses, with PL2^{pro} playing a major role and dominating over the activity of PL1^{pro}. This conclusion is also supported by the conservation of PL2^{pro} in all coronaviruses (Ziebuhr et al. 2001; Snijder et al. 2003).

IBV encodes only one proteolytically active PL^{pro}, which is PL2^{pro}. The IBV PL1^{pro} domain, although being conserved, has lost its proteolytic activity in the course of evolution because of the accumulation of active site mutations (Ziebuhr et al. 2001). Apparently, IBV does not en-

---

[2] Note that similar designations (nsp or ns) are occasionally used for some of the group-specific *nonstructural* proteins encoded in the 3'-structural protein regions of coronaviruses (Brown and Brierley, 1995).

code a counterpart of the nsp1 protein of other coronaviruses. Thus there are only two cleavage sites in this region of pp1a/pp1ab, nsp2|nsp3 and nsp3|nsp4, which are both processed by PL2^{pro} (Lim and Liu 1998; Lim et al. 2000). In SARS-CoV, only one PL^{pro} is conserved (Marra et al. 2003; Rota et al. 2003). The domain occupies a position in pp1a/pp1ab that corresponds to that of the PL2^{pro} domains of other coronaviruses and therefore is considered an ortholog of coronavirus PL2^{pro}s (Snijder et al. 2003). Obviously, the SARS-CoV PL2^{pro} must be responsible for the processing of all three sites identified in this region and, indeed, the activity of PL2^{pro} at the nsp2|nsp3 site was demonstrated recently (Thiel et al. 2003). The arrangement of the N-terminal domains of SARS-CoV nsp3 differs from that of other coronaviruses (Ziebuhr et al. 2001; Snijder et al. 2003). Thus, the conserved ADRP domain ("X" in Fig. 1) resides immediately downstream of the acidic domain (Ac) in nsp3, a position that is occupied by PL1^{pro} in other coronaviruses. Further downstream, another domain of unknown function has been identified in the region separating the ADRP and PL2^{pro} domains. It has been termed "SARS-CoV unique domain" (SUD) (Snijder et al. 2003) (Fig. 1).

The sequence similarity between coronaviral PL^{pro}s and the prototypic cellular proteinases is very low. A closer relationship seems to exist between the active sites of coronavirus PL^{pro}s and the leader proteinase (L^{pro}) of the picornavirus foot-and-mouth-disease virus (FMDV) (Gorbalenya et al. 1991). Crystal structure analysis revealed that the active site of L^{pro} also diverged profoundly from its cellular homologs, which explains some of the unique biochemical properties of this enzyme, such as salt sensitivity and narrow pH optimum (Guarné et al. 1998, 2000). It remains to be studied whether the sequence affinity between L^{pro} and coronavirus PL^{pro}s is associated with common structural and functional features.

Only very few amino acids are absolutely conserved among coronavirus PL^{pro}s (Herold et al. 1999). Furthermore, there are only very few PL1^{pro} versus PL2^{pro} lineage-specific residues, which do not provide sufficient evidence for clustering the PL1^{pro} and PL2^{pro} domains into two separate groups. Despite this divergency at the sequence level, coronavirus PL^{pro}s share a number of common properties. Thus they all (1) process sites that are located in the N-terminal half of the replicase polyproteins, far upstream of the conserved ORF1b-encoded domains (Fig. 1), (2) cleave sites that have at least one small residue (Gly, Ala) at the scissile bond (Dong and Baker 1994; Hughes et al. 1995; Bonilla et al. 1997; Herold et al. 1998; Lim and Liu 1998; Lim et al. 2000; Ziebuhr et al. 2001; Kanjanahaluethai et al. 2003), (3) have a catalytic dyad consisting of Cys (followed by Trp or Tyr) and a downstream His (Baker et al. 1993; Herold et al. 1998; Lim and Liu 1998), and (4) employ variants of the papainlike $\alpha+\beta$ fold (Gorbalenya et al. 1991; Herold et al. 1999). Molecular modeling suggests that the α and β domains are connected by a transcription factor-like domain that includes a zinc-binding domain (ZBD) essential for proteolytic activity (Herold et al. 1999) (Fig. 1). It seems likely that the domain also has other functions, for example, in sg mRNA transcription. This hypothesis is based on (1) the sequence similarity with cellular transcription factors (Herold et al. 1999) and (2) the fact that the related ZBD-containing EAV nsp1 papainlike proteinase has a clearly established role in arterivirus sg mRNA synthesis (Tijms et al. 2001).

The presence of two PL^{pro}s in most coronavirus replicases suggests that these enzymes originated from the duplication of a PL^{pro} domain in one of the ancestors of the contemporary coronaviruses. Surprisingly, however, phylogenetic trees inferred from multiple sequence comparisons of coronavirus PL^{pro}s revealed that only the $PL1^{pro}$ and $PL2^{pro}$ domains of the most closely related coronaviruses were clustered together (Ziebuhr et al. 2001). Therefore, multiple independent gene duplications in different coronaviruses cannot be excluded entirely. Alternatively and much more probably, the above result can be interpreted to reflect homoplasy events that, subsequent to the initial gene duplication, have driven a parallel evolution of the two coronavirus PL^{pro} paralogs, while other regions of the replicase diverged much more profoundly (Ziebuhr et al. 2001). Often, such homoplasy events are driven by common substrates. Thus the identification of a common cleavage site that is processed by both $PL1^{pro}$ and $PL2^{pro}$ in HCoV-229E may indicate that, in this virus and probably also other coronaviruses, the conservation of overlapping substrate specificities was an important driving force of evolution. The underlying selective advantage that led to the conservation of such a partial redundancy of two proteinase domains in most coronaviruses remains to be investigated. Conservation of overlapping substrate specificities also appears to affect the cleavage site structures. Thus a comparison of PL^{pro} cleavage sites of SARS-CoV and IBV, which both employ only one PL^{pro} activity, with the corresponding cleavage sites of HCoV-229E, which employs two PL^{pro} domains, revealed a much better conservation of the IBV/SARS-CoV $PL2^{pro}$ sites compared with the HCoV $PL1^{pro}/PL2^{pro}$ sites (Thiel et al. 2003).

3.2.2
Main Proteinase

The coronavirus main proteinase, 3CL^{pro}, is encoded by ORF1a and resides in nsp5 (Fig. 1). In the polyprotein, it is flanked by hydrophobic domains. The ~33-kDa proteinase releases itself from pp1a/pp1ab at flanking sites and directs the proteolytic processing of all downstream domains of pp1a/pp1ab (Fig. 1). In total, 3CL^{pro} cleaves at 11 conserved sites to produce 13 processing end products and, probably, multiple intermediates. Because of the central role in the expression of the major replicative proteins, 3CL^{pro} is also called "main" proteinase (M^{pro}).

Coronavirus 3CL^{pro}s represent a highly diverged branch of two-β-barrel proteinases (Gorbalenya et al. 1989a,c). In contrast to what the name suggests, coronavirus 3CL^{pro}s also deviate significantly from the picornavirus 3C and other +RNA viral 3C-like proteinases. Characterization of a roniviral 3CL^{pro} has indicated that the 3C-like proteinases of potyviruses may represent the closest relatives of coronavirus 3CL^{pro}s (outside the *Nidovirales* order) (Cowley et al. 2000; Gorbalenya 2001; Ziebuhr et al. 2003). In common with the prototypic picornavirus 3C proteinases (Allaire et al. 1994; Matthews et al. 1994; Mosimann et al. 1997), coronavirus 3C-like proteinases have a chymotrypsin-like, two-β-barrel fold that is formed by 12 antiparallel β-strands (Allaire et al. 1994; Matthews et al. 1994; Mosimann et al. 1997; Anand et al. 2002, 2003). However, both the size and orientation of secondary structure elements vary considerably between the two groups of enzymes, making reliable structural alignments difficult, if not impossible. Furthermore, in contrast to 3C proteinases but in common with other nidovirus 3C-like proteinases (Barrette-Ng et al. 2002; Ziebuhr et al. 2003), coronavirus 3CL^{pro}s have a C-terminal extension, which is called domain III to distinguish it from the β-barrel domains I and II. Domain III of the TGEV 3CL^{pro} comprises 103 amino acids and consists of 5 α-helices that adopt a unique structure that currently has no homologs in the database (Anand et al. 2002) (Figs. 2 and 3). The structure of the coronavirus 3CL^{pro} domain III differs from the corresponding domain of the arterivirus nsp4 proteinase, which comprises only 49 residues and consists of 2 short pairs of β-strands and 2 α-helices (Barrette-Ng et al. 2002).

The differences between picornavirus and coronavirus chymotrypsin-like proteinases also extend to the catalytic residues. Thus, whereas the vast majority of picornavirus enzymes employ a catalytic triad, Cys-

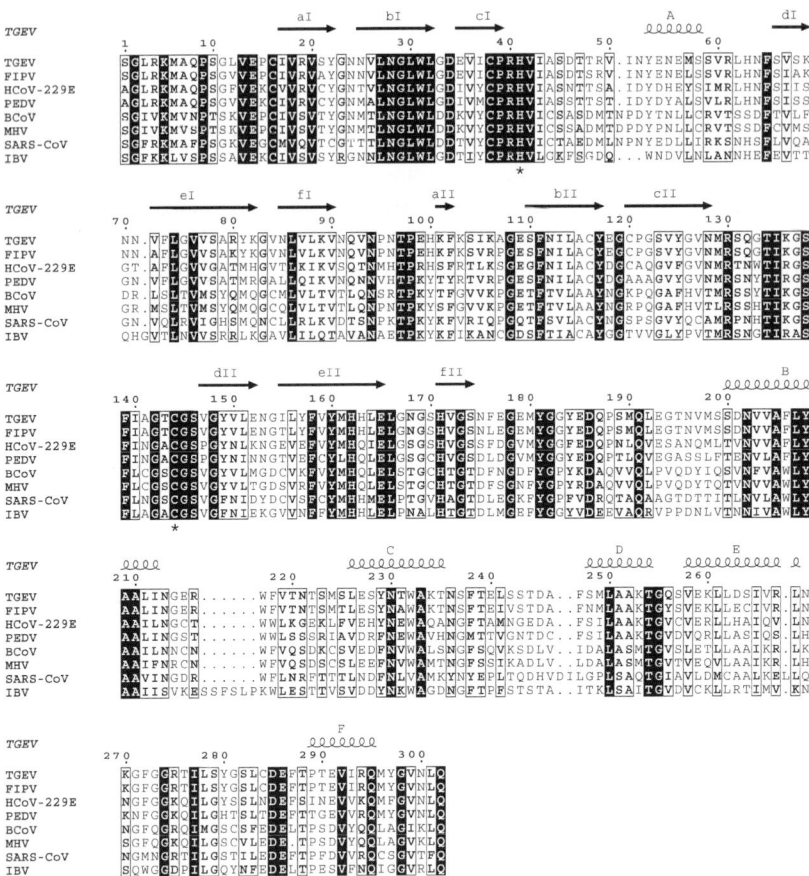

Fig. 2. Sequence comparison of coronavirus 3C-like main proteinases. The alignment was generated with the ClustalW program (version 1.82) (http://www.ebi.ac.uk/clustalw/) and used as input for the ESPript program (version 2.1) (http://prodes.toulouse.inra.fr/ESPript/cgi-bin/ESPript.cgi). The 3CL^{pro} sequences of transmissible gastroenteritis virus (*TGEV*, strain Purdue 46), feline infectious peritonitis virus (*FIPV*, strain 79-1146), human coronavirus 229E (*HCoV-229E*), porcine epidemic diarrhea virus (PEDV, strain CV777) bovine coronavirus (*BCoV*, isolate LUN), mouse hepatitis virus (*MHV*, strain A59), avian infectious peritonitis virus (IBV, strain Beaudette), and SARS coronavirus (*SARS-CoV*, isolate Frankfurt 1) were derived from the replicative polyproteins of the respective viruses whose sequences are deposited at the DDBJ/EMBL/GenBank database (accession numbers: TGEV, AJ271965; FIPV, AF326575; HCoV, X69721; PEDV, AF353511; BCoV, AF391542; MHV, NC 001846; IBV, M95169; SARS-CoV, AY291315). The β-strands and α-helices as revealed by the TGEV 3CL^{pro} crystal structure (Anand et al. 2002; PDB 1LVO) are shown *above* the sequence alignment. Catalytic Cys and His residues are indicated by *asterisks*

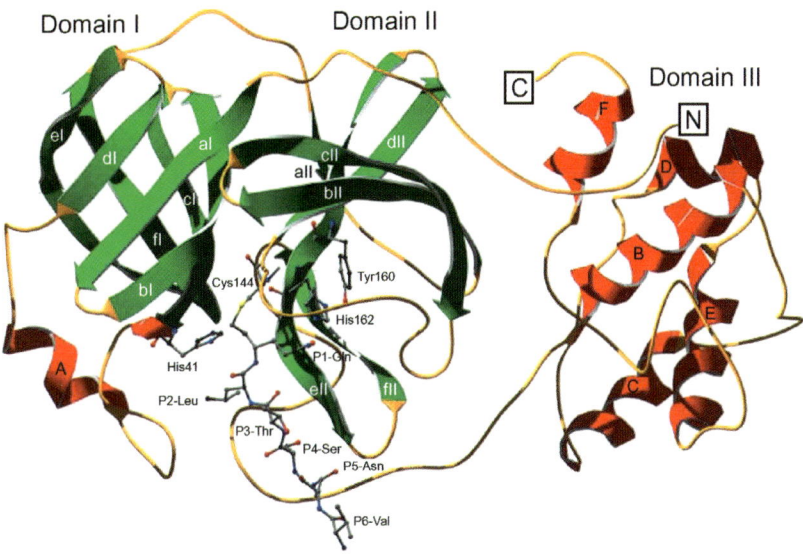

Fig. 3. Structure of monomer B of TGEV 3CL^{pro} with a hexapeptidyl chloromethyl ketone inhibitor bound to the active site (Anand et al. 2002, 2003). 3CL^{pro} domains I, II, and III are indicated. α-Helices are shown in *red* and are *labeled A to F*. β-Strands are shown in *green* and are *labeled a to f*, followed by an indication of the domain to which they belong. Shown in ball-and-stick representation are the substrate analog inhibitor (residues P1 to P6), the catalytic residues (Cys144 and His41), and the S1 subsite His162 residue interacting with Tyr160 and the P1 Gln side chain of the substrate (see text for details). N- and C termini are labeled *N* and *C*

His-Asp(Glu) (Allaire et al. 1994; Matthews et al. 1994; Mosimann et al. 1997; Seipelt et al. 1999), which is reminiscent of the charge-relay system of chymotrypsin-like serine proteinases, the coronavirus 3CL^{pro}s use a catalytic dyad consisting of Cys (nucleophile) and His (general base) (Figs. 2 and 3). Mutation analyses performed with recombinant enzymes from different coronavirus species had consistently failed to identify a third catalytic residue, suggesting that coronavirus 3CL^{pro}s may lack a counterpart to the catalytic Asp(Glu) of other chymotrypsin-like proteinases (Liu and Brown 1995; Lu and Denison 1997; Ziebuhr et al. 1997). This hypothesis was confirmed by crystal structure analyses of the TGEV (Anand et al. 2002), HCoV-229E (Anand et al. 2003), and SARS-CoV 3CL^{pro} enzymes (PDB acc: 1Q2W). Thus, for example, in the TGEV 3CL^{pro} structure, a buried water molecule was found in the place that is normally occupied by the third member of the triad (Asp or Glu).

The water was hydrogen-bonded to His41[3] $N^{\delta 1}$, His163 $N^{\delta 1}$, and Asp186 $O^{\delta 1}$. An equivalent water molecule is also found in the HCoV 3CL^{pro} structure. Here, it is stabilized by His41 $N^{\delta 1}$, Gln163 $N^{\delta 1}$, and Asp186 $O^{\delta 1}$. The TGEV 3CL^{pro} structure also suggested that, after the attack of the active-site Cys144 nucleophile on the carbonyl carbon of the scissile bond, the developing oxyanion is stabilized by hydrogen bonds donated by the main chain amides of Gly142, Thr143, and Cys144, which together form the "oxyanion hole."

The substrate specificity of coronavirus 3CL^{pro}s resembles that of many other 3C and 3C-like proteinases (Blom et al. 1996; Ryan and Flint 1997) in so far as all the coronavirus 3CL^{pro} sites share a Gln residue at the P1 position, whereas small residues (Ala, Ser, and Gly) are conserved at the P1' position (Ziebuhr et al. 2000). Larger residues, such as Asn (which is found at the P1' position of all coronavirus nsp8|nsp9 sites), result in significantly reduced cleavage efficiencies (Ziebuhr and Siddell 1999; Hegyi and Ziebuhr 2002; Fan et al. 2003). Leu is strongly preferred at the P2 position of coronavirus 3CL^{pro} substrates, although other hydrophobic residues, such as Ile, Val, Phe, and Met, are occasionally also found at this position. At the P4 position, small residues, Val, Thr, Ser, Pro, and Ala, are favored. The structural basis for the pronounced specificity of coronavirus 3CL^{pro}s was elucidated recently by structure analysis of a hexapeptidyl chloromethyl ketone inhibitor bound to the active site of the TGEV 3CL^{pro} (Anand et al. 2003). Because the sequence of the inhibitor was derived from the P6–P1 region of a natural cleavage site (Val-Asn-Ser-Thr-Leu-Gln) of TGEV 3CL^{pro}, the structure most likely represents the binding mode of coronavirus 3CL^{pro} substrates in general. It was found that the P region of 3CL^{pro} substrates binds in a shallow groove at the surface of the proteinase, between domains I and II (Fig. 3). Residues P5 to P3 form an antiparallel β-sheet with residues 164–167 of strand eII and residues 189–191 of the loop linking domains II and III. Deletion of the loop region abolishes the proteolytic activity of 3CL^{pro}, supporting the functional significance of the interaction between the substrate and this loop region (Anand et al. 2002).

The conserved Gln side chain at the P1 position of 3CL^{pro} substrates interacts with the imidazole of His162 (Fig. 3), at the bottom of the S1 subsite, which is formed by the main-chain atoms of Ile51, Leu164, Glu165, and His171 (Anand et al. 2003). The neutral state of His162 over a broad pH range appears to be maintained by (1) stacking onto the

---

[3] Amino acid residues of coronavirus 3CL^{pro}s are numbered from Ser(Ala)1 to Gln302.

phenyl ring of Phe139 and (2) accepting a hydrogen bond from the hydroxyl group of the buried Tyr160. This interpretation is supported by mutagenesis data obtained for bacterially expressed HCoV-229E and feline infectious peritonitis (FIPV) 3CL^{pro}s (Ziebuhr et al. 1997; Hegyi et al. 2002). Tyr160 is part of the conserved coronavirus 3CL^{pro} signature, Tyr-X-His, whereas Gly(Ala)-X-His is found at the equivalent sequence position in most 3C and 3C-like proteinases (Gorbalenya et al. 1989a; Gorbalenya and Snijder 1996). Accordingly, stabilization of histidine in the neutral tautomeric state needs to be ensured by other residues (Bergmann et al. 1997; Mosimann et al. 1997).

The hydrophobic S2 subsite of the proteinase, which accommodates the conserved Leu residue and, in few cases, other hydrophobic residues, is formed by the side chains of Leu164, Ile51, Thr47, His41, and Tyr53 (Anand et al. 2003). The fact that, in the structure, the P3 side chain of the substrate analog was oriented toward bulk solvent explains why there is no specificity for any particular side chain at the P3 position of coronavirus 3CL^{pro} cleavage sites (Ziebuhr et al. 2000). The S4 site is rather congested (Anand et al. 2003), explaining the conservation of small residues, such as Ser, Thr, Val, or Pro, at this position of coronavirus 3CL^{pro} substrates. On the basis of the TGEV 3CL^{pro}–inhibitor structure, it has been proposed that the relatively small P1' residues (Ser, Ala, or Gly) may be accommodated by a S1' subsite that involves Leu27, His41, and Thr47 (Anand et al. 2003).

It is generally believed that most of the pp1a/pp1ab cleavages are mediated *in trans* by the fully processed form of 3CL^{pro} (nsp5). The *trans* activity of 3CL^{pro} has been well characterized, both biochemically and structurally (Ziebuhr et al. 1995; Grötzinger et al. 1996; Lu et al. 1996; Heusipp et al. 1997a,b; Tibbles et al. 1999; Ziebuhr and Siddell 1999; Anand et al. 2002, 2003; Hegyi and Ziebuhr 2002; Fan et al. 2003). However, it is not clear whether 3CL^{pro} cleaves itself from pp1a/pp1ab *in cis* or *in trans*. Also, it is not clear whether 3CL^{pro} can cleave downstream pp1a/pp1ab sites *in cis*. Thus, on the one hand, there is biochemical and structural evidence to suggest that 3CL^{pro} self-processing occurs *in trans* (Lu et al. 1996; Anand et al. 2002). Furthermore, in MHV-infected cells, 3CL^{pro} was found to be part of a rather stable 150-kDa processing intermediate (nsp4–10 or nsp4–11), which also argues against a rapid, co-translational release of 3CL^{pro} *in cis* (Schiller et al. 1998). On the other hand, a number of MHV and IBV 3CL^{pro}-containing precursors were shown to require microsomal membranes for efficient autocatalytic release of 3CL^{pro} from the flanking TM2 (nsp4) and TM3 (nsp6) domains (Tibbles et al. 1996; Piñon et al. 1997), indicating that the flanking domains (when properly folded) affect the activity of 3CL^{pro}. In other words, interdomain interactions in pp1ab may modulate the structure (and activity) of the enzyme, for example, to render 3CL^{pro} competent for *cis* cleavages at flanking sites or even further downstream sites. In fact, one might expect that at least some of the pp1a/pp1ab cleavages need to occur *in cis* early in infection, when the concentration of 3CL^{pro} is low and intermolecular reactions are less likely to occur. Otherwise, if there were no *cis* cleavages at all, pp1a/pp1ab should operate initially as an extremely large polyprotein that is only processed at its N-terminus by PL^{pro} cleavages. Structure information for larger 3CL^{pro} precursors will be required to answer the question of whether or not 3CL^{pro} adopts alternative conformations in its fully processed form and larger precursor molecules. Notably, reorientation of secondary structure elements after intramolecular release is believed to occur in picornavirus 3C proteinases (Khan et al. 1999), illustrating the significance of this question.

At present, structure information is only available for the fully processed coronavirus 3CL^{pro} (Anand et al. 2002, 2003). Both the crystal structures and dynamic light scattering data show that 3CL^{pro} forms dimers (Anand et al. 2002, 2003). The two molecules in the dimer are oriented perpendicular to one another (Fig. 4). The contact interface mainly involves conserved residues of the N-terminus of one molecule and domain II of the other molecule (and vice versa). The N-terminal amino acid residues are squeezed in between domains II and III of the parent

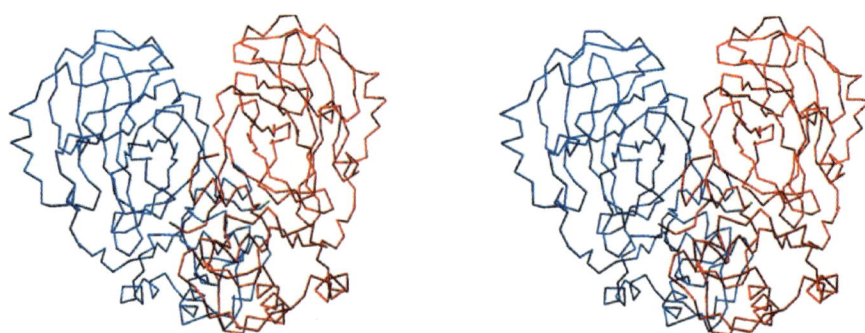

Fig. 4. Coronavirus main proteinases form dimers (Anand et al. 2002). Stereo representation of a Cα plot of a TGEV 3CL^{pro} dimer (PDB accession number: 1LVO). Monomers A and B are shown in *blue* and *red*, respectively. The monomers are oriented perpendicular to one another. Dimerization mainly involves interactions of the N terminus with domain II of the other dimer (see text for details). The N termini of monomers A and B are shown in *green* and *brown*, respectively

monomer and domain II of the other monomer, where they make a number of very specific interactions that appear tailor-made to bind this segment with high affinity. Apparently, this mechanism allows the active site to remain competent for binding and cleaving other sites in the polyprotein after autocleavage of 3CL^{pro}. In addition, the exact placement of the N-terminus seems to have a structural role for the mature 3CL^{pro}, because deletion of residues 1 to 5 leads to a dramatic decrease in proteolytic activity (Anand et al. 2003). It has been speculated that the tight interaction of the N-terminus with domains II and III may help to maintain the loop connecting domains II and III in the orientation required to bind the P3–P5 residues of the substrate (Anand et al. 2002, 2003). The presumed indirect role of domain III in proteolysis may explain the results from previous mutagenesis studies that consistently reported a dramatic loss of *trans*-cleavage activity with C-terminally truncated forms of HCoV-229E, TGEV, MHV, and IBV 3CL^{pro}s (Lu and Denison 1997; Ziebuhr et al. 1997; Ng and Liu 2000; Anand et al. 2002).

Genetic data also point to a (direct or indirect) role of domain III in RNA synthesis. Thus characterization of temperature-sensitive (*ts*) MHV mutants revealed that substitution of the MHV 3CL^{pro} Phe219 residue, which is part of the loop connecting α-helices B and C in domain III (Fig. 2), with Leu causes an RNA-minus phenotype at the restrictive temperature (Siddell et al. 2001). Further characterization of the *ts* mutant, Alb *ts*16, showed that both plus- and minus-strand synthesis was not greatly affected when the temperature was shifted late in infection. However, when the temperature was shifted to the nonpermissive temperature early, at a time when the rate of MHV RNA synthesis increases rapidly, no increase of plus-strand synthesis was observed with Alb *ts*16. Furthermore, inhibition of minus-strand synthesis (by inhibition of protein synthesis) was found to cause a decline of plus-strand synthesis after 30–60 min. The data can be interpreted to indicate that the defect in 3CL^{pro} activity interferes with minus-strand synthesis and reduces it to a low level that merely ensures the replenishment of minus strands being lost because of turnover. Alternatively, the mutation may cause a defect in the activity of 3CL^{pro} that blocks the formation of plus-strand polymerase activity (or prevents its conversion from the minus strand-synthesizing precursor). It remains to be determined whether the observed *ts* phenotype is caused by specific defects in the proteolytic activity of 3CL^{pro} or whether another, nonproteolytic function of domain III is affected. Thus, for example, protein-protein interactions involving domain III—as proposed to be mediated by the C-terminal domain of the EAV nsp4 proteinase (Barrette-Ng et al. 2002)—may be affected.

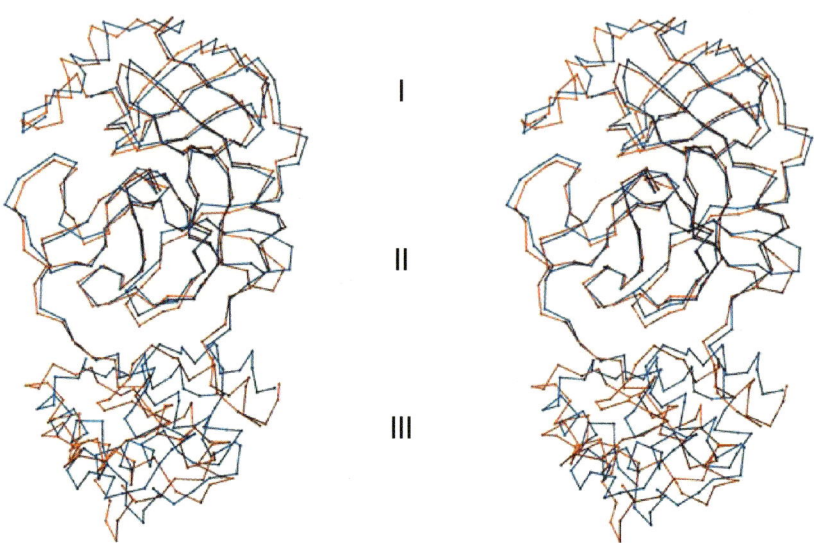

Fig. 5. Differential orientation of the C-terminal domains III of TGEV and SARS-CoV 3C-like main proteinases (PDB 1LVO and 1Q2 W). Superimposition (stereo image) of TGEV (*orange*) and SARS-CoV (*blue*) 3CL^{pro}s shows little variation between the structures of the N-terminal β-barrel domains I and II. The orientation (rather than the structure) of the respective C-terminal domains of TGEV and SARS-CoV 3CL^{pro} differs slightly in the two proteins, resulting in less perfect superimposition

Comparison of coronavirus main proteinase structures shows that domains I and II superimpose much better than the C-terminal domains III (Fig. 5). This is mainly due to a slightly different orientation of domain III in relation to domains I and II rather than differences in the domain III structures themselves.

3.3
Helicase

RNA helicases represent the second most conserved subunit of the RNA synthesis machinery of +RNA viruses and are involved in diverse steps of the viral life cycle (Buck 1996; Kadaré and Haenni 1997). They utilize the energy derived from hydrolysis of nucleoside triphosphates (NTPs) to unwind double-stranded (ds) RNA. Conservation of specific sequence motifs allows helicases to be classified into three large superfamilies (SFs), termed SF1, SF2, and SF3, as well as several small families (Gorbalenya et al. 1989b; Gorbalenya and Koonin 1993). The coronavirus helicase resides in nsp13 and has been classified as belonging to SF1 (Gorbalenya et al. 1989b, c) (Fig. 1). Nsp13 and its homologs in other nidoviruses have a putative zinc-binding domain (ZBD) at their N-terminus (Gorbalenya et al. 1989c), which is known to be required for the enzymatic activities of coronavirus and arterivirus helicases (Seybert, van Dinten, Posthuma, Snijder, Gorbalenya, and Ziebuhr, unpublished data). EAV reverse genetics data have shown that the ZBD and a downstream segment ("hinge spacer") that links ZBD to the C-terminal helicase domain have distinct functions in arterivirus replication, sg mRNA transcription, and virion morphogenesis (van Dinten et al. 2000). It is tempting to suggest that coronavirus helicases may have similarly diverse functions. Biochemical characterization of a recombinant form of HCoV-229E nsp13 demonstrated both nucleic acid-stimulated NTPase and duplex-unwinding activities (Seybert et al. 2000a). Similar data have subsequently been obtained for two arterivirus nsp10 helicases and the SARS-CoV nsp13 helicase (Seybert et al. 2000b; Bautista et al. 2002; Tanner et al. 2003; Thiel et al. 2003).

Coronavirus (and arterivirus) helicases were shown to unwind their dsRNA substrates with $5'$-to-$3'$ polarity, that is, they move in a $5'$-to-$3'$ direction on the strand to which they initially bind (Seybert et al. 2000a, b). Obviously, this stands in contrast to the $3'$-to-$5'$ polarity of the SF2 helicases of flavi-, pesti-, and hepaciviruses (Kadaré and Haenni 1997; Kwong et al. 2000) and may indicate fundamental differences in biological functions between the two groups of enzymes. For example, the $5'$-to-$3'$ polarity of the coronavirus nsp13 helicase activity argues against a role in the separation of secondary structures in the RNA template during minus-strand synthesis (as has been suggested for RNA viral SF2 helicases), because this would require a helicase with $3'$-to-$5'$ polarity.

Interestingly, coronavirus nsp13 is one of the few helicases that have no marked preference for RNA or DNA substrates. Thus they have been found to unwind partial-duplex DNA substrates with high efficacy (Seybert et al. 2000; Thiel et al. 2003). This property allows DNA-based assays to be used in the characterization of coronavirus helicases (for example, in mutagenesis studies and high-throughput tests of potential inhibitors). Because coronaviruses replicate in the cytoplasm and the helicase has not been found to localize to the nucleus (Sims et al. 2000; Bost et al. 2001), a biological significance of the DNA-unwinding activity of nsp13 seems unlikely, although it cannot be excluded entirely at the present stage. It should be mentioned in this context that the hepatitis C virus (HCV) NS3 helicase also has DNA duplex-unwinding activity, which, however, has been proposed to affect the structure of host cell DNA (Pang et al. 2002).

Duplex unwinding by coronavirus helicases is an energy-dependent process that derives its energy from NTP hydrolysis (Seybert et al. 2000a; Seybert and Ziebuhr 2001). Coronavirus helicases appear to be highly promiscuous with respect to the NTP cofactor used. Thus all standard NTPs and dNTPs were found to be hydrolyzed by coronavirus helicases (Seybert et al. 2000a; Seybert and Ziebuhr 2001; Tanner et al. 2003). Finally, coronavirus helicases possess RNA 5'-triphosphatase activity that may be involved in the formation of the 5' RNA cap structure of coronavirus plus-strand RNAs (Ivanov et al. 2004; Ivanov and Ziebuhr 2004).

3.4
RNA-Dependent RNA Polymerase

As discussed above for other coronavirus pp1a/pp1ab proteins, the RdRp domain also differs substantially from its homologs in other +RNA viruses. Coronavirus RdRps and their nidovirus relatives have been classified as an outgroup of SF1 RdRps (Koonin 1991). The coronavirus RdRp domain comprising the finger, palm, and thumb subdomains occupies the C-terminal two-thirds of nsp12 (Gorbalenya et al. 1989c). Recent data suggest that replication complex association of the RdRp may occur through interactions of the nsp12 segment 411–448 (located upstream of the RdRp core domain in nsp12) with ORF1a-encoded proteins, such as nsp5 (3CL^{pro}), nsp8, and nsp9 (Brockway et al. 2003). Consistent with the presumed RdRp activity of nsp12, a mutation in nsp12 (His868 to Arg) was found to cause an RNA-negative phenotype in an MHV ts mutant, Alb ts22 (Siddell et al. 2001). Thus, when infected cultures of Alb ts22 were shifted to the restrictive temperature at 40°C, both plus- and minus-strand RNA synthesis ceased immediately. Even at the permissive temperature, the ts mutant synthesized 4–5 times less RNA compared with revertants. The defect of this mutant in RNA synthesis can easily be explained by the fact that His868 is part of the predicted thumb subdomain of the MHV RdRp that, in other RNA polymerases, has been implicated in polymerase activity (Burns et al. 1989; Mills et al. 1989; Plotch et al. 1989; Hansen et al. 1997).

The Cys/His-rich nsp10 that immediately precedes RdRp in pp1ab (Fig. 1) has also been implicated in RNA synthesis. An MHV ts mutant, Alb ts6, encoding a mutant form of nsp10 (Gln65 to Glu), was shown to have a defect in minus-strand RNA synthesis (Siddell et al. 2001). Thus, when the temperature was shifted to 40°C, minus-strand synthesis stopped immediately but plus-strand synthesis continued at the same level as was occurring at the time of temperature shift. Plus-strand RNA synthesis gradually declined over 3–4 h (starting at 30–60 min after the shift to 40°C) because the minus strands produced at the permissive temperature were turned over (Wang and Sawicki 2001) and, because of the defect in their synthesis, were not replenished at the restrictive temperature.

Nsp10 and nsp12 (RdRp) are adjacent domains in pp1ab (Fig. 1). Peptide cleavage data have shown that, most likely because of a replacement of the conserved P2 Leu residue, the nsp10|nsp12 cleavage site is less efficiently cleaved than other SARS-CoV 3CL^{pro} sites (Fan et al. 2003). Also, the nsp10|nsp12 sites of other coronaviruses have the P2 position occupied by noncanonical residues. It is thus tempting to speculate that the nsp10|nsp12 site has to be cleaved more slowly than other sites, probably to attain a specific activity mediated by an nsp10–nsp12-containing intermediate. The IBV nsp10 has been reported to form dimers. It localizes to membranes near the site of viral RNA synthesis (Ng and Liu 2002).

4
Subcellular Localization of the Coronavirus Replicase

Genome replication and transcription of virtually all +RNA viruses takes place at intracellular membranes that are derived from various cellular organelles including, for example, the endoplasmic reticulum, lysosomes and endosomes, intermediate compartment and *trans*-Golgi network, peroxisomes, mitochondria, and chloroplasts (Russo et al. 1983; Froshauer et al. 1988; Peränen and Kääriäinen 1991; De Graaff et al. 1993; Peränen et al. 1995; Restrepo-Hartwig and Ahlquist 1996; Schaad et al. 1997; van der Meer et al. 1998; Mackenzie et al. 1999; Restrepo-Hartwig and Ahlquist 1999; Miller et al. 2001). The viral replication complex, which consists of multiple viral but also cellular subunits (see the chapter by Shi and Lai, this volume), is associated with these membranes and, in many cases, also directs their synthesis and/or modification (Peränen and Kääriäinen 1991; Cho et al. 1994; Schlegel et al. 1996; Teterina et al. 1997; Snijder et al. 2001; Egger et al. 2002). Typically, multiple vesicles or membrane invaginations (spherules) on cellular organelles are induced to which the replication complex is attached by specific structural elements, such as hydrophobic domains (van Kuppeveld et al. 1995; Snijder et al. 2001) amphipathic helices (Datta and Dasgupta 1994), palmitate side chains (Laakkonen et al. 1996), and C-terminal membrane insertion sequences (Schmidt-Mende et al. 2001). As a result, replication takes place in a membrane-protected (and, thus, nuclease resistant) microenvironment that contains (and sequesters) the protein functions required for viral RNA synthesis. This strategy is believed to improve template specificity by retaining negative strands for template use and to repress host defenses that may be induced by double-stranded RNA (Schwartz et al. 2002).

Association of the viral replication/transcription complex with intracellular membranes has also been established for coronaviruses (Sethna and Brian 1997). Thus TGEV genome- and subgenome-length minus strands, which are the templates for viral genome RNA replication and subgenomic mRNA transcription, respectively (Sethna et al. 1989; Sawicki and Sawicki 1990; Schaad and Baric 1994; Sawicki et al. 2001), were predominantly found in nuclease-resistant membranous complexes. In contrast, positive-strand RNAs proved to be much more susceptible to nuclease digestion, indicating that plus-strand RNAs, which also act as mRNAs, are mainly in solution or part of easily dissociable complexes in the cytosol (Sethna and Brian 1997).

Immunofluorescence (IF) studies provided clear evidence that the vast majority of coronavirus replicase subunits localize to perinuclear membrane compartments (Heusipp et al. 1997a; Bi et al. 1998; Denison et al. 1999; Shi et al. 1999; van der Meer et al. 1999; Ziebuhr and Siddell 1999; Bost et al. 2000; Sims et al. 2000; Bost et al. 2001; Xu et al. 2001; Ng and Liu 2002). Whereas most ORF1a-encoded replicase components remain tightly associated with membranes throughout the viral life cycle, at least some of the ORF1b-encoded subunits seem to be only temporarily present in the complex, probably when still part of the polyprotein. Thus, for example, partial detachment from the membrane-bound complexes was reported for MHV nsp12 and nsp13 later in infection (van der Meer et al. 1999; Bost et al. 2001; Xu et al. 2001). Also, the most C-terminal IBV pp1ab processing products show, in contrast to all other IBV pp1a/pp1ab proteins tested, a diffuse, cytoplasmic staining pattern in IF experiments (van der Meer et al. 1999; Bost et al. 2001; Xu et al. 2001). The membrane-bound replicase proteins overlap to a large extent with the site of viral RNA synthesis (Denison et al. 1999; Shi et al. 1999; van der Meer et al. 1999; Bost et al. 2001; Gosert et al. 2002; Ng and Liu 2002). There is some controversy regarding the intracellular compartment at which viral RNA synthesis takes place and, in particular, the cellular origin of the membranes employed. In a recent EM study (Gosert et al. 2002), virus-induced double membrane vesicles (DMVs) were reported to be the site of MHV-A59 replication and transcription in HeLa-MHVR (Gallagher 1996) and 17CL-1 cells. These DMVs have a diameter of 200–350 nm and consist of a double membrane that, occasionally, is fused into a trilayer. At the time of maximum RNA synthesis, both genome- and subgenome-length positive-strand RNA was detected on DMVs by in situ hybridization, and also the results of BrUTP labeling suggest that DMVs are the site of viral RNA synthesis. The subcellular origin of the DMVs has not been determined to date. However, a previous IF study (Shi et al. 1999) using MHV-A59-infected 17CL-1 and HeLa-MHVR cells suggested that N-terminal pp1a/pp1ab proteins and newly synthesized RNA colocalize with ER- or Golgi-derived membranes, depending on the cell type studied.

In clear contrast to these results, another study revealed that, in MHV-A59-infected L cells at 5 h p.i., the C-terminal pp1a region (CT1a), 3CL^{pro} (nsp5), RdRp (nsp12), helicase (nsp13), and the N protein are associated with virus-induced, late endosomal/lysosomal membranes, which were confirmed to be the site of RNA synthesis (van der Meer et al. 1999). In IF experiments, the sites of maximum CT1a accumulation overlapped only partially with those of nsp5, nsp12, and nsp13. A thorough EM study suggested that the low (albeit significant) degree of colocalization of CT1a and nsp12 is probably due to the existence of two distinct types of membrane structures that are closely adjacent to each other but have different morphologies and protein compositions. Thus CT1a was found to be associated mainly with endosomes, whereas the majority of nsp12 was associated with multilayered membranes, probably originating from invaginations on continuous membrane sheets. The latter structures were morphologically reminiscent of endocytic carrier vesicles (ECVs) or multivesicular bodies (MVBs). However, the fact that many of these structures had membrane continuities to late endosomes argues against typical ECVs and rather favors the idea that both the multivesicular (carrying the bulk of CT1a) and multilayered (carrying the bulk of nsp12) structures represent different subdomains of the same endocytic compartment. Most intriguingly, it has also been found (van der Meer et al. 1999) that CT1a- and nsp12-positive membranes appear to be secreted. Similar observations have also been reported recently for endosome-derived cytoplasmic vacuoles carrying the alphavirus replication complex (Kujala et al. 2001). The functional significance of this phenomenon is currently unclear but may have parallels in the regulated lysosomal secretion systems employed by, for example, lymphocytes (Stinchcombe and Griffiths 1999).

The existence of two closely associated but physically distinct membrane compartments was also shown by iodixanol gradient centrifugation of intracellular membranes isolated from MHV-A59-infected DBT cells (Sims et al. 2000). The ORF1a-encoded proteins nsp2 (p65) and nsp8 (p22) cofractionated with membranes with a buoyant density of 1.05–1.09 g/ml. In contrast, nsp13, the N protein, nsp1 (p28), and newly synthesized RNA were detected in another membrane fraction of 1.12–1.13 g/ml. Both membrane fractions were LAMP-1 positive, confirming previous conclusions on the endosomal/lysosomal origin of the MHV replication compartment. Interestingly, later in infection, there appears to be a translocation of nsp13 and the N protein to the ER/*cis*-Golgi compartment, resulting in colocalization of these two proteins with the M protein at the site of virion assembly (Bost et al. 2001). The combined data suggest a multipartite structure of the coronavirus replication complex, with the N protein playing a specific role in RNA synthesis as suggested earlier (Compton et al. 1987; Baric et al. 1988). Apparently, the coronavirus replication complex undergoes structural rearrangements at the transition from maximum RNA synthesis to virion assembly at later time points (8–12 h p.i.). If this is confirmed, the localization of nsp13 at the site of assembly may correspond with a specific role of nsp13 in virion biogenesis. Such an activity has also been proposed for the related arterivirus nsp10 helicase (van Dinten et al. 1999, 2000; Seybert et al. 2000b).

To date, the mechanisms by which components of the coronavirus replication complex are integrated in or attached to intracellular membranes have not been elucidated in detail. However, it seems very likely that the strongly hydrophobic domains, TM1 to TM3 (see Fig. 1), that are present in nsp3, nsp4, and nsp6 (Gorbalenya et al. 1989c; Ziebuhr et al. 2001) play a major role in this process. This hypothesis is supported by arterivirus data showing that homologous hydrophobic domains present in EAV nsp2 and nsp3 are necessary and sufficient to trigger the synthesis of the membrane structures carrying the arterivirus replication complex (Pedersen et al. 1999; Snijder et al. 2001). The fact that several MHV pp1a/pp1ab processing products including nsp3 (Gosert et al. 2002) and nsp4–10(11) (Schiller et al. 1998), which contain TM1 and TM2/TM3, respectively, are integral membrane proteins strongly suggests a scaffold function for these proteins. There is also biochemical evidence indicating that the majority of ORF1a-encoded proteins and, to a lesser extent, ORF1b-encoded proteins are tightly bound in the complex (Gosert et al. 2002). The precise protein-protein and protein-RNA interactions stabilizing this complex remain to be characterized.

5
Concluding Remarks

Although much has been learned about coronavirus replicase organization, localization, proteolytic processing, and some of the viral replicative enzymes (e.g., proteinases and helicases), there are still major gaps in our knowledge. Given the availability of full-length clones of coronaviruses, directed genetic analysis is now possible (Almazán et al. 2000; Yount et al. 2000; Casais et al. 2001; Thiel et al. 2001a; Yount et al. 2002, 2003). In vivo studies as well as biochemical and structural information should yield important new information on the molecular details of coronaviral RNA synthesis. In this context, it will be of particular interest to define the proteins that are responsible for the unique features of coronavirus RNA synthesis, for example, the production of an extensive set of 5'- and 3'-coterminal subgenomic RNAs and the synthesis and maintenance of RNA genomes of this unique size. Studies on coronavirus replicases and their homologs on closely related viruses may also help to determine the structural and functional constraints that have driven the evolution of nidoviruses and enable them to infect a broad range of vertebrate and invertebrate hosts. Furthermore, the relationship of the recently identified coronavirus RNA processing activities with cellular proteins may reveal interesting insights into similarities and differences (or even an interplay) between coronaviral and cellular RNA metabolism pathways. In the long term, the unique structural properties of coronavirus replicative enzymes may allow the development of very selective enzyme inhibitors and possibly even drugs suitable to combat coronavirus infections.

References

Allaire M, Chernaia MM, Malcolm BA, James MN (1994) Picornaviral 3C cysteine proteinases have a fold similar to chymotrypsin-like serine proteinases. Nature 369:72–76

Almazán F, González JM, Pénzes Z, Izeta A, Calvo E, Plana-Durán J, Enjuanes L (2000) Engineering the largest RNA virus genome as an infectious bacterial artificial chromosome. Proc Natl Acad Sci USA 97:5516–5521

Anand K, Palm GJ, Mesters JR, Siddell SG, Ziebuhr J, Hilgenfeld R (2002) Structure of coronavirus main proteinase reveals combination of a chymotrypsin fold with an extra alpha-helical domain. EMBO J 21:3213–3224

Anand K, Ziebuhr J, Wadhwani P, Mesters JR, Hilgenfeld R (2003) Coronavirus main proteinase (3CL^{pro}) structure: basis for design of anti-SARS drugs. Science 300:1763–1767

Baker SC, Shieh CK, Soe LH, Chang MF, Vannier DM, Lai MM (1989) Identification of a domain required for autoproteolytic cleavage of murine coronavirus gene A polyprotein. J Virol 63:3693–3699

Baker SC, Yokomori K, Dong S, Carlisle R, Gorbalenya AE, Koonin EV, Lai MM (1993) Identification of the catalytic sites of a papain-like cysteine proteinase of murine coronavirus. J Virol 67:6056–6063

Baric RS, Nelson GW, Fleming JO, Deans RJ, Keck JG, Casteel N, Stohlman SA (1988) Interactions between coronavirus nucleocapsid protein and viral RNAs: implications for viral transcription. J Virol 62:4280–4287

Barrette-Ng IH, Ng KK, Mark BL, Van Aken D, Cherney MM, Garen C, Kolodenko Y, Gorbalenya AE, Snijder EJ, James MN (2002) Structure of arterivirus nsp4. The smallest chymotrypsin-like proteinase with an alpha/beta C-terminal extension and alternate conformations of the oxyanion hole. J Biol Chem 277:39960–39966

Bautista EM, Faaberg KS, Mickelson D, McGruder ED (2002) Functional properties of the predicted helicase of porcine reproductive and respiratory syndrome virus. Virology 298:258–270

Bergmann EM, Mosimann SC, Chernaia MM, Malcolm BA, James MN (1997) The refined crystal structure of the 3C gene product from hepatitis A virus: specific proteinase activity and RNA recognition. J Virol 71:2436–2448

Bi W, Piñon JD, Hughes S, Bonilla PJ, Holmes KV, Weiss SR, Leibowitz JL (1998) Localization of mouse hepatitis virus open reading frame 1A derived proteins. J Neurovirol 4:594–605

Blom N, Hansen J, Blaas D, Brunak S (1996) Cleavage site analysis in picornaviral polyproteins: discovering cellular targets by neural networks. Protein Sci 5:2203–2216

Bonilla PJ, Gorbalenya AE, Weiss SR (1994) Mouse hepatitis virus strain A59 RNA polymerase gene ORF 1a: heterogeneity among MHV strains. Virology 198:736–740

Bonilla PJ, Hughes SA, Weiss SR (1997) Characterization of a second cleavage site and demonstration of activity in trans by the papain-like proteinase of the murine coronavirus mouse hepatitis virus strain A59. J Virol 71:900–909

Bost AG, Carnahan RH, Lu XT, Denison MR (2000) Four proteins processed from the replicase gene polyprotein of mouse hepatitis virus colocalize in the cell periphery and adjacent to sites of virion assembly. J Virol 74:3379–3387

Bost AG, Prentice E, Denison MR (2001) Mouse hepatitis virus replicase protein complexes are translocated to sites of M protein accumulation in the ERGIC at late times of infection. Virology 285:21–29

Boursnell ME, Brown TD, Foulds IJ, Green PF, Tomley FM, Binns MM (1987) Completion of the sequence of the genome of the coronavirus avian infectious bronchitis virus. J Gen Virol 68:57–77

Bredenbeek PJ, Pachuk CJ, Noten AF, Charite J, Luytjes W, Weiss SR, Spaan WJ (1990) The primary structure and expression of the second open reading frame of the polymerase gene of the coronavirus MHV-A59; a highly conserved polymerase is expressed by an efficient ribosomal frameshifting mechanism. Nucleic Acids Res 18:1825–1832

Brierley I, Boursnell ME, Binns MM, Bilimoria B, Blok VC, Brown TD, Inglis SC (1987) An efficient ribosomal frame-shifting signal in the polymerase-encoding region of the coronavirus IBV. EMBO J 6:3779–3785

Brierley I, Digard P, Inglis SC (1989) Characterization of an efficient coronavirus ribosomal frameshifting signal: requirement for an RNA pseudoknot. Cell 57:537–547

Brockway SM, Clay CT, Lu XT, Denison MR (2003) Characterization of the expression, intracellular localization, and replication complex association of the putative mouse hepatitis virus RNA-dependent RNA polymerase. J Virol 77:10515–10527

Buck KW (1996) Comparison of the replication of positive-stranded RNA viruses of plants and animals. Adv Virus Res 47:159–251

Bügl H, Fauman EB, Staker BL, Zheng F, Kushner SR, Saper MA, Bardwell JC, Jakob U (2000) RNA methylation under heat shock control. Mol Cell 6:349–360

Burns CC, Lawson MA, Semler BL, Ehrenfeld E (1989) Effects of mutations in poliovirus 3Dpol on RNA polymerase activity and on polyprotein cleavage. J Virol 63:4866–4874

Casais R, Thiel V, Siddell SG, Cavanagh D, Britton P (2001) Reverse genetics system for the avian coronavirus infectious bronchitis virus. J Virol 75:12359–12369

Cavanagh D (1997) Nidovirales: a new order comprising Coronaviridae and Arteriviridae. Arch Virol 142:629–633

Cho MW, Teterina N, Egger D, Bienz K, Ehrenfeld E (1994) Membrane rearrangement and vesicle induction by recombinant poliovirus 2C and 2BC in human cells. Virology 202:129–145

Chouljenko VN, Lin XQ, Storz J, Kousoulas KG, Gorbalenya AE (2001) Comparison of genomic and predicted amino acid sequences of respiratory and enteric bovine coronaviruses isolated from the same animal with fatal shipping pneumonia. J Gen Virol 82:2927–2933

Compton SR, Rogers DB, Holmes KV, Fertsch D, Remenick J, McGowan JJ (1987) In vitro replication of mouse hepatitis virus strain A59. J Virol 61:1814–1820

Cowley JA, Dimmock CM, Spann KM, Walker PJ (2000) Gill-associated virus of *Penaeus monodon* prawns: an invertebrate virus with ORF1a and ORF1b genes related to arteri- and coronaviruses. J Gen Virol 81:1473–1484

Culver GM, Consaul SA, Tycowski KT, Filipowicz W, Phizicky EM (1994) tRNA splicing in yeast and wheat germ. A cyclic phosphodiesterase implicated in the metabolism of ADP-ribose 1″,2″-cyclic phosphate. J Biol Chem 269:24928–24934

Datta U, Dasgupta A (1994) Expression and subcellular localization of poliovirus VPg-precursor protein 3AB in eukaryotic cells: evidence for glycosylation in vitro. J Virol 68:4468–4477

De Graaff M, Coscoy L, Jaspars EM (1993) Localization and biochemical characterization of alfalfa mosaic virus replication complexes. Virology 194:878–881

de Vries AAF, Horzinek MC, Rottier PJM, de Groot RJ (1997) The genome organization of the Nidovirales: similarities and differences between arteri-, toro-, and coronaviruses. Sem Virol 8:33–47

den Boon JA, Snijder EJ, Chirnside ED, de Vries AA, Horzinek MC, Spaan WJ (1991) Equine arteritis virus is not a togavirus but belongs to the coronaviruslike superfamily. J Virol 65:2910–2920

Denison MR, Hughes SA, Weiss SR (1995) Identification and characterization of a 65-kDa protein processed from the gene 1 polyprotein of the murine coronavirus MHV-A59. Virology 207:316–320

Denison MR, Spaan WJ, van der Meer Y, Gibson CA, Sims AC, Prentice E, Lu XT (1999) The putative helicase of the coronavirus mouse hepatitis virus is processed from the replicase gene polyprotein and localizes in complexes that are active in viral RNA synthesis. J Virol 73:6862–6871

Dong S, Baker SC (1994) Determinants of the p28 cleavage site recognized by the first papain-like cysteine proteinase of murine coronavirus. Virology 204:541–549

Dougherty WG, Semler BL (1993) Expression of virus-encoded proteinases: functional and structural similarities with cellular enzymes. Microbiol Rev 57:781–822

Egger D, Wölk B, Gosert R, Bianchi L, Blum HE, Moradpour D, Bienz K (2002) Expression of hepatitis C virus proteins induces distinct membrane alterations including a candidate viral replication complex. J Virol 76:5974–5984

Eleouet JF, Rasschaert D, Lambert P, Levy L, Vende P, Laude H (1995) Complete sequence (20 kilobases) of the polyprotein-encoding gene 1 of transmissible gastroenteritis virus. Virology 206:817–822

Fan K, Wei P, Feng Q, Chen S, Huang C, Ma L, Lai B, Pei J, Liu Y, Chen J, Lai L (2003) Biosynthesis, purification, and substrate specificity of severe acute respiratory syndrome coronavirus 3C-like proteinase. J Biol Chem 279:1637–1642

Filipowicz W, Pogacic V (2002) Biogenesis of small nucleolar ribonucleoproteins. Curr Opin Cell Biol 14:319–327

Froshauer S, Kartenbeck J, Helenius A (1988) Alphavirus RNA replicase is located on the cytoplasmic surface of endosomes and lysosomes. J Cell Biol 107:2075–2086

Gallagher TM (1996) Murine coronavirus membrane fusion is blocked by modification of thiols buried within the spike protein. J Virol 70:4683–4690

Goldbach R (1987) Genome similarities between plant and animal RNA viruses. Microbiol Sci 4:197–202

Gorbalenya AE, Donchenko AP, Blinov VM, Koonin EV (1989a) Cysteine proteases of positive strand RNA viruses and chymotrypsin-like serine proteases. A distinct protein superfamily with a common structural fold. FEBS Lett 243:103–114

Gorbalenya AE, Koonin EV, Donchenko AP, Blinov VM (1989b) Two related superfamilies of putative helicases involved in replication, recombination, repair and expression of DNA and RNA genomes. Nucleic Acids Res 17:4713–4730

Gorbalenya AE, Koonin EV, Donchenko AP, Blinov VM (1989c) Coronavirus genome: prediction of putative functional domains in the non-structural polyprotein by comparative amino acid sequence analysis. Nucleic Acids Res 17:4847–4861

Gorbalenya AE, Koonin EV, Lai MM (1991) Putative papain-related thiol proteases of positive-strand RNA viruses. Identification of rubi- and aphthovirus proteases and delineation of a novel conserved domain associated with proteases of rubi-, alpha- and coronaviruses. FEBS Lett 288:201–205

Gorbalenya AE, Koonin EV (1993) Helicases: amino acid sequence comparisons and structure-function relationships. Curr Opin Struct Biol 3:419–429

Gorbalenya AE, Snijder EJ (1996) Viral cysteine proteinases. Persp Drug Discov Des 6:64–86
Gorbalenya AE (2001) Big nidovirus genome. When count and order of domains matter. Adv Exp Med Biol 494:1-17
Gosert R, Kanjanahaluethai A, Egger D, Bienz K, Baker SC (2002) RNA replication of mouse hepatitis virus takes place at double-membrane vesicles. J Virol 76:3697–3708
Grötzinger C, Heusipp G, Ziebuhr J, Harms U, Süss J, Siddell SG (1996) Characterization of a 105-kDa polypeptide encoded in gene 1 of the human coronavirus HCV 229E. Virology 222:227–235
Guarné A, Tormo J, Kirchweger R, Pfistermueller D, Fita I, Skern T (1998) Structure of the foot-and-mouth disease virus leader protease: a papain-like fold adapted for self-processing and eIF4G recognition. EMBO J 17:7469–7479
Guarné A, Hampoelz B, Glaser W, Carpena X, Tormo J, Fita I, Skern T (2000) Structural and biochemical features distinguish the foot-and-mouth disease virus leader proteinase from other papain-like enzymes. J Mol Biol 302:1227–240
Hansen JL, Long AM, Schultz SC (1997) Structure of the RNA-dependent RNA polymerase of poliovirus. Structure 5:1109–1122
Hegyi A, Friebe A, Gorbalenya AE, Ziebuhr J (2002) Mutational analysis of the active centre of coronavirus 3C-like proteases. J Gen Virol 83:581–593
Hegyi A, Ziebuhr J (2002) Conservation of substrate specificities among coronavirus main proteases. J Gen Virol 83:595–599
Herold J, Raabe T, Schelle-Prinz B, Siddell SG (1993) Nucleotide sequence of the human coronavirus 229E RNA polymerase locus. Virology 195:680–691
Herold J, Siddell SG (1993) An 'elaborated' pseudoknot is required for high frequency frameshifting during translation of HCV 229E polymerase mRNA. Nucleic Acids Res 21:5838–5842
Herold J, Gorbalenya AE, Thiel V, Schelle B, Siddell SG (1998) Proteolytic processing at the amino terminus of human coronavirus 229E gene 1-encoded polyproteins: identification of a papain-like proteinase and its substrate. J Virol 72:910–918
Herold J, Siddell SG, Gorbalenya AE (1999) A human RNA viral cysteine proteinase that depends upon a unique Zn^{2+}-binding finger connecting the two domains of a papain-like fold. J Biol Chem 274:14918–14925
Heusipp G, Grötzinger C, Herold J, Siddell SG, Ziebuhr J (1997a) Identification and subcellular localization of a 41 kDa, polyprotein 1ab processing product in human coronavirus 229E-infected cells. J Gen Virol 78:2789–2794
Heusipp G, Harms U, Siddell SG, Ziebuhr J (1997b) Identification of an ATPase activity associated with a 71-kilodalton polypeptide encoded in gene 1 of the human coronavirus 229E. J Virol 71:5631–5634
Hughes SA, Bonilla PJ, Weiss SR (1995) Identification of the murine coronavirus p28 cleavage site. J Virol 69:809–813
Ivanov KA, Thiel V, Dobbe JC, van der Meer Y, Snijder EJ, Ziebuhr J (2004) Multiple enzymatic activities associated with severe acute respiratory syndrome coronavirus helicase. J Virol 78:5619–5632
Ivanov KA, Ziebuhr J (2004) Human coronavirus nonstructural protein 13: characterization of duplex-unwinding, (deoxy)nucleoside triphosphatase, and RNA 5′-triphosphatase activities. J Virol 78:7833–7838

Kadaré G, Haenni AL (1997) Virus-encoded RNA helicases. J Virol 71:2583–2590
Kanjanahaluethai A, Baker SC (2000) Identification of mouse hepatitis virus papain-like proteinase 2 activity. J Virol 74:7911–7921
Kanjanahaluethai A, Jukneliene D, Baker SC (2003) Identification of the murine coronavirus MP1 cleavage site recognized by papain-like proteinase 2. J Virol 77:7376–7382
Khan AR, Khazanovich-Bernstein N, Bergmann EM, James MN (1999) Structural aspects of activation pathways of aspartic protease zymogens and viral 3C protease precursors. Proc Natl Acad Sci USA 96:10968–10975
Kim JC, Spence RA, Currier PF, Lu X, Denison MR (1995) Coronavirus protein processing and RNA synthesis is inhibited by the cysteine proteinase inhibitor E64d. Virology 208:1–8
Kiss T (2001) Small nucleolar RNA-guided post-transcriptional modification of cellular RNAs. EMBO J 20:3617–3622
Kocherhans R, Bridgen A, Ackermann M, Tobler K (2001) Completion of the porcine epidemic diarrhoea coronavirus (PEDV) genome sequence. Virus Genes 23:137–144
Koonin EV (1991) The phylogeny of RNA-dependent RNA polymerases of positive-strand RNA viruses. J Gen Virol 72:2197–2206
Koonin EV, Dolja VV (1993) Evolution and taxonomy of positive-strand RNA viruses: implications of comparative analysis of amino acid sequences. Crit Rev Biochem Mol Biol 28:375–430
Kräusslich HG, Wimmer E (1988) Viral proteinases. Annu Rev Biochem 57:701–754
Kujala P, Ikäheimonen A, Ehsani N, Vihinen H, Auvinen P, Kääriäinen L (2001) Biogenesis of the Semliki Forest virus RNA replication complex. J Virol 75:3873–3884
Kwong AD, Kim JL, Lin C (2000) Structure and function of hepatitis C virus NS3 helicase. Curr Top Microbiol Immunol 242:171–196
Laakkonen P, Ahola T, Kääriäinen L (1996) The effects of palmitoylation on membrane association of Semliki forest virus RNA capping enzyme. J Biol Chem 271:28567–28571
Lai MM, Patton CD, Baric RS, Stohlman SA (1983) Presence of leader sequences in the mRNA of mouse hepatitis virus. J Virol 46:1027–1033
Lai MM, Cavanagh D (1997) The molecular biology of coronaviruses. Adv Virus Res 48:1–10
Laneve P, Altieri F, Fiori ME, Scaloni A, Bozzoni I, Caffarelli E (2003) Purification, cloning, and characterization of XendoU, a novel endoribonuclease involved in processing of intron-encoded small nucleolar RNAs in *Xenopus laevis*. J Biol Chem 278:13026–13032
Lee HJ, Shieh CK, Gorbalenya AE, Koonin EV, La Monica N, Tuler J, Bagdzhadzhyan A, Lai MM (1991) The complete sequence (22 kilobases) of murine coronavirus gene 1 encoding the putative proteases and RNA polymerase. Virology 180:567–582
Lemm JA, Rümenapf T, Strauss EG, Strauss JH, Rice CM (1994) Polypeptide requirements for assembly of functional Sindbis virus replication complexes: a model for the temporal regulation of minus- and plus-strand RNA synthesis. EMBO J 13:2925–2934

Lim KP, Liu DX (1998) Characterization of the two overlapping papain-like proteinase domains encoded in gene 1 of the coronavirus infectious bronchitis virus and determination of the C-terminal cleavage site of an 87-kDa protein. Virology 245:303–312

Lim KP, Ng LF, Liu DX (2000) Identification of a novel cleavage activity of the first papain-like proteinase domain encoded by open reading frame 1a of the coronavirus avian infectious bronchitis virus and characterization of the cleavage products. J Virol 74:1674–1685

Liu C, Xu HY, Liu DX (2001) Induction of caspase-dependent apoptosis in cultured cells by the avian coronavirus infectious bronchitis virus. J Virol 75:6402–6409

Liu DX, Brown TD (1995) Characterisation and mutational analysis of an ORF 1a-encoding proteinase domain responsible for proteolytic processing of the infectious bronchitis virus 1a/1b polyprotein. Virology 209:420–427

Lu X, Lu Y, Denison MR (1996) Intracellular and in vitro-translated 27-kDa proteins contain the 3C-like proteinase activity of the coronavirus MHV-A59. Virology 222:375–382

Lu Y, Denison MR (1997) Determinants of mouse hepatitis virus 3C-like proteinase activity. Virology 230:335–342

Mackenzie JM, Jones MK, Westaway EG (1999) Markers for trans-Golgi membranes and the intermediate compartment localize to induced membranes with distinct replication functions in flavivirus-infected cells. J Virol 73:9555–9567

Marra MA, Jones SJ, Astell CR, Holt RA, Brooks-Wilson A, Butterfield YS, Khattra J, Asano JK, Barber SA, Chan SY, Cloutier A, Coughlin SM, Freeman D, Girn N, Griffith OL, Leach SR, Mayo M, McDonald H, Montgomery SB, Pandoh PK, Petrescu AS, Robertson AG, Schein JE, Siddiqui A, Smailus DE, Stott JM, Yang GS, Plummer F, Andonov A, Artsob H, Bastien N, Bernard K, Booth TF, Bowness D, Czub M, Drebot M, Fernando L, Flick R, Garbutt M, Gray M, Grolla A, Jones S, Feldmann H, Meyers A, Kabani A, Li Y, Normand S, Stroher U, Tipples GA, Tyler S, Vogrig R, Ward D, Watson B, Brunham RC, Krajden M, Petric M, Skowronski DM, Upton C, Roper RL (2003) The genome sequence of the SARS-associated coronavirus. Science 300:1399–1404

Martzen MR, McCraith SM, Spinelli SL, Torres FM, Fields S, Grayhack EJ, Phizicky EM (1999) A biochemical genomics approach for identifying genes by the activity of their products. Science 286:1153–1155

Matthews DA, Smith WW, Ferre RA, Condon B, Budahazi G, Sisson W, Villafranca JE, Janson CA, McElroy HE, Gribskov CL, et al. (1994) Structure of human rhinovirus 3C protease reveals a trypsin-like polypeptide fold, RNA-binding site, and means for cleaving precursor polyprotein. Cell 77:761–771

Miller DJ, Schwartz MD, Ahlquist P (2001) Flock house virus RNA replicates on outer mitochondrial membranes in *Drosophila* cells. J Virol 75:11664–11676

Mills DR, Priano C, DiMauro P, Binderow BD (1989) Q beta replicase: mapping the functional domains of an RNA-dependent RNA polymerase. J Mol Biol 205:751–764

Mosimann SC, Cherney MM, Sia S, Plotch S, James MN (1997) Refined X-ray crystallographic structure of the poliovirus 3C gene product. J Mol Biol 273:1032–1047

Nasr F, Filipowicz W (2000) Characterization of the *Saccharomyces cerevisiae* cyclic nucleotide phosphodiesterase involved in the metabolism of ADP-ribose 1″,2″-cyclic phosphate. Nucleic Acids Res 28:1676–1683

Ng LF, Liu DX (2000) Further characterization of the coronavirus infectious bronchitis virus 3C-like proteinase and determination of a new cleavage site. Virology 272:27–39

Ng LF, Liu DX (2002) Membrane association and dimerization of a cysteine-rich, 16-kilodalton polypeptide released from the C-terminal region of the coronavirus infectious bronchitis virus 1a polyprotein. J Virol 76:6257–6267

Pang PS, Jankowsky E, Planet PJ, Pyle AM (2002) The hepatitis C viral NS3 protein is a processive DNA helicase with cofactor enhanced RNA unwinding. EMBO J 21:1168–1176

Pedersen KW, van der Meer Y, Roos N, Snijder EJ (1999) Open reading frame 1a-encoded subunits of the arterivirus replicase induce endoplasmic reticulum-derived double-membrane vesicles which carry the viral replication complex. J Virol 73:2016–2026

Penzes Z, González JM, Calvo E, Izeta A, Smerdou C, Mendez A, Sánchez CM, Sola I., Almazán F, Enjuanes L (2001) Complete genome sequence of transmissible gastroenteritis coronavirus PUR46-MAD clone and evolution of the Purdue virus cluster. Virus Genes 23:105–118

Peränen J, Kääriäinen L (1991) Biogenesis of type I cytopathic vacuoles in Semliki Forest virus-infected BHK cells. J Virol 65:1623–1627

Peränen J, Laakkonen P, Hyvönen M, Kääriäinen L (1995) The alphavirus replicase protein nsP1 is membrane-associated and has affinity to endocytic organelles. Virology 208:610–620

Piñon JD, Mayreddy RR, Turner JD, Khan FS, Bonilla PJ, Weiss SR (1997) Efficient autoproteolytic processing of the MHV-A59 3C-like proteinase from the flanking hydrophobic domains requires membranes. Virology 230:309–322

Piñon JD, Teng H, Weiss SR (1999) Further requirements for cleavage by the murine coronavirus 3C-like proteinase: identification of a cleavage site within ORF1b. Virology 263:471–484

Plotch SJ, Palant O, Gluzman Y (1989) Purification and properties of poliovirus RNA polymerase expressed in *Escherichia coli*. J Virol 63:216–225

Restrepo-Hartwig M, Ahlquist P (1999) Brome mosaic virus RNA replication proteins 1a and 2a colocalize and 1a independently localizes on the yeast endoplasmic reticulum. J Virol 73:10303–10309

Restrepo-Hartwig MA, Ahlquist P (1996) Brome mosaic virus helicase- and polymerase-like proteins colocalize on the endoplasmic reticulum at sites of viral RNA synthesis. J Virol 70:8908–8916

Rota PA, Oberste MS, Monroe SS, Nix WA, Campagnoli R, Icenogle JP, Penaranda S, Bankamp B, Maher K, Chen MH, Tong S, Tamin A, Lowe L, Frace M, DeRisi JL, Chen Q, Wang D, Erdman DD, Peret TC, Burns C, Ksiazek TG, Rollin PE, Sanchez A, Liffick S, Holloway B, Limor J, McCaustland K, Olsen-Rasmussen M, Fouchier R, Gunther S, Osterhaus AD, Drosten C, Pallansch MA, Anderson LJ, Bellini WJ (2003) Characterization of a novel coronavirus associated with severe acute respiratory syndrome. Science 300:1394–1399

Ruan YJ, Wei CL, Ee AL, Vega VB, Thoreau H, Su ST, Chia JM, Ng P, Chiu KP, Lim L, Zhang T, Peng CK, Lin EO, Lee NM, Yee SL, Ng LF, Chee RE, Stanton LW, Long PM, Liu ET (2003) Comparative full-length genome sequence analysis of 14 SARS coronavirus isolates and common mutations associated with putative origins of infection. Lancet 361:1779–1785

Russo M, Di Franco A, Martelli GP (1983) The fine structure of *Cymbidium* ringspot virus infections in host tissues. III. Role of peroxisomes in the genesis of multivesicular bodies. J Ultrastruct Res 82:52–63

Ryan MD, Flint M (1997) Virus-encoded proteinases of the picornavirus supergroup. J Gen Virol 78:699–723

Sawicki D, Wang T, Sawicki S (2001) The RNA structures engaged in replication and transcription of the A59 strain of mouse hepatitis virus. J Gen Virol 82:385–396

Sawicki SG, Sawicki DL (1990) Coronavirus transcription: subgenomic mouse hepatitis virus replicative intermediates function in RNA synthesis. J Virol 64:1050–1056

Schaad MC, Baric RS (1994) Genetics of mouse hepatitis virus transcription: evidence that subgenomic negative strands are functional templates. J Virol 68:8169–8179

Schaad MC, Jensen PE, Carrington JC (1997) Formation of plant RNA virus replication complexes on membranes: role of an endoplasmic reticulum-targeted viral protein. EMBO J 16:4049–4059

Schiller JJ, Kanjanahaluethai A, Baker SC (1998) Processing of the coronavirus MHV-JHM polymerase polyprotein: identification of precursors and proteolytic products spanning 400 kilodaltons of ORF1a. Virology 242:288–302

Schlegel A, Giddings TH, Jr., Ladinsky MS, Kirkegaard K (1996) Cellular origin and ultrastructure of membranes induced during poliovirus infection. J Virol 70:6576–6588

Schmidt-Mende J, Bieck E, Hügle T, Penin F, Rice CM, Blum HE, Moradpour D (2001) Determinants for membrane association of the hepatitis C virus RNA-dependent RNA polymerase. J Biol Chem 276:44052–44063

Schwartz M, Chen J, Janda M, Sullivan M, den Boon J, Ahlquist P (2002) A positive-strand RNA virus replication complex parallels form and function of retrovirus capsids. Mol Cell 9:505–514

Seipelt J, Guarne A, Bergmann E, James M, Sommergruber W, Fita I, Skern T (1999) The structures of picornaviral proteinases. Virus Res 62:159–168

Sethna PB, Hung SL, Brian DA (1989) Coronavirus subgenomic minus-strand RNAs and the potential for mRNA replicons. Proc Natl Acad Sci USA 86:5626–5630

Sethna PB, Brian DA (1997) Coronavirus genomic and subgenomic minus-strand RNAs copartition in membrane-protected replication complexes. J Virol 71:7744–7749

Seybert A, Hegyi A, Siddell SG, Ziebuhr J (2000a) The human coronavirus 229E superfamily 1 helicase has RNA and DNA duplex-unwinding activities with $5'$-to-$3'$ polarity. RNA 6:1056–1068

Seybert A, van Dinten LC, Snijder EJ, Ziebuhr J (2000b) Biochemical characterization of the equine arteritis virus helicase suggests a close functional relationship between arterivirus and coronavirus helicases. J Virol 74:9586–9593

Seybert A, Ziebuhr J (2001) Guanosine triphosphatase activity of the human coronavirus helicase. Adv Exp Med Biol 494:255–260

Shi ST, Schiller JJ, Kanjanahaluethai A, Baker SC, Oh JW, Lai MM (1999) Colocalization and membrane association of murine hepatitis virus gene 1 products and de novo-synthesized viral RNA in infected cells. J Virol 73:5957–5969

Siddell S, Sawicki D, Meyer Y, Thiel V, Sawicki S (2001) Identification of the mutations responsible for the phenotype of three MHV RNA-negative ts mutants. Adv Exp Med Biol 494:453–458

Siddell SG. (1995). The Coronaviridae: an introduction. In "The Coronaviridae" (Siddell SG, ed.), pp. 1–10. Plenum Press, New York.

Sims AC, Ostermann J, Denison MR (2000) Mouse hepatitis virus replicase proteins associate with two distinct populations of intracellular membranes. J Virol 74:5647–5654

Snijder EJ, den Boon JA, Bredenbeek PJ, Horzinek MC, Rijnbrand R, Spaan WJ (1990a) The carboxyl-terminal part of the putative Berne virus polymerase is expressed by ribosomal frameshifting and contains sequence motifs which indicate that toro- and coronaviruses are evolutionarily related. Nucleic Acids Res 18:4535–4542

Snijder EJ, Horzinek MC (1993) Toroviruses: replication, evolution and comparison with other members of the coronavirus-like superfamily. J Gen Virol 74:2305–2316

Snijder EJ, Meulenberg JJ (1998) The molecular biology of arteriviruses. J Gen Virol 79:961–979

Snijder EJ, van Tol H, Roos N, Pedersen KW (2001) Non-structural proteins 2 and 3 interact to modify host cell membranes during the formation of the arterivirus replication complex. J Gen Virol 82:985–994

Snijder EJ, Bredenbeek PJ, Dobbe JC, Thiel V, Ziebuhr J, Poon LL, Guan Y, Rozanov M, Spaan WJ, Gorbalenya AE (2003) Unique and conserved features of genome and proteome of SARS-coronavirus, an early split-off from the coronavirus group 2 lineage. J Mol Biol 331:991–1004

Spaan W, Delius H, Skinner M, Armstrong J, Rottier P, Smeekens S, van der Zeijst BA, Siddell SG (1983) Coronavirus mRNA synthesis involves fusion of non-contiguous sequences. EMBO J 2:1839–1844

Stinchcombe JC, Griffiths GM (1999) Regulated secretion from hemopoietic cells. J Cell Biol 147:1–6

Strauss JH, Strauss EG (1988) Evolution of RNA viruses. Annu Rev Microbiol 42:657–683

Tanner JA, Watt RM, Chai YB, Lu LY, Lin MC, Peiris JS, Poon LL, Kung HF, Huang JD (2003) The severe acute respiratory syndrome (SARS) coronavirus NTPase/helicase belongs to a distinct class of $5'$ to $3'$ viral helicases. J Biol Chem 278:39578–39582

Teng H, Piñon JD, Weiss SR (1999) Expression of murine coronavirus recombinant papain-like proteinase: efficient cleavage is dependent on the lengths of both the substrate and the proteinase polypeptides. J Virol 73:2658–2666

Teterina NL, Bienz K, Egger D, Gorbalenya AE, Ehrenfeld E (1997) Induction of intracellular membrane rearrangements by HAV proteins 2C and 2BC. Virology 237:66–77

Thiel V, Herold J, Schelle B, Siddell SG (2001a) Infectious RNA transcribed in vitro from a cDNA copy of the human coronavirus genome cloned in vaccinia virus. J Gen Virol 82:1273–1281

Thiel V, Herold J, Schelle B, Siddell SG (2001b) Viral replicase gene products suffice for coronavirus discontinuous transcription. J Virol 75:6676–6681

Thiel V, Ivanov KA, Putics A, Hertzig T, Schelle B, Bayer S, Weissbrich B, Snijder EJ, Rabenau H, Doerr HW, Gorbalenya AE, Ziebuhr J (2003) Mechanisms and enzymes involved in SARS coronavirus genome expression. J Gen Virol 84:2305–2315

Tibbles KW, Brierley I, Cavanagh D, Brown TD (1996) Characterization in vitro of an autocatalytic processing activity associated with the predicted 3C-like proteinase domain of the coronavirus avian infectious bronchitis virus. J Virol 70:1923–1930

Tibbles KW, Cavanagh D, Brown TD (1999) Activity of a purified His-tagged 3C-like proteinase from the coronavirus infectious bronchitis virus. Virus Res. 60:137–145

Tijms MA, van Dinten LC, Gorbalenya AE, Snijder EJ (2001) A zinc finger-containing papain-like protease couples subgenomic mRNA synthesis to genome translation in a positive-stranded RNA virus. Proc Natl Acad Sci USA 98:1889–1894

van der Meer Y, van Tol H, Krijnse Locker J, Snijder EJ (1998) ORF1a-encoded replicase subunits are involved in the membrane association of the arterivirus replication complex. J Virol 72:6689–6698

van der Meer Y, Snijder EJ, Dobbe JC, Schleich S, Denison MR, Spaan WJ, Krijnse Locker J (1999) Localization of mouse hepatitis virus nonstructural proteins and RNA synthesis indicates a role for late endosomes in viral replication. J Virol 73:7641–7657

van Dinten LC, Rensen S, Gorbalenya AE, Snijder EJ (1999) Proteolytic processing of the open reading frame 1b-encoded part of arterivirus replicase is mediated by nsp4 serine protease and Is essential for virus replication. J Virol 73:2027–2037

van Dinten LC, van Tol H, Gorbalenya AE, Snijder EJ (2000) The predicted metal-binding region of the arterivirus helicase protein is involved in subgenomic mRNA synthesis, genome replication, and virion biogenesis. J Virol 74:5213–5223

van Kuppeveld FJ, Galama JM, Zoll J, Melchers WJ (1995) Genetic analysis of a hydrophobic domain of coxsackie B3 virus protein 2B: a moderate degree of hydrophobicity is required for a *cis*-acting function in viral RNA synthesis. J Virol 69:7782–7790.

Vasiljeva L, Merits A, Golubtsov A, Sizemskaja V, Kaariainen L, Ahola T (2003) Regulation of the sequential processing of Semliki Forest virus replicase polyprotein. J Biol Chem 278:41636–41645

Wang T, Sawicki SG (2001) Mouse hepatitis virus minus-strand templates are unstable and turnover during viral replication. Adv Exp Med Biol 494:491–497

Xu HY, Lim KP, Shen S, Liu DX (2001) Further identification and characterization of novel intermediate and mature cleavage products released from the ORF 1b region of the avian coronavirus infectious bronchitis virus 1a/1b polyprotein. Virology 288:212–222

Yount B, Curtis KM, Baric RS (2000) Strategy for systematic assembly of large RNA and DNA genomes: transmissible gastroenteritis virus model. J Virol 74:10600–10611

Yount B, Denison MR, Weiss SR, Baric RS (2002) Systematic assembly of a full-length infectious cDNA of mouse hepatitis virus strain A59. J Virol 76:11065–11078

Yount B, Curtis KM, Fritz EA, Hensley LE, Jahrling PB, Prentice E, Denison MR, Geisbert TW, Baric RS (2003) Reverse genetics with a full-length infectious cDNA of severe acute respiratory syndrome coronavirus. Proc Natl Acad Sci USA 100:12995–13000

Ziebuhr J, Herold J, Siddell SG (1995) Characterization of a human coronavirus (strain 229E) 3C-like proteinase activity. J Virol 69:4331–4338

Ziebuhr J, Heusipp G, Siddell SG (1997) Biosynthesis, purification, and characterization of the human coronavirus 229E 3C-like proteinase. J Virol 71:3992–3997

Ziebuhr J, Siddell SG (1999) Processing of the human coronavirus 229E replicase polyproteins by the virus-encoded 3C-like proteinase: identification of proteolytic products and cleavage sites common to pp1a and pp1ab. J Virol 73:177–185

Ziebuhr J, Snijder EJ, Gorbalenya AE (2000) Virus-encoded proteinases and proteolytic processing in the *Nidovirales*. J Gen Virol 81:853–879

Ziebuhr J, Thiel V, Gorbalenya AE (2001) The autocatalytic release of a putative RNA virus transcription factor from its polyprotein precursor involves two paralogous papain-like proteases that cleave the same peptide bond. J Biol Chem 276:33220–33232

Ziebuhr J, Bayer S, Cowley JA, Gorbalenya AE (2003) The 3C-like proteinase of an invertebrate nidovirus links coronavirus and potyvirus homologs. J Virol 77:1415–1426

Zuo Y, Deutscher MP (2001) Exoribonuclease superfamilies: structural analysis and phylogenetic distribution. Nucleic Acids Res 29:1017–1026

Viral and Cellular Proteins Involved in Coronavirus Replication

S. T. Shi · M. M. C. Lai (✉)

Department of Molecular Microbiology and Immunology,
University of Southern California, Keck School of Medicine, 2011 Zonal Avenue,
Los Angeles, CA 90033, USA
michlai@hsc.usc.edu

1	Introduction	96
2	Viral Proteins in Coronavirus Replication	99
2.1	The Polymerase Gene Products	100
2.1.1	RNA-Dependent RNA Polymerase	102
2.1.2	Helicase	104
2.1.3	Proteases	105
2.1.4	Other Polymerase Gene Proteins	107
2.2	The N Protein	108
3	Cellular Proteins in Coronavirus Replication	109
3.1	HNRNP A1	110
3.2	PTB	112
3.3	PABP	114
3.4	Mitochondrial Aconitase	115
3.5	Other Cellular Proteins	116
3.6	Proposed Functions of Cellular Proteins	117
4	Perspectives	118
	References	119

Abstract As the largest RNA virus, coronavirus replication employs complex mechanisms and involves various viral and cellular proteins. The first open reading frame of the coronavirus genome encodes a large polyprotein, which is processed into a number of viral proteins required for viral replication directly or indirectly. These proteins include the RNA-dependent RNA polymerase (RdRp), RNA helicase, proteases, metal-binding proteins, and a number of other proteins of unknown function. Genetic studies suggest that most of these proteins are involved in viral RNA replication. In addition to viral proteins, several cellular proteins, such as heterogeneous nuclear ribonucleoprotein (hnRNP) A1, polypyrimidine-tract-binding (PTB) protein, poly(A)-binding protein (PABP), and mitochondrial aconitase (m-aconitase), have been identified to interact with the critical *cis*-acting elements of coronavirus replication. Like many other RNA viruses, coronavirus may subvert these cellular proteins from cellular RNA processing or translation machineries to play a role in viral replication.

1
Introduction

Studies of diverse groups of positive-stranded RNA viruses reveal that they employ common strategies for replication, although the precise nature of these proteins varies for each virus (Pogue et al. 1994). In general, the formation of viral translation and RNA replication complexes require multiple viral and cellular proteins. By analogy with the phage $Q\beta$, which recruits four host (bacterial) proteins to be an integral part of the replicase complex together with the viral polymerase (Blumenthal and Carmichael 1979), it is likely that replication complexes of positive-stranded RNA viruses consist of both virus- and host-encoded proteins. In addition, viral and cellular proteins interact with various *cis*-acting elements on viral RNAs and play essential roles in the regulation of viral replication. They may mediate the cross talk between the 5' and 3' ends of the viral RNA and bring other distant *cis*-acting elements close together to carry out complex processes, such as subgenomic RNA transcription, coupling between translation and RNA replication, and asymmetric production of excess genomic positive- over negative-strand RNAs. The switch between translation and replication in poliovirus has been shown to involve the cellular protein poly(rC)-binding protein (PCBP), which upregulates viral translation, and the viral protein 3CD, which represses viral translation and promotes negative-strand synthesis (Gamarnik and Andino 1998). Identification of the roles of viral and cellular proteins should provide valuable insights into the mechanisms of viral replication.

The replication of the genome is considered as the most fundamental aspect of the biology of positive-stranded RNA viruses. Like all other positive-stranded RNA viruses, coronavirus replicates its genome through the synthesis of a complementary negative-strand RNA using the genomic RNA as a template. The negative-strand RNA, in turn, serves as the template for synthesizing more progeny positive-strand RNAs. Analysis of the structure of mouse hepatitis virus (MHV) defective-interfering (DI) RNAs indicates that approximately 470 nucleotides (nt) at the 5' terminus, 436 nt at the 3' terminus, and about 135 internal nt are required for coronavirus DI RNA replication and suggests that these sequences contain signals necessary for viral RNA replication

(Kim et al. 1993; Kim and Makino 1995b; Lin and Lai 1993; Lin et al. 1996). Both of the 5′ and 3′ ends of the genome are necessary for positive-strand synthesis (Kim et al. 1993; Lin and Lai 1993), whereas the *cis*-acting signals for the synthesis of negative-strand RNA exist within the last 55 nt and the poly(A) tail at the 3′ end of the MHV genome (Lin et al. 1994). One unique feature of coronaviruses is the expression of their genetic information by transcription of a 3′ coterminal nested set of subgenomic mRNAs that contain a common 5′ leader sequence derived from the 5′ end of the RNA genome. The interaction between the leader sequence and an intergenic (IG) sequence upstream of each open reading frame (ORF), also named transcription-regulating sequence (TRS), is required for the transcription of subgenomic mRNAs (Chang et al. 1994; Liao and Lai 1994; Zhang and Lai 1995b). Logically, these *cis*-acting sequences for viral genomic RNA replication and subgenomic RNA transcription serve as ideal signals to recruit viral factors and possibly cellular proteins for the formation of the RNA replication and transcription complex.

Apart from the findings that continuous synthesis of viral proteins is a prerequisite for the synthesis of both positive- and negative-strand RNA and subgenomic mRNAs (Perlman et al. 1986; Sawicki and Sawicki 1986), little information is currently available concerning the identities and functions of the viral proteins that participate in coronavirus replication. Because of the unparalleled size of the coronavirus RNA genome, genetic approaches to the analysis of replicase gene function have been limited to date. Nevertheless, studies of the temperature-sensitive mutants of coronavirus demonstrate the importance of ORF 1 polyprotein (also known as the polymerase or replicase protein) in coronavirus RNA synthesis and suggest that different domains of this polyprotein are involved in different steps of viral RNA synthesis (Baric et al. 1990a; Fu and Baric 1994; Leibowitz et al. 1982; Schaad et al. 1990). Evolutionarily, the virus genome is composed of relatively constant replicative genes that are indispensable for viral replication and more flexible genes coding for virion structural proteins and various accessory proteins (Koonin and Dolja 1993). Despite the high mutation frequency that is typical of RNA viruses, viral proteins mediating the replication and expression of virus genomes contain arrays of conserved sequence motifs. Proteins with such motifs include RdRp, putative RNA helicase, chymotrypsin-like and papain-like proteases, and metal-binding proteins, all of which are present in the coronavirus ORF 1 polyprotein as shown by sequence comparisons (Bonilla et al. 1994; Bredenbeek et al. 1990; Gorbalenya et al. 1989b; Lee et al. 1991). Strategically located as the

5'-most gene in the viral genome, the coronavirus ORF 1 is translated into a large polyprotein immediately upon virus entry and processed by viral proteases into functional proteins, which are responsible for RNA replication and transcription. The processing scheme of the coronavirus ORF 1 polyprotein has been largely delineated by a number of recent studies. As a result, the functions of the domains that have not been identified before are beginning to emerge. In addition to the proteins with apparent enzymatic activities required for viral RNA synthesis, a number of other coronavirus proteins have also been implicated in viral replication.

Many studies have shown that viruses use cellular proteins for multiple purposes in their replication cycles, including the attachment and entry into the cells, the initiation and regulation of RNA replication/ transcription, the translation of their mRNAs, and the assembly of progeny virions. Because many aspects of the replication cycles of different types of viruses are unique, the cellular proteins used by different types of viruses also differ. Nevertheless, viruses typically subvert the normal components of cellular RNA processing or translation machineries to play an integral or regulatory role in the replication/transcription and translation of viral RNA (Lai 1998). These cellular proteins include, but are not limited to:

1. Heterogeneous nuclear ribonucleoproteins and other RNA processing factors: hnRNP A1 (Black et al. 1995, 1996; Li et al. 1997; Shi et al. 2000; Wang et al. 1997) and other hnRNP type A/B proteins (Bilodeau et al. 2001; Caputi et al. 1999; Shi et al. 2003), hnRNP C (Gontarek et al. 1999; Sokolowski and Schwartz 2001; Spangberg et al. 2000), hnRNP E (PCBP) (Gamarnik and Andino 1997; Parsley et al. 1997), hnRNP H (Caputi and Zahler 2002), hnRNP I (PTB) (Black et al. 1995, 1996; Chung and Kaplan 1999; Gontarek et al. 1999; Hellen et al. 1994; Ito and Lai 1997; Li et al. 1999; Wu-Baer et al. 1996), hnRNP L (Gutierrez-Escolano et al. 2000; Hahm et al. 1998), HuR (Spangberg et al. 2000), and Lsm1p-related protein (Diez et al. 2000).
2. Translation factors: elongation factors EF-1α (Blackwell and Brinton 1997; Harris et al. 1994; Joshi et al. 1986), -β and -γ (Das et al. 1998), EF-Tu (Blumenthal and Carmichael 1979), and eukaryotic initiation factor eIF-3 (Osman and Buck 1997; Quadt et al. 1993).
3. Noncanonical translation factors: hnRNP A1, PTB, and La antigen (Meerovitch et al. 1993; Pardigon and Strauss 1996; Svitkin et al. 1996).
4. Cytoskeletal or chaperone proteins: tubulin (Huang et al. 1993; Moyer et al. 1990; Moyer et al. 1986), actin (De et al. 1991), and heat shock protein (Oglesbee et al. 1996).

These cellular proteins typically bind to viral RNAs or polymerase to form replication or translation complexes (Lai 1998). Remarkably, most of them can interact with RNAs of several different viruses or bind to viral RNA in one virus but associate with viral polymerase in another.

Coronavirus RNA synthesis, including replication of viral genome and transcription of subgenomic mRNAs, has been shown to be regulated by several viral RNA elements, including 5′-untranslated region (UTR), *cis*- and *trans*-acting leader RNAs (Liao and Lai 1994; Zhang et al. 1994; Zhang and Lai 1995b), IG sequence (Makino et al. 1991), and 3′-UTR (Lin et al. 1996). Biochemical evidence suggests that these regulatory sequences likely interact with each other either directly or indirectly, probably through protein-RNA and protein-protein interactions involving both viral and cellular proteins (Zhang and Lai 1995b). Indeed, hnRNP A1 (Huang and Lai 2001; Li et al. 1997; Shi et al. 2000), PTB (Huang and Lai 1999; Li et al. 1999), PABP (Spagnolo and Hogue 2000), and mitochondrial aconitase (Nanda and Leibowitz 2001), have been identified as binding specifically to the known *cis*-acting regulatory sequences. The functional importance of hnRNP A1 (Shi et al. 2000) and PTB (Huang and Lai 1999) in viral RNA synthesis has also been established, further supporting the notion that cellular proteins play an integral or regulatory role in viral replication.

Viruses invariably rely on cellular architecture as an important structural element of their replication machineries. The replication complexes of numerous positive-stranded RNA viruses have been found to be membrane associated (Bienz et al. 1994; Chambers et al. 1990; Froshauer et al. 1988; Miller et al. 2001; Schwartz et al. 2002; van Dinten et al. 1996). Thus, many cellular membrane proteins are expected to serve as scaffolds to provide support for the formation of viral replication complexes, for localized protein translation, and for viral assembly. Very little is currently known about these cellular factors. In this chapter, we focus on the proteins that are the integral parts of the replication complexes. Left out are the cellular factors involved in other aspects of viral replication, such as virus entry and virus assembly.

2
Viral Proteins in Coronavirus Replication

Although the mechanism of coronavirus RNA replication is still controversial, the consensus is that coronavirus RNA replication is directed by *cis*-acting sequences present on the viral RNAs with the help of *trans*-

acting factors encoded by the virus. Indeed, continuous protein synthesis is required for RNA synthesis, due to the fact that the application of inhibitors of protein synthesis at any time during the viral life cycle inhibits viral RNA synthesis (Perlman et al. 1986; Sawicki and Sawicki 1986). A similar observation has been made with an inhibitor of cysteine protease, which inhibits the processing of the MHV ORF 1 (termed the polymerase or the replicase gene) polyprotein (Kim et al. 1995), suggesting that continuous production of the polymerase gene products is required for viral RNA synthesis. The precise nature of many of these products, however, is largely unknown.

2.1
The Polymerase Gene Products

The coronavirus polymerase gene accounts for approximately two-thirds of the genome. It contains two overlapping ORFs, ORF 1a and ORF 1b, which overlap by 76 nt (Fig. 1). The expression of the downstream

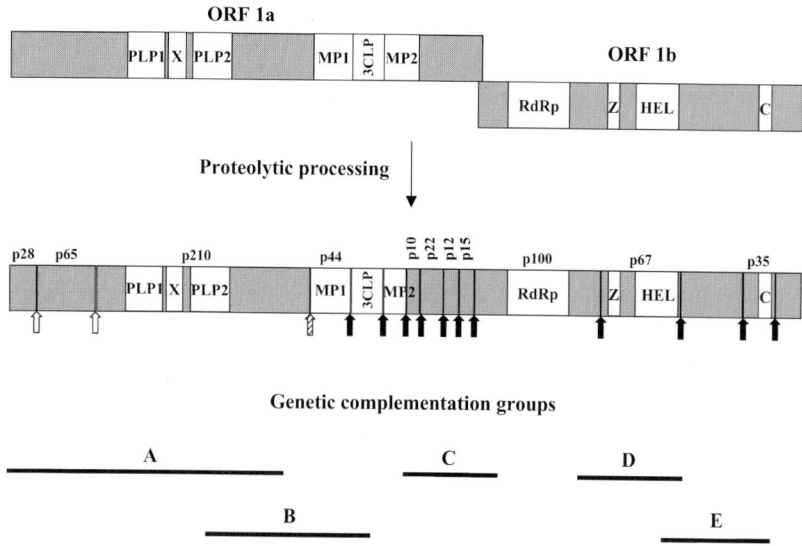

Fig. 1. The domain structure and processing scheme of the MHV polymerase gene products and the approximate location of genetic complementation groups (Baric et al. 1990a). *PLP*, papain-like protease; *3CLP*, 3C-like protease; *MP*, membrane protein; *RdRp*, RNA-dependent RNA polymerase; *Z*, zinc-binding domain; *HEL*, helicase; *C*, conserved domain. The *open*, *hatched*, and *closed arrows* indicate the PLP1, PLP2, and 3CLP cleavage sites, respectively

ORF 1b is mediated by a ribosomal frameshift event that is aided by the formation of a pseudoknot structure within the overlapping region (Bredenbeek et al. 1990; Brierley et al. 1987; Herold and Siddell 1993). To date, the full-length product of ORF 1 has not been detected in coronavirus-infected cells, most probably because it is cotranslationally and auto-proteolytically processed into numerous intermediates and mature nonstructural proteins. Based on the primary sequences of several different coronaviruses, the degree of amino acid identity for this gene product is greater than that is observed for any other coronavirus gene products. A combination of computer-based motif prediction and experimental analysis has identified a number of functional domains in the ORF 1 polyprotein (Fig. 1) (Gorbalenya et al. 1991; Lee et al. 1991). ORF 1a contains the papain-like cysteine proteases (PLPs), a chymotrypsin/picornaviral 3C-like protease (3CLP), and membrane-associated proteins (MP). The more conserved ORF 1b includes domains for an RdRp, a zinc-finger nucleic acid-binding domain (metal-binding domain), and a nucleoside triphosphate (NTP)-binding/helicase domain. Both the synthesis and the processing of the ORF 1 polyprotein have been shown to be essential throughout infection to sustain RNA synthesis and virus replication (Denison et al. 1995b; Kim et al. 1995; Shi et al. 1999).

The importance of the polymerase gene products in viral replication has been established by the study of temperature-sensitive (ts) mutants, which are a practical tool for investigating the roles of viral proteins in replication. The ts mutants are grouped into two categories, RNA^- and RNA^+, based on the ability of these mutants to support viral RNA synthesis at the restrictive temperature (Leibowitz et al. 1982; Robb and Bond 1979). Complementation analysis of ts mutants suggests that at least five RNA^- complementation groups are encoded in the MHV genome (Baric et al. 1990a; Koolen et al. 1983; Leibowitz et al. 1982; Martin et al. 1988; Schaad et al. 1990). All of the RNA^- complementation groups are mapped within the ORF 1 region, suggesting that the coronavirus ORF 1 encodes all of the proteins required for viral RNA replication. Different complementation groups within MHV ORF 1 have been demonstrated to affect distinct steps of RNA synthesis, including the synthesis of leader RNA, negative-strand RNA, and positive-strand RNA, suggesting that different steps of RNA synthesis require different viral proteins (Baric et al. 1990b). Among the five RNA^- complementation groups, A, B, C, D, and E, identified by Baric et al. (Fig. 1) (Baric et al. 1990a), groups A and B are defective in the synthesis of all viral RNAs, whereas the rest of the groups are only defective in certain steps of viral RNA synthesis. The group C mutants encode a function required early in viral transcription to synthesize negative-strand RNA, whereas the group E mutants are blocked at a later stage in the virus growth cycle. The group D mutants are incapable of subgenomic mRNA transcription. Taken together, at least four cistrons are required for positive-strand RNA synthesis whereas the group C cistron functions during negative-strand RNA synthesis. A comparison of three disparate panels of MHV ORF 1 mutants, one for JHM (Robb and Bond 1979) and two for A59 (Koolen et al. 1983; Schaad et al. 1990), concluded that there are at least eight genetically complementable, *trans*-acting functions encoded by ORF 1 (Stalcup et al. 1998).

Genetic recombination analysis revealed that the five RNA⁻ complementation groups of MHV are arranged in alphabetical order in the 5' to 3' direction, with some overlaps between the group A/B and D/E mutants (Fig. 1) (Baric et al. 1990a, b). Group A most likely includes the PLP1 and PLP2 domains, whereas group B encompasses the 3CLP domain. Group C spans the ORF 1a/ORF 1b junction, including the site of ribosomal frameshifting and the N-terminal part of the putative RdRp. Group D is mapped approximately in the middle part of the ORF 1b, possibly encoding the C-terminal part of the putative RdRp and the helicase domain. Group E is located at the C terminus of ORF 1b, about 20–22 kb from the 5' end of the genome (Fu and Baric 1994). Further characterization of the ts mutants showed that one group C mutant carries a mutation in the 5' end of ORF 1b encoding the putative RdRp, which is the only mutation found in a domain with an assigned function. Because most of the mutations in other ts mutants have not been identified, it is still not possible to correlate all the genetic defects with the processed products of the ORF 1 polyprotein.

Studies of the localization and interactions of MHV replicase proteins in infected cells have also provided critical insights into the possible roles of these proteins during viral replication. The localization of polymerase gene products, including PLP1 and PLP2, 3CLP, RdRp, and helicase, to cytoplasmic foci active in viral RNA synthesis has been well documented, suggesting that they may participate in the formation and function of the viral replication complexes (Denison et al. 1999; Shi et al. 1999; van der Meer et al. 1999).

2.1.1
RNA-Dependent RNA Polymerase

The RdRp is the most conserved domain of all RNA viruses and is certainly the most fundamental component of the viral replication machinery. It functions as the catalytic subunit of the viral replicase required for the replication of all positive-stranded RNA viruses (Buck 1996). The vast majority of RdRps, including the coronavirus RdRp, have been identified solely on the basis of sequence similarity. Most viral RNA polymerases contain a signature GDD motif, which is considered to be the most characteristic sequence of the RdRps of positive-stranded RNA viruses. In coronavirus, an SDD motif is detected instead of GDD; the effect of this substitution on the activity of coronavirus RdRp is not clear (Gorbalenya et al. 1989b). Based on sequence analysis, the coronavirus RdRp is encoded by the 5' end of the ORF 1b gene, synthesized as part of the gene 1 polyprotein, and processed by cysteine proteases into an approximately 100-kDa protein (Fig. 1) (Gorbalenya et al. 1989b; Lee et al. 1991). The viral proteins that contain the putative RNA polymerase domain have been detected by immunofluorescence or immunoprecipitation in cells infected with MHV (Shi et al. 1999; van der Meer et al. 1999), IBV (Liu et al. 1994), and HCoV-229E(Grotzinger et al. 1996) but it is not known whether they represent the functional RdRp.

Earlier studies on transmissible gastroenteritis virus (TGEV), bovine coronavirus (BCV), and MHV demonstrated viral polymerase activities in membrane fractions of virus-infected cells (Brayton et al. 1982, 1984; Dennis and Brian 1982; Mahy et al. 1983). Two temporally and enzymatically distinct RdRp activities have been detected in MHV-infected cells (Brayton et al. 1982), suggesting that the enzyme represents two different species of RNA polymerase that perform different roles in virus-specific RNA synthesis. The early polymerase is most likely responsible for negative-strand RNA synthesis, whereas the late polymerase is responsible for the positive-stranded RNA synthesis (Brayton et al. 1984). It is unknown whether the protein components of these two complexes are different or whether the same polymerase is modified by other viral or cellular proteins to perform distinct functions. Because coronaviruses are known to have a unique mechanism of subgenomic RNA synthesis quite distinct from that of genome replication, it is possible that the viruses could have more than one RNA polymerase. After the initial detection of polymerase activities in the fractions of coronavirus-infected cells, several in vitro RNA synthesis systems were also reported (Baker and Lai 1990; Compton et al. 1987; Leibowitz and DeVries 1988). The nature of the polymerases in these systems, however, has not been characterized.

The catalytic activity of the coronavirus RdRp has so far not been demonstrated biochemically. In fact, only a handful of viral RdRps, such as $Q\beta$ replicase subunit II (Landers et al. 1974), poliovirus 3D pol protein (Neufeld et al. 1991; Rothstein et al. 1988; Van Dyke and Flanegan 1980), hepatitis C virus NS5B protein (Behrens et al. 1996; Lohmann et al. 1997; Yuan et al. 1997), dengue virus NS5 protein (Tan et al. 1996), and tobacco vein mottling virus (TVMV) nuclear inclusion protein NIb (Hong and Hunt 1996), have been shown to possess RNA replicating activities in vitro. It is likely that the extremely hydrophobic nature of the coronavirus RdRps prevents the purification and biochemical characterization of this protein. Thus, the precise role of coronavirus RdRps in viral RNA synthesis has not been established.

2.1.2
Helicase

The RNA helicase is the second most conserved component of the RNA virus replication machinery (Gorbalenya et al. 1988, 1989a; Gorbalenya and Koonin 1989; Koonin and Dolja 1993). Nearly all double-stranded and positive-stranded RNA viruses are predicted to encode putative helicases (Gorbalenya and Koonin 1989). RNA helicases are a diverse class of enzymes that use the energy of NTP hydrolysis to unwind duplex RNA. There is extensive genetic evidence suggesting a key function for helicases in the life cycle of positive-stranded RNA viruses (Buck 1996; Kadare and Haenni 1997). They are involved in virtually every aspect of RNA metabolism, including transcription, splicing, translation, export, ribosome biogenesis, mitochondrial gene expression, and the regulation of mRNA stability (de la Cruz et al. 1999; Linder and Daugeron 2000; Lohman and Bjornson 1996; Schmid and Linder 1992). The idea of involvement of RNA helicase in RNA replication came from the observation that helicase mutants of BMV are defective in template recruitment for RNA replication and the synthesis of negative-strand or subgenomic RNA (Ahola et al. 2000).

The RNA helicase domains of coronaviruses are encoded by ORF 1b and processed by 3CLP (Denison et al. 1999). They have been proposed to represent a separate phylogenetic lineage of the RNA virus superfamily 1 (SF1) helicases, which include the majority of putative RNA virus helicases (Gorbalenya and Koonin 1989; Kadare and Haenni 1997; Koonin and Dolja 1993). The putative MHV RNA helicase, which is processed from the ORF 1b polyprotein by 3CLP, has been detected in MHV-infected cells throughout the viral life cycle (Denison et al. 1999). Numerous attempts to detect the predicted RNA duplex-unwinding activity of these proteins have failed until recently when duplex-unwinding activity was observed for the human coronavirus (HCoV) helicase, providing valuable insights into the functions of this protein in viral replication (Seybert et al. 2000). Biochemical characterization revealed that this helicase has both RNA and DNA duplex-unwinding activities with a 5′ to 3′ polarity, in contrast to the previously characterized RNA virus SF2 helicases. A zinc finger/nucleic acid-binding domain, which has been found in numerous cellular helicases (Fig. 1) (Gorbalenya and Koonin 1993), is also present in the coronavirus ORF 1b, upstream of the helicase domain, but it is not known whether it contributes to the activity of the coronavirus helicase.

Although there is no direct evidence indicating the involvement of the helicase in coronavirus RNA replication and transcription, the helicase was localized to the perinuclear sites where active viral RNA synthesis was observed (Denison et al. 1999). It was further detected by biochemical analysis in membrane fractions that contain viral RNAs, suggesting that helicase is a component of the viral replication complex (Bost et al. 2000, 2001; Denison et al. 1999; Sims et al. 2000). Furthermore, because double-stranded replicative intermediates are believed to be the predominant RNA structures in coronavirus RNA synthesis, it is tempting to speculate that, in analogy to models described for the DNA replisome (Baker and Bell 1998), the coronavirus helicase cooperates with the RdRp by providing the single-stranded RNA template for processive RNA synthesis. It is noteworthy that the vaccinia virus NPH-II RNA helicase was recently shown to be a highly processive enzyme that unwinds long duplex RNA structures, supporting the hypothesis that at least some viral RNA helicases might be directly involved in RNA replication (Jankowsky et al. 2000).

2.1.3
Proteases

The coronavirus replicase is translated from the genomic RNA as a large precursor polyprotein, which is then processed by viral proteases to generate functional replicase proteins. Whereas the RdRp and RNA helicase play direct roles in viral RNA synthesis, the proteases are involved in viral replication through the processing of viral polyproteins into mature products critical for the appropriate localization, assembly, and function of the replicase complex. They also play an important regulatory role in the generation of specific protein functions at certain stages of the viral life cycle. This controlled proteolysis is thought to be determined mainly by the substrate specificity of the proteases and the accessibility of cleavage sites in the context of specific intermediate products (van Dinten et al. 1997, 1999; Ziebuhr and Siddell 1999). Sequence analysis of coronavirus genomic RNA led to the prediction of two or three protease domains in ORF 1a: one or two PLPs and a 3CLP (Gorbalenya et al. 1991; Lee et al. 1991). All of these proteases have been shown to function during viral replication and drive the processing of the MHV ORF 1 replicase polyprotein into at least 15 products (Fig. 1) (Baker et al. 1989, 1993; Bonilla et al. 1994, 1995; Bost et al. 2000; Denison et al. 1992, 1995a, 1999; Gao et al. 1996; Lu et al. 1995, 1996, 1998; Lu and Denison 1997; Pinon et al. 1999; Schiller et al. 1998; Shi et al. 1999). Comparable, but distinct, proteolytic processing pathways have also been reported for some other coronaviruses, most notably IBV (Liu et al. 1994, 1998; Liu and Brown 1995) and HCoV-229E (Ziebuhr et al. 2000).

The coronavirus ORF 1 polyprotein can be divided into an N-terminal region that is processed by one or two PLPs and a C-terminal region that is processed by the 3CLP (Ziebuhr et al. 2000). The N-terminal region of the polyprotein spans from the initiator Met to the N terminus of the hydrophobic domain MP1 (Fig. 1). All coronaviruses, except IBV, encode two paralogous and sequentially positioned PLP1 and PLP2 that flank a conserved X domain from both sides (Fig. 1) (Gorbalenya et al. 1991; Lee et al. 1991). At least three proteins, p28, p65, and p210 (also known as p240), are produced from this region of the ORF 1a polyprotein in MHV (Denison and Perlman 1987; Denison et al. 1995a; Schiller et al. 1998). The MHV p210 protein is autocatalytically released through cleavages mediated by PLP1 at the N-terminal site (Bonilla et al. 1995, 1997) and PLP2 at the C-terminal site (Kanjanahaluethai and Baker 2000). PLP1 also cleaves the p28-p65 junction (Baker et al. 1989, 1993; Dong and Baker 1994; Hughes et al. 1995), which, except for IBV, is conserved in all coronaviruses (Herold et al. 1998). Accordingly, a PLP1-mediated cleavage at this site, resulting in the production of a small N-terminal protein (p9, p28 equivalent), was also detected in HCoV-infected cells (Herold et al. 1998). The single IBV PLP corresponds to the PLP2 domain of other coronaviruses. It is part of a p195 protein, which is cleaved to produce an N-terminal product, p87 (Lim and Liu 1998; Lim et al. 2000).

Coronavirus PLPs contain a transcription factor-like zinc finger (Herold et al. 1999), suggesting that they might also be directly involved in coronavirus RNA synthesis. This hypothesis is strongly supported by a recent report showing the equine arteritis virus (EAV) nonstructural protein 1, which is likely a distant homolog of the coronavirus PLPs, to be a transcriptional factor indispensable for subgenomic mRNA synthesis (Tijms et al. 2001).

The C-terminal part of the ORF 1 polyprotein encompasses all of the major conserved domains starting from the hydrophobic domain MP1 and extending to the C terminus of the replicase polyprotein. The 3CLP, flanked on either side by membrane-spanning regions MP1 and MP2 (Bonilla et al. 1994; Lee et al. 1991; Lu et al. 1995), is believed to be the principal viral protease responsible for the processing events leading to the formation of the viral replicase complex. At least 12 processing products, including the 3CLP itself, RdRp, and helicase, are generated by 3CLP-mediated cleavage (Fig. 1) (Gorbalenya et al. 1991; Lee et al. 1991). Treatment of infected cells with E64d, a known inhibitor of the 3CLP, results in the inhibition of viral RNA replication in these cells (Kim et al. 1995), demonstrating the importance of the action of the 3CLP in the events leading to viral replication. The importance of 3CLP cleavages was demonstrated with an infectious clone of the related arterivirus EAV (van Dinten et al. 1999). Introduction of mutations into the candidate ORF 1b 3CLP cleavage sites had drastic effects on RNA synthesis and virus replication. 3CLP has also been localized to the site of viral RNA synthesis by immunofluorescence staining and biochemical fractionation studies (Bost et al. 2000, 2001; Denison et al. 1999; Shi et al. 1999; Sims et al. 2000).

2.1.4
Other Polymerase Gene Proteins

Apart from the RdRp, helicase, and proteases, the identities of many of the ORF 1 products have not been established. Thus, their roles in viral replication remain unknown. By immunofluorescence staining and confocal microscopy, several studies have shown that a number of ORF 1a products, p65, p10, p22, p12, and p15, and an ORF 1b product, p35, are associated with the site of viral RNA synthesis (Fig. 1) (Bost et al. 2000; Shi et al. 1999). However, biochemical studies revealed two distinct but tightly associated membrane populations, only one of which appears to be a site for viral RNA synthesis (Sims et al. 2000). p28, helicase, 3CLP, and nucleocapsid (N) protein cosegregated with the viral RNA and, therefore, are likely to be the components of the viral replication complexes, whereas p65 and p22 are present in different membrane fractions and may serve roles during infection that are distinct from viral RNA transcription or replication (Sims et al. 2000).

The hydrophobic domains, MP1 and MP2, within the ORF 1a polyprotein were postulated to mediate the association of the coronavirus replicase with cellular membrane structures. MP1 has indeed been detected in microsomal membranes (Pinon et al. 1997), but its role in membrane association and coronavirus replication is largely speculative. A recent study on the related arterivirus demonstrated that the EAV nonstructural proteins (nsp) 2 and 3, which contain one or two hydrophobic regions, induce the formation of double-membrane structures where EAV RNA synthesis takes place (Snijder et al. 2001). Similarly, the membrane proteins of coronavirus may serve to alter the cell architecture so that it is more favorable for viral replication.

2.2
The N Protein

The coronavirus N protein associates with the genomic RNA to form a helical nucleocapsid. In addition to its role as a major structural component of virions, N may also be involved in viral RNA replication and translation control. In an in vitro replication system for MHV, it was demonstrated that antibodies against the N protein could inhibit RNA synthesis (Compton et al. 1987). Optimal replication of the bovine coronavirus (BCV) DI RNA also requires the translation of most, if not all, of the N protein *in cis* (Chang and Brian 1996). Structural analysis of DI RNAs shows that the presence of gene 1 and N gene is sufficient for viral RNA replication (Kim and Makino 1995a). In addition, the MHV N protein was detected in membrane fractions containing viral RNA (Sims et al. 2000) and colocalized with putative replicase proteins in virus-infected cells, providing further support that N may be involved in RNA replication (Denison et al. 1999; van der Meer et al. 1999). However, a mutational study of an infectious cDNA clone of EAV, a close relative of coronavirus, reported that all structural proteins, including N, are dispensable for genome replication and subgenomic mRNA transcription (Molenkamp et al. 2000). The coronavirus replicase gene products were also shown to be sufficient for discontinuous subgenomic mRNA transcription with a partial cDNA clone representing the 5' and 3' ends of the HCoV-229E genome, the HCoV-229E replicase gene, and a reporter gene located downstream of a regulatory element for coronavirus mRNA transcription (Thiel et al. 2001). The RNA replication levels observed in these systems are much lower than those containing the wild-type full-length viral genome, indicating that factors other than the replicase polyprotein are required for efficient RNA replication.

Because the N protein has the ability to interact with viral RNA, it most likely functions in viral RNA synthesis by binding to RNA and forming a ribonucleoprotein (RNP) complex. The N protein binds to the leader RNA sequences present at the 5′ end of genomic RNA and all six subgenomic mRNAs in MHV-infected cells (Baric et al. 1988; Nelson et al. 2000). Biochemical analysis measured a dissociation constant of 14 nM for bacterially expressed MHV N-binding to the leader RNA (Nelson et al. 2000). The MHV negative-stranded RNA was also immunoprecipitated by the anti-N monoclonal antibody. These data indicate that the MHV N protein is associated with MHV-specific RNAs and RNA intermediates and may play an important functional role during MHV transcription and replication. Furthermore, the N-leader-RNA-containing RNP complexes were also immunoprecipitated from BCV-infected cells (Cologna et al. 2000). The interactions between the N protein and the RNA encompassing the N ORF may also contribute to the formation of the N-RNA complexes that are present in coronavirus-infected cells (Cologna et al. 2000).

The N protein of MHV is also involved in positive translational control (Tahara et al. 1993, 1998). It stimulated translation of a chimeric reporter mRNA containing an intact MHV 5′-untranslated region and the chloramphenicol acetyltransferase (CAT)-coding sequence. Preferential translation of viral mRNA in MHV-infected cells is stimulated in part by the interaction between the N protein and a 12-nt tract at the 3′ end of the leader.

Other coronavirus proteins, including structural protein hemagglutinin-esterase (HE) (Luytjes et al. 1988; Yokomori et al. 1991) and nonstructural proteins NS2 (Schwarz et al. 1990), NS4 and NS5 (Yokomori and Lai 1991), are not essential for coronavirus replication. However, it is not clear whether any of these proteins can modulate viral replication.

3
Cellular Proteins in Coronavirus Replication

Coronavirus replication involves not only the viral proteins, but also cellular proteins, which are subverted from the normal functions of the host to play roles in the viral replication cycle. No coronavirus proteins in the infected cell extract could be cross-linked to the viral RNA in vitro, suggesting that viral proteins may interact with viral RNA only indirectly through cellular proteins. Several cellular proteins have been shown to bind to the regulatory elements of MHV RNA, including the 5′

and 3' ends of the genomic RNA and the 3' end of the negative-strand RNA and IG sites. So far, only a handful of them have been identified, among which hnRNP A1 and PTB are the only two proteins found to interact with regions other than the 3' end of the coronavirus genome. These proteins are likely to serve as mediators to bring the *cis*-regulatory regions together to form viral replication complexes. They may also help recruit and stabilize the RdRp to the initiation sites of viral RNA synthesis.

3.1
HNRNP A1

UV cross-linking experiments using cytoplasmic extracts of uninfected cells and the IG sequence showed that three different cellular proteins bind to IG of the template RNA (Zhang and Lai 1995a). Deletion analyses and site-directed mutagenesis of IG further demonstrated a correlation between protein binding and transcription efficiency, suggesting that these RNA-binding proteins are involved in the regulation of coronavirus mRNA transcription. One of these proteins was identified by partial peptide sequencing to be hnRNP A1 (Li et al. 1997). hnRNP A1 is an RNA-binding protein that contains two RNA-binding domains (RBDs) and a glycine-rich domain responsible for protein-protein interaction. It is predominantly a nuclear protein but also shuttles between the nucleus and the cytoplasm (Pinol-Roma and Dreyfuss 1992). A 38-amino acid sequence, termed M9, located near the C terminus of hnRNP A1 between amino acids 268 and 305 has been determined to be the signal that mediates shuttling (Michael et al. 1995; Siomi and Dreyfuss 1995; Weighardt et al. 1995). The nuclear hnRNP A1 is known to be involved in pre-mRNA splicing and transport of cellular RNAs (Dreyfuss et al. 1993), whereas the cytoplasmic hnRNP A1 is capable of high-affinity binding to AU-rich elements and thus modulating mRNA turnover and translation (Hamilton et al. 1993, 1997; Henics et al. 1994). Another function of hnRNP A1 in the cytoplasm is to promote ribosome binding by a cap-mediated mechanism and to prevent spurious initiations at aberrant translation start sites (Svitkin et al. 1996).

hnRNP A1 binds MHV negative-strand leader and IG sequences (Furuya and Lai 1993; Li et al. 1997), which are critical elements for the discontinuous viral RNA transcription (Fig. 2). Site-directed mutagenesis of the IG sequences demonstrated that the extent of binding of hnRNP A1 to the IG sequences correlated with the efficiency of transcription from the IG site (Furuya and Lai 1993; Li et al. 1997; Zhang and Lai

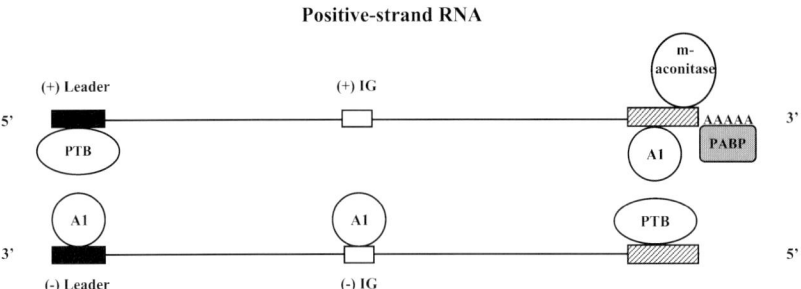

Fig. 2. Schematic drawings of the cellular proteins that interact with coronavirus RNA. hnRNP A1 interacts with the negative-strand leader and IG sequences as well as the positive-strand 3′-UTR, whereas PTB interacts with the positive-strand leader and the complementary sequence of 3′-UTR. These two proteins bind to sequences that are complementary to each other at both the 5′ and 3′ ends of coronavirus RNAs. The poly(A) tail and the 3′-most 42 nt of the genomic RNA serve as binding domains for PABP and m-aconitase, respectively

1995b). Immunostaining of hnRNP A1 showed that hnRNP A1 relocated to the cytoplasm of MHV-infected cells, where viral RNA synthesis occurs (Li et al. 1997). hnRNP A1 also interacts with the MHV N protein (Wang and Zhang 1999), which also binds to the MHV RNA directly (Baric et al. 1988; Stohlman et al. 1988). Furthermore, hnRNP A1 mediates the formation of a ribonucleoprotein complex containing the MHV negative-strand leader and IG sequences (Zhang et al. 1999), suggesting that it may serve as a protein mediator for distant RNA regions to interact with each other to form a transcription initiation complex. Remarkably, hnRNP A1 has also been shown to bind the positive-stranded 3′-UTR and may play a role in negative-strand RNA synthesis (Fig. 2) (Huang and Lai 2001).

The functional importance of hnRNP A1 in coronavirus RNA replication was shown in cells stably expressing the wild-type hnRNP A1 or a dominant-negative mutant of hnRNP A1, which lacks the C-terminal nuclear localization domain (Shi et al. 2000). Viral RNA synthesis was accelerated by the overexpression of hnRNP A1 but delayed by the expression of the mutant hnRNP A1 in the cytoplasm. Thus, the truncation mutant of hnRNP A1 interferes with viral RNA replication in a dominant-negative fashion. In addition to the general inhibition of viral RNA synthesis, the hnRNP A1 mutant also caused a preferential inhibition of the replication of DI RNAs, suggesting that the inhibition of MHV replication by the hnRNP A1 mutant was most likely a direct effect on viral RNA synthesis rather than an indirect effect on other aspects of cellular or viral functions. Because hnRNP A1 binds directly to the *cis*-acting MHV RNA sequences critical for MHV RNA transcription (Li et al. 1997) and replication (Huang and Lai 2001), it is most likely that hnRNP A1 may participate in the formation of the transcription/replication complex.

However, a mouse erythroleukemia cell line, CB3, that lacks detectable hnRNP A1 expression (Ben-David et al. 1992) can still support efficient MHV replication (Shen and Masters 2001). Because hnRNP A1 protein is involved in a variety of important cellular functions, it is conceivable that other cellular gene products may substitute for the function of hnRNP A1 in both uninfected and virus-infected CB3 cells. Indeed, a number of CB3 cellular proteins comparable to hnRNP A1 in size were found to interact with the MHV negative-strand leader RNA. All of these proteins were identified to be hnRNP A1-related proteins, including hnRNP A/B, hnRNP A2/B1, and hnRNP A3 (Shi et al. 2003). These hnRNPs have primary sequence structure, biochemical properties, and function similar to those of hnRNP A1 (Dreyfuss et al. 1993; Ma et al. 2002; Mayeda et al. 1994). They also have binding specificity and affinity similar to MHV RNA compared with hnRNP A1 (Shi et al. 2003). One of these proteins, hnRNP A2/B1, can substitute for hnRNP A1 in regulating the splicing of cellular (Mayeda et al. 1994) and viral (Bilodeau et al. 2001; Caputi et al. 1999) pre-mRNAs. Together, these multiple hnRNP A1-related proteins may perform similar functions in MHV replication.

3.2
PTB

PTB, which is also known as hnRNP I, binds to the UC-rich RNA sequences typically found near the 3' end of introns. Similar to hnRNP A1, PTB shuttles between the nucleus and cytoplasm and plays a role in the regulation of alternative splicing of pre-mRNAs and translation of cellular and viral RNAs (Kaminski et al. 1995; Svitkin et al. 1996; Valcarcel and Gebauer 1997). Studies of picornaviruses revealed that PTB plays a role in internal ribosome entry site (IRES)-mediated translation by mechanisms distinct from those governing the cap-dependent translation of most eukaryotic mRNAs (Jackson and Kaminski 1995). PTB was found to be associated with the IRES elements of encephalomyocarditis virus and foot-and-mouth-disease virus and to stimulate translation initiated from these IRES elements (Kaminski et al. 1995; Niepmann 1996; Niepmann et al. 1997).

UV cross-linking and immunoprecipitation studies using cellular extracts and a recombinant PTB established that PTB binds specifically to the MHV positive-strand leader RNA (Fig. 2) (Li et al. 1999), which is required for MHV RNA synthesis (Kim et al. 1993; Liao and Lai 1994) and regulates translation (Tahara et al. 1994). The PTB-binding sites were mapped to the UCUAA pentanucleotide repeats within the leader RNA; deletion of these binding sites significantly inhibits RNA transcription (Li et al. 1999). Interestingly, PTB also interacts with the complementary strand of the 3'-UTR (c3'-UTR) (Fig. 2) (Huang and Lai 1999). A strong PTB-binding site was mapped to nt 53–149, and another weak binding site was mapped to nt 270–307 on c3'-UTR. Partial substitutions of the PTB-binding nucleotides reduced PTB binding in vitro. Furthermore, DI RNAs harboring these mutations showed substantially reduced ability to synthesize subgenomic mRNA. Remarkably, the binding of PTB to nt 53–149 caused a conformational change in the neighboring RNA region. Partial deletions within the PTB-binding sequence completely abolished the PTB-induced conformational change in the mutant RNA even when the RNA retained partial PTB-binding activity. Correspondingly, the MHV DI RNAs containing these deletions lost their ability to transcribe mRNAs. Thus, the conformational change in the c3'-UTR caused by PTB binding may play a role in mRNA transcription.

It is interesting to note that hnRNP A1 and PTB bind to the precisely complementary sites on the negative- and positive-stranded RNA, respectively, of the leader region of MHV RNA, and also the 5'- and 3'-ends of both the positive- and negative-strand RNAs (Fig. 2) (Huang and Lai 2001; Huang and Lai 1999; Li et al. 1997, 1999;). Furthermore, hnRNP A1 and PTB together mediate the formation of an RNP complex involving the 5'- and 3' end fragments of MHV RNA in vitro (Huang and Lai 2001). The interaction between hnRNP A1 and PTB have also been detected in a splicing complex in uninfected cells (Bothwell et al. 1991). All of these findings support the notion that hnRNP A1 and PTB may be involved in the formation of a ribonucleoprotein complex, which functions in MHV RNA synthesis.

Most coronavirus mRNAs are capped at the 5' end and translated by a cap-dependent mechanism. The binding of PTB to the coronavirus leader RNA, which regulates MHV RNA translation (Tahara et al. 1994), suggests a possible role of PTB in coronavirus mRNA translation as well. Surprisingly, PTB was found to have no direct effect on the cap-dependent MHV RNA translation (Choi and Lai, unpublished data). It is, however, still possible that PTB may affect the IRES-mediated translation of coronavirus ORF 5b, which encodes the envelope (E) protein (Lai and Cavanagh 1997; Thiel and Siddell 1994). The ORF 5b IRES has been shown to serve as a binding site for cellular proteins (Jendrach et al. 1999), although it is not known whether PTB is among these proteins.

3.3
PABP

The 3'-UTRs of coronavirus RNA are necessary for the synthesis of negative-strand viral RNA (Lin et al. 1994) and both genomic and subgenomic positive-strand RNA synthesis (Kim et al. 1993; Lin and Lai 1993; Lin et al. 1996). They contain structures that are conserved among divergent coronaviruses (Hsue et al. 2000; Hsue and Masters 1997; Liu et al. 2001). It is possible that these secondary structural elements serve as binding sites for cellular proteins and function in viral replication. Indeed, the mutations at the 3' end of the viral genomic RNA that abolished the binding of cellular proteins also inhibited both negative-strand and positive-strand RNA synthesis, although the correlation between protein binding and RNA synthesis was not absolute (Liu et al. 1997; Yu and Leibowitz 1995a).

A number of cellular proteins have been found to interact with multiple sites within the 3' end of positive-strand MHV RNA (Huang and Lai 2001; Liu et al. 1997; Spagnolo and Hogue 2000; Yu and Leibowitz 1995a, b). Several cellular proteins have also been shown to interact with the BCV 3'-UTR [287 nt plus poly(A) tail] (Huang and Lai 2001; Liu et al. 1997; Spagnolo and Hogue 2000; Yu and Leibowitz 1995a, b). Competition with the MHV 3'-UTR [301 nt plus poly(A) tail] suggests that the interactions are conserved for the two viruses (Huang and Lai 2001; Liu et al. 1997; Spagnolo and Hogue 2000; Yu and Leibowitz 1995a, b). Proteins with molecular masses of 99, 95, 73, 40–50, and 30 kDa were detected, among which the 73-kDa protein was identified to be poly(A)-binding protein (PABP) by immunoprecipitation experiments. PABP is known to interact specifically with poly(A), which is an important *cis*-acting signal for coronavirus RNA replication (Fig. 2) (Lin et al. 1994). RNAs with shortened poly(A) tails exhibited less in vitro PABP binding. Furthermore, binding of PABP to the 3'-UTR of the DI RNA replicons correlated with the ability of the DI RNA to replicate, suggesting that the interaction between PABP and the poly(A) tail may affect coronavirus RNA replication (Huang and Lai 2001; Liu et al. 1997; Spagnolo and Hogue 2000; Yu and Leibowitz 1995a, b).

PABP is a highly abundant cytoplasmic protein (Gorlach et al. 1994) that binds the 3' poly(A) tail on eukaryotic mRNAs and helps promote both efficient translation initiation and mRNA stability. It interacts with the translation factor eukaryotic initiation factor (eIF) 4G (Imataka et al. 1998; Le et al. 1997; Tarun and Sachs 1996; Tarun et al. 1997), which is part of the eIF4F triple complex that binds mRNA cap structures during translation, and PABP-interacting protein (PAIP-1), a protein with homology to eIF-4G (Craig et al. 1998). This interaction, known as the closed-loop model of translation initiation, mediates the cross talk between the 5' and 3' ends of mRNAs (Gallie 1998; Sachs et al. 1997). Because coronavirus RNA is capped and polyadenylated like the host mRNAs, PABP is likely involved in the translation of the coronavirus genome upon virus entry into the cell. Because translation is required for efficient coronavirus RNA replication, it is conceivable that PABP can indirectly modulate RNA synthesis through its effect on translation. It is also possible that the PABP-poly(A) interaction may play a more direct role in coronavirus RNA replication in view of the apparent requirement for both the 5' and 3' ends, including the poly(A) tail, of the coronavirus genome for DI RNA replication and mRNA transcription (Kim et al. 1993; Lai 1998; Liao and Lai 1994; Lin et al. 1994, 1996). Indeed, hnRNP A1 and PTB together have been shown to mediate the interaction between the 5' and 3' ends of MHV RNA (Huang and Lai 2001). PABP may be another cellular factor that facilitates a similar interaction of the ends.

3.4
Mitochondrial Aconitase

The 3'-most 42 nt of the MHV genomic RNA has been shown to interact with host factors and form at least three RNA-protein complexes (Nanda and Leibowitz 2001). Four proteins of approximately 90, 70, 58, and 40 kDa were resolved from these complexes, and the 90-kDa protein was identified as mitochondrial aconitase (m-aconitase), which catalyzes stereospecific interconversion of citrate into isocitrate through a *cis*-aconitase intermediate in the Krebs cycle (Beinert and Kennedy 1993). UV cross-linking studies indicate that the highly purified m-aconitase binds specifically to the MHV 3' protein-binding element despite the absence of a consensus RNA-binding domain (Fig. 2) (Burd and Dreyfuss 1994). Colocalization of m-aconitase with the MHV N protein was observed in virus-infected cells, suggesting a possible interaction of m-aconitase with the MHV replication complexes (Nanda and Leibowitz 2001).

A cytoplasmic homolog of m-aconitase, cytoplasmic aconitase (c-aconitase), also known as iron regulatory protein 1, is a well-recognized RNA-binding protein (Kennedy et al. 1992). The binding properties of m-aconitase and the functional relevance of RNA binding appear to parallel those of c-aconitase. c-Aconitase is a bifunctional protein, which has been shown to interact with iron-responsive elements located in the 5′-UTR of ferritin mRNA and the 3′-UTR of transferrin receptor (TfR) mRNA and to function to coordinate posttranscriptional regulation of cellular iron metabolism (Hentze and Kuhn 1996; Kuhn and Hentze 1992). Similarly, m-aconitase can function as a posttranscriptional regulator as well (Beinert and Kennedy 1993; Klausner et al. 1993). A link between cellular iron status and m-aconitase expression has also been established (Kim et al. 1996; Schalinske et al. 1998). Increasing the intracellular level of m-aconitase of MHV-infected cells by iron supplementation resulted in increased RNA-binding activity of cell extracts and increased virus production as well as viral protein synthesis at early hours of infection (Nanda and Leibowitz 2001). It is possible that the binding of m-aconitase to the 3′-UTR increases the stability of the viral mRNAs and hence augments the translation of viral proteins, similar to the role of IRP in regulating TfR (Kuhn and Hentze 1992).

3.5
Other Cellular Proteins

Accumulating evidence indicates the presence of additional cellular proteins that interact with coronavirus RNA. The 3′-UTRs of murine and bovine coronaviruses were reported to contain bulged stem-loop (Hsue et al. 2000; Hsue and Masters 1997) and pseudoknot (Williams et al. 1995) structures, which are essential for viral replication. These motifs are potential binding sites for the proteins shown to interact with the 3′-UTR. Indeed, a number of cellular proteins have been shown to interact with different regions within the 3′-UTR of MHV (Liu et al. 1997; Nanda and Leibowitz 2001; Yu and Leibowitz 1995a, b). The 3′-most 42-nt sequence interacts with at least four proteins 90, 70, 58, and 40 kDa in size, among which the 90-kDa protein was identified as m-aconitase (Nanda and Leibowitz 2001). A distinct host cellular protein-binding element was also mapped within a 26-nt sequence at positions 154–129 from the 3′ end of the MHV-JHM genome (Liu et al. 1997). The resulting RNA-protein complex contains six host cellular proteins with one protein of 120-kDa molecular mass, two poorly resolved species approximately 55 kDa in size, a second pair of poorly resolved 40-kDa proteins, and a minor component of 25 kDa. This region contains multiple stem-loop and hairpin-loop structures, which are shown by mutational analysis to be important for efficient MHV replication (Liu et al. 2001). In the study that identified PABP, several other proteins with molecular masses of 99, 95, 40–50, and 30 kDa were also shown to interact with the 3'-UTRs of both BCV and MHV (Spagnolo and Hogue 2000). These cellular proteins have the potential to regulate viral RNA synthesis through their binding to the 3' ends of the coronavirus genomes; however, their identities and functions remain to be determined.

3.6
Proposed Functions of Cellular Proteins

The *cis*-acting signals for viral RNA replication or transcription often consist of multiple distant sequences on the viral RNA. In many cases, there appears to be a cross talk between the 5' and 3' ends of viral RNAs so that the 3' end sequence often can regulate RNA synthesis or translation initiated from the 5' end of the RNA. The 5'- and 3'-UTRs of both positive- and negative-sense RNA and the IG sequences are thought to contain important sequence and structural elements that function in the initiation and regulation of RNA replication, transcription, and translation. The 3' end of the MHV RNA has been shown to regulate mRNA synthesis starting from an upstream internal promoter (Lin et al. 1996). The poly(A) tail is also involved in coronavirus RNA synthesis (Huang and Lai 2001; Liu et al. 1997; Spagnolo and Hogue 2000; Yu and Leibowitz 1995a, b). Furthermore, there is an apparent interaction between the leader and IG sequences, which regulates the synthesis of coronavirus subgenomic mRNAs (Lai and Cavanagh 1997; Zhang et al. 1994). When no sequence complementarity exists between the 5' and 3' ends, RNA-protein and protein-protein interactions must be involved. hnRNP A1 and PTB have the ability to interact with each other, thus allowing different RNA regions to interact (Fig. 3A). By analogy to translation regulation, the binding of PABP to the 3' end of the coronavirus genome may also facilitate the cross talk between the 3' end and the other upstream *cis*-acting sequences. Furthermore, because most of the viral RdRps do not appear to bind directly to the *cis*-acting regulatory or promoter sequences on the RNA, their ability to initiate RNA synthesis at specific sites probably depends on their interactions with the cellular proteins that bind directly to the viral RNA template. These cellular proteins may serve as a platform on which other proteins, both viral and

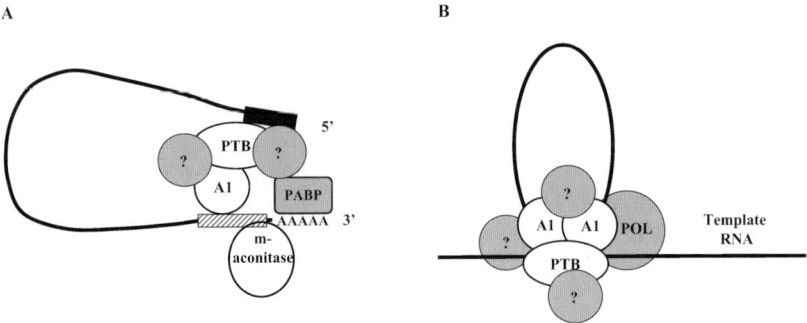

Fig. 3A, B. Proposed functions of cellular proteins in coronavirus replication. **A** Interactions between distant RNA elements are mediated by hnRNP A1, PTB, and PABP. **B** Formation of coronavirus replication/transcription complexes through the recruitment of additional viral and cellular proteins by hnRNP A1 and PTB

cellular, subsequently bind to form functional replication and transcription complexes (Fig. 3B).

Together, cellular proteins play important roles in coronavirus replication. Identification of these proteins and analysis of their functions in virus replication are critical to furthering our understanding of virus-host interactions and will provide clues to unveil the replication strategies of other positive-stranded RNA viruses.

4
Perspectives

Although an increasing body of literature supports the importance of various viral and cellular proteins in coronavirus replication, our current understanding of the roles of these proteins is still limited. The availability of the reverse genetics approach for coronaviruses is expected to greatly accelerate the understanding of coronavirus replication as well as the functional importance of viral and cellular factors in coronavirus replication. In addition, the growing knowledge of the properties of the individual protein products of the coronavirus ORF 1 should help in understanding the makeup of the replication machinery. The recent advances in gene knockout by RNA interference (RNAi) in mammalian cells will likely be a valuable tool in establishing the functional relevance of these cellular proteins. Nevertheless, the ultimate unraveling of the viral and cellular proteins involved in coronavirus replication is expected to come after the purification of coronavirus RdRp and the reconstitution of virus replication in vitro.

References

Ahola T, den Boon JA, Ahlquist P (2000) Helicase and capping enzyme active site mutations in brome mosaic virus protein 1a cause defects in template recruitment, negative-strand RNA synthesis, and viral RNA capping. J Virol 74:8803–8811

Baker SC, Lai MM (1990) An in vitro system for the leader-primed transcription of coronavirus mRNAs. EMBO J 9:4173–4179

Baker SC, Shieh CK, Soe LH, Chang MF, Vannier DM, Lai MM (1989) Identification of a domain required for autoproteolytic cleavage of murine coronavirus gene A polyprotein. J Virol 63:3693–3699

Baker SC, Yokomori K, Dong S, Carlisle R, Gorbalenya AE, Koonin EV, Lai MM (1993) Identification of the catalytic sites of a papain-like cysteine proteinase of murine coronavirus. J Virol 67:6056–6063

Baker TA, Bell SP (1998) Polymerases and the replisome: machines within machines. Cell 92:295–305

Baric RS, Fu K, Schaad MC, Stohlman SA (1990a) Establishing a genetic recombination map for murine coronavirus strain A59 complementation groups. Virology 177:646–656

Baric RS, Nelson GW, Fleming JO, Deans RJ, Keck JG, Casteel N, Stohlman SA (1988) Interactions between coronavirus nucleocapsid protein and viral RNAs: implications for viral transcription. J Virol 62:4280–4287

Baric RS, Schaad MC, Wei T, Fu KS, Lum K, Shieh C, Stohlman SA (1990b) Murine coronavirus temperature sensitive mutants. Adv Exp Med Biol 276:349–356

Behrens SE, Tomei L, De Francesco R (1996) Identification and properties of the RNA-dependent RNA polymerase of hepatitis C virus. EMBO J 15:12–22

Beinert H, Kennedy MC (1993) Aconitase, a two-faced protein: enzyme and iron regulatory factor. FASEB J 7:1442–1449

Ben-David Y, Bani MR, Chabot B, De Koven A, Bernstein A (1992) Retroviral insertions downstream of the heterogeneous nuclear ribonucleoprotein A1 gene in erythroleukemia cells: evidence that A1 is not essential for cell growth. Mol Cell Biol 12:4449–4455

Bienz K, Egger D, Pfister T (1994) Characteristics of the poliovirus replication complex. Arch Virol Suppl 9:147–157

Bilodeau PS, Domsic JK, Mayeda A, Krainer AR, Stoltzfus CM (2001) RNA splicing at human immunodeficiency virus type 1 3′ splice site A2 is regulated by binding of hnRNP A/B proteins to an exonic splicing silencer element. J Virol 75:8487–8497

Black AC, Luo J, Chun S, Bakker A, Fraser JK, Rosenblatt JD (1996) Specific binding of polypyrimidine tract binding protein and hnRNP A1 to HIV-1 CRS elements. Virus Genes 12:275–285

Black AC, Luo J, Watanabe C, Chun S, Bakker A, Fraser JK, Morgan JP, Rosenblatt JD (1995) Polypyrimidine tract-binding protein and heterogeneous nuclear ribonucleoprotein A1 bind to human T-cell leukemia virus type 2 RNA regulatory elements. J Virol 69:6852–6858

Blackwell JL, Brinton MA (1997) Translation elongation factor-1 alpha interacts with the 3' stem-loop region of West Nile virus genomic RNA. J Virol 71:6433–6444

Blumenthal T, Carmichael GG (1979) RNA replication: function and structure of Qbeta-replicase. Annu Rev Biochem 48:525–548

Bonilla PJ, Gorbalenya AE, Weiss SR (1994) Mouse hepatitis virus strain A59 RNA polymerase gene ORF 1a: heterogeneity among MHV strains. Virology 198:736–740

Bonilla PJ, Hughes SA, Pinon JD, Weiss SR (1995) Characterization of the leader papain-like proteinase of MHV-A59: identification of a new in vitro cleavage site. Virology 209:489–497

Bonilla PJ, Hughes SA, Weiss SR (1997) Characterization of a second cleavage site and demonstration of activity in trans by the papain-like proteinase of the murine coronavirus mouse hepatitis virus strain A59. J Virol 71:900–909

Bost AG, Carnahan RH, Lu XT, Denison MR (2000) Four proteins processed from the replicase gene polyprotein of mouse hepatitis virus colocalize in the cell periphery and adjacent to sites of virion assembly. J Virol 74:3379–3387

Bost AG, Prentice E, Denison MR (2001) Mouse hepatitis virus replicase protein complexes are translocated to sites of M protein accumulation in the ERGIC at late times of infection. Virology 285:21–29

Bothwell AL, Ballard DW, Philbrick WM, Lindwall G, Maher SE, Bridgett MM, Jamison SF, Garcia-Blanco MA (1991) Murine polypyrimidine tract binding protein. Purification, cloning, and mapping of the RNA binding domain. J Biol Chem 266:24657–24663

Brayton PR, Lai MM, Patton CD, Stohlman SA (1982) Characterization of two RNA polymerase activities induced by mouse hepatitis virus. J Virol 42:847–853

Brayton PR, Stohlman SA, Lai MM (1984) Further characterization of mouse hepatitis virus RNA-dependent RNA polymerases. Virology 133:197–201

Bredenbeek PJ, Pachuk CJ, Noten AF, Charite J, Luytjes W, Weiss SR, Spaan WJ (1990) The primary structure and expression of the second open reading frame of the polymerase gene of the coronavirus MHV-A59; a highly conserved polymerase is expressed by an efficient ribosomal frameshifting mechanism. Nucleic Acids Res 18:1825–1832

Brierley I, Boursnell ME, Binns MM, Bilimoria B, Blok VC, Brown TD, Inglis SC (1987) An efficient ribosomal frame-shifting signal in the polymerase-encoding region of the coronavirus IBV. EMBO J 6:3779–3785

Buck KW (1996) Comparison of the replication of positive-stranded RNA viruses of plants and animals. Adv Virus Res 47:159–251

Burd CG, Dreyfuss G (1994) Conserved structures and diversity of functions of RNA-binding proteins. Science 265:615–621

Caputi M, Mayeda A, Krainer AR, Zahler AM (1999) hnRNP A/B proteins are required for inhibition of HIV-1 pre-mRNA splicing. EMBO J 18:4060–4067

Caputi M, Zahler AM (2002) SR proteins and hnRNP H regulate the splicing of the HIV-1 tev-specific exon 6D. EMBO J 21:845–855

Chambers TJ, Hahn CS, Galler R, Rice CM (1990) Flavivirus genome organization, expression, and replication. Annu Rev Microbiol 44:649–688

Chang RY, Brian DA (1996) *cis* Requirement for N-specific protein sequence in bovine coronavirus defective interfering RNA replication. J Virol 70:2201–2207

Chang RY, Hofmann MA, Sethna PB, Brian DA (1994) A *cis*-acting function for the coronavirus leader in defective interfering RNA replication. J Virol 68:8223–8231

Chung RT, Kaplan LM (1999) Heterogeneous nuclear ribonucleoprotein I (hnRNP-I/PTB) selectively binds the conserved 3′ terminus of hepatitis C viral RNA. Biochem Biophys Res Commun 254:351–362

Cologna R, Spagnolo JF, Hogue BG (2000) Identification of nucleocapsid binding sites within coronavirus-defective genomes. Virology 277:235–249

Compton SR, Rogers DB, Holmes KV, Fertsch D, Remenick J, McGowan JJ (1987) In vitro replication of mouse hepatitis virus strain A59. J Virol 61:1814–1820

Craig AW, Haghighat A, Yu AT, Sonenberg N (1998) Interaction of polyadenylate-binding protein with the eIF4G homologue PAIP enhances translation. Nature 392:520–523

Das T, Mathur M, Gupta AK, Janssen GM, Banerjee AK (1998) RNA polymerase of vesicular stomatitis virus specifically associates with translation elongation factor-$1\alpha\beta\gamma$ for its activity. Proc Natl Acad Sci USA 95:1449–1454

De BP, Lesoon A, Banerjee AK (1991) Human parainfluenza virus type 3 transcription in vitro: role of cellular actin in mRNA synthesis. J Virol 65:3268–3275

de la Cruz J, Kressler D, Linder P (1999) Unwinding RNA in *Saccharomyces cerevisiae*: DEAD-box proteins and related families. Trends Biochem Sci 24:192–198

Denison M, Perlman S (1987) Identification of putative polymerase gene product in cells infected with murine coronavirus A59. Virology 157:565–568

Denison MR, Hughes SA, Weiss SR (1995a) Identification and characterization of a 65-kDa protein processed from the gene 1 polyprotein of the murine coronavirus MHV-A59. Virology 207:316–320

Denison MR, Kim JC, Ross T (1995b) Inhibition of coronavirus MHV-A59 replication by proteinase inhibitors. Adv Exp Med Biol 380:391–397

Denison MR, Spaan WJ, van der Meer Y, Gibson CA, Sims AC, Prentice E, Lu XT (1999) The putative helicase of the coronavirus mouse hepatitis virus is processed from the replicase gene polyprotein and localizes in complexes that are active in viral RNA synthesis. J Virol 73:6862–6871

Denison MR, Zoltick PW, Hughes SA, Giangreco B, Olson AL, Perlman S, Leibowitz JL, Weiss SR (1992) Intracellular processing of the N-terminal ORF 1a proteins of the coronavirus MHV-A59 requires multiple proteolytic events. Virology 189:274–284

Dennis DE, Brian DA (1982) RNA-dependent RNA polymerase activity in coronavirus-infected cells. J Virol 42:153–164

Diez J, Ishikawa M, Kaido M, Ahlquist P (2000) Identification and characterization of a host protein required for efficient template selection in viral RNA replication. Proc Natl Acad Sci USA 97:3913–3918

Dong S, Baker SC (1994) Determinants of the p28 cleavage site recognized by the first papain-like cysteine proteinase of murine coronavirus. Virology 204:541–549

Dreyfuss G, Matunis MJ, Pinol-Roma S, Burd CG (1993) hnRNP proteins and the biogenesis of mRNA. Annu Rev Biochem 62:289–321

Froshauer S, Kartenbeck J, Helenius A (1988) Alphavirus RNA replicase is located on the cytoplasmic surface of endosomes and lysosomes. J Cell Biol 107:2075–2086

Fu K, Baric RS (1994) Map locations of mouse hepatitis virus temperature-sensitive mutants: confirmation of variable rates of recombination. J Virol 68:7458–7466

Furuya T, Lai MMC (1993) Three different cellular proteins bind to complementary sites on the 5′-end-positive and 3′ end-negative strands of mouse hepatitis virus RNA. J Virol 67:7215–7222

Gallie DR (1998) A tale of two termini: a functional interaction between the termini of an mRNA is a prerequisite for efficient translation initiation. Gene 216:1–11

Gamarnik AV, Andino R (1997) Two functional complexes formed by KH domain containing proteins with the 5′ noncoding region of poliovirus RNA. RNA 3:882–892

Gamarnik AV, Andino R (1998) Switch from translation to RNA replication in a positive-stranded RNA virus. Genes Dev 12:2293–2304

Gao HQ, Schiller JJ, Baker SC (1996) Identification of the polymerase polyprotein products p72 and p65 of the murine coronavirus MHV-JHM. Virus Res 45:101–109

Gontarek RR, Gutshall LL, Herold KM, Tsai J, Sathe GM, Mao J, Prescott C, Del Vecchio AM (1999) hnRNP C and polypyrimidine tract-binding protein specifically interact with the pyrimidine-rich region within the 3′NTR of the HCV RNA genome. Nucleic Acids Res 27:1457–1463

Gorbalenya AE, Blinov VM, Donchenko AP, Koonin EV (1989a) An NTP-binding motif is the most conserved sequence in a highly diverged monophyletic group of proteins involved in positive strand RNA viral replication. J Mol Evol 28:256–268

Gorbalenya AE, Koonin EV (1989) Viral proteins containing the purine NTP-binding sequence pattern. Nucleic Acids Res 17:8413–8440

Gorbalenya AE, Koonin EV (1993) Helicases: amino acid comparisons and structure-function relationships. Curr Opin Struct Biol 3:419–429

Gorbalenya AE, Koonin EV, Donchenko AP, Blinov VM (1988) A novel superfamily of nucleoside triphosphate-binding motif containing proteins which are probably involved in duplex unwinding in DNA and RNA replication and recombination. FEBS Lett 235:16–24

Gorbalenya AE, Koonin EV, Donchenko AP, Blinov VM (1989b) Coronavirus genome: prediction of putative functional domains in the non-structural polyprotein by comparative amino acid sequence analysis. Nucleic Acids Res 17:4847–4861

Gorbalenya AE, Koonin EV, Lai MM (1991) Putative papain-related thiol proteases of positive-strand RNA viruses. Identification of rubi- and aphthovirus proteases and delineation of a novel conserved domain associated with proteases of rubi-, alpha- and coronaviruses. FEBS Lett 288:201–205

Gorlach M, Burd CG, Dreyfuss G (1994) The mRNA poly(A)-binding protein: localization, abundance, and RNA-binding specificity. Exp Cell Res 211:400–407

Grotzinger C, Heusipp G, Ziebuhr J, Harms U, Suss J, Siddell SG (1996) Characterization of a 105-kDa polypeptide encoded in gene 1 of the human coronavirus HCV 229E. Virology 222:227–235

Gutierrez-Escolano AL, Brito ZU, del Angel RM, Jiang X (2000) Interaction of cellular proteins with the 5' end of Norwalk virus genomic RNA. J Virol 74:8558–8562

Hahm B, Kim YK, Kim JH, Kim TY, Jang SK (1998) Heterogeneous nuclear ribonucleoprotein L interacts with the 3' border of the internal ribosomal entry site of hepatitis C virus. J Virol 72:8782–8788

Hamilton BJ, Burns CM, Nichols RC, Rigby WFC (1997) Modulation of AUUUA response element binding by heterogeneous nuclear ribonucleoprotein A1 in human T lymphocytes. The roles of cytoplasmic location, transcription, and phosphorylation. J Biol Chem 272:28732–28741

Hamilton BJ, Nagy E, Malter JS, Arrick BA, Rigby WFC (1993) Association of heterogeneous nuclear ribonucleoprotein A1 and C proteins with reiterated AUUUA sequences. J Biol Chem 268:8881–8887

Harris KS, Xiang W, Alexander L, Lane WS, Paul AV, Wimmer E (1994) Interaction of poliovirus polypeptide 3CDpro with the 5' and 3' termini of the poliovirus genome. Identification of viral and cellular cofactors needed for efficient binding. J Biol Chem 269:27004–27014

Hellen CU, Pestova TV, Litterst M, Wimmer E (1994) The cellular polypeptide p57 (pyrimidine tract-binding protein) binds to multiple sites in the poliovirus 5' nontranslated region. J Virol 68:941–950

Henics T, Sanfridson A, Hamilton BJ, Nagy E, Rigby WFC (1994) Enhanced stability of interleukin-2 mRNA in MLA 144 cells. Possible role of cytoplasmic AU-rich sequence-binding proteins. J Biol Chem 269:5377–5383

Hentze MW, Kuhn LC (1996) Molecular control of vertebrate iron metabolism: mRNA-based regulatory circuits operated by iron, nitric oxide, and oxidative stress. Proc Natl Acad Sci USA 93:8175–8182

Herold J, Gorbalenya AE, Thiel V, Schelle B, Siddell SG (1998) Proteolytic processing at the amino terminus of human coronavirus 229E gene 1-encoded polyproteins: identification of a papain-like proteinase and its substrate. J Virol 72:910–918

Herold J, Siddell SG (1993) An 'elaborated' pseudoknot is required for high frequency frameshifting during translation of HCV 229E polymerase mRNA. Nucleic Acids Res 21:5838–5842

Herold J, Siddell SG, Gorbalenya AE (1999) A human RNA viral cysteine proteinase that depends upon a unique Zn^{2+}-binding finger connecting the two domains of a papain-like fold. J Biol Chem 274:14918–14925

Hong Y, Hunt AG (1996) RNA polymerase activity catalyzed by a potyvirus-encoded RNA-dependent RNA polymerase. Virology 226:146–151

Hsue B, Hartshorne T, Masters PS (2000) Characterization of an essential RNA secondary structure in the 3' untranslated region of the murine coronavirus genome. J Virol 74:6911–6921

Hsue B, Masters PS (1997) A bulged stem-loop structure in the 3' untranslated region of the genome of the coronavirus mouse hepatitis virus is essential for replication. J Virol 71:7567–7578

Huang P, Lai MM (2001) Heterogeneous nuclear ribonucleoprotein a1 binds to the 3′-untranslated region and mediates potential 5′-3′ end cross talks of mouse hepatitis virus RNA. J Virol 75:5009–5017

Huang P, Lai MMC (1999) Polypyrimidine tract-binding protein binds to the complementary strand of the mouse hepatitis virus 3′ untranslated region, thereby altering RNA conformation. J Virol 73:9110–9116

Huang YT, Romito RR, De BP, Banerjee AK (1993) Characterization of the in vitro system for the synthesis of mRNA from human respiratory syncytial virus. Virology 193:862–867

Hughes SA, Bonilla PJ, Weiss SR (1995) Identification of the murine coronavirus p28 cleavage site. J Virol 69:809–813

Imataka H, Gradi A, Sonenberg N (1998) A newly identified N-terminal amino acid sequence of human eIF4G binds poly(A)-binding protein and functions in poly(A)-dependent translation. EMBO J 17:7480–7489

Ito T, Lai MMC (1997) Determination of the secondary structure of and cellular protein binding to the 3′-untranslated region of the hepatitis C virus RNA genome. J Virol 71:8698–8706

Jackson RJ, Kaminski A (1995) Internal initiation of translation in eukaryotes: the picornavirus paradigm and beyond. RNA 1:985–1000

Jankowsky E, Gross CH, Shuman S, Pyle AM (2000) The DExH protein NPH-II is a processive and directional motor for unwinding RNA. Nature 403:447–451

Jendrach M, Thiel V, Siddell S (1999) Characterization of an internal ribosome entry site within mRNA 5 of murine hepatitis virus. Arch Virol 144:921–933

Joshi RL, Ravel JM, Haenni AL (1986) Interaction of turnip yellow mosaic virus Val-RNA with eukaryotic elongation factor EF-1α. Search for a function. EMBO J 5:1143–1148

Kadare G, Haenni AL (1997) Virus-encoded RNA helicases. J Virol 71:2583–2590

Kaminski A, Hunt SL, Patton JG, Jackson RJ (1995) Direct evidence that polypyrimidine tract binding protein (PTB) is essential for internal initiation of translation of encephalomyocarditis virus RNA. RNA 1:924–938

Kanjanahaluethai A, Baker SC (2000) Identification of mouse hepatitis virus papain-like proteinase 2 activity. J Virol 74:7911–7921

Kennedy MC, Mende-Mueller L, Blondin GA, Beinert H (1992) Purification and characterization of cytosolic aconitase from beef liver and its relationship to the iron-responsive element binding protein. Proc Natl Acad Sci USA 89:11730–11734

Kim HY, LaVaute T, Iwai K, Klausner RD, Rouault TA (1996) Identification of a conserved and functional iron-responsive element in the 5′-untranslated region of mammalian mitochondrial aconitase. J Biol Chem 271:24226–24230

Kim JC, Spence RA, Currier PF, Lu X, Denison MR (1995) Coronavirus protein processing and RNA synthesis is inhibited by the cysteine proteinase inhibitor E64d. Virology 208:1–8

Kim KH, Makino S (1995a) Two murine coronavirus genes suffice for viral RNA synthesis. J Virol 69:2313–2321

Kim YN, Jeong YS, Makino S (1993) Analysis of *cis*-acting sequences essential for coronavirus defective interfering RNA replication. Virology 197:53–63

Kim YN, Makino S (1995b) Characterization of a murine coronavirus defective interfering RNA internal *cis*-acting replication signal. J Virol 69:4963-4971

Klausner RD, Rouault TA, Harford JB (1993) Regulating the fate of mRNA: the control of cellular iron metabolism. Cell 72:19-28

Koolen MJ, Osterhaus AD, Van Steenis G, Horzinek MC, Van der Zeijst BA (1983) Temperature-sensitive mutants of mouse hepatitis virus strain A59: isolation, characterization and neuropathogenic properties. Virology 125:393-402

Koonin EV, Dolja VV (1993) Evolution and taxonomy of positive-strand RNA viruses: implications of comparative analysis of amino acid sequences. Crit Rev Biochem Mol Biol 28:375-430

Kuhn LC, Hentze MW (1992) Coordination of cellular iron metabolism by post-transcriptional gene regulation. J Inorg Biochem 47:183-195

Lai MMC (1998) Cellular factors in the transcription and replication of viral RNA genomes: a parallel to DNA-dependent RNA transcription. Virology 244:1-12

Lai MMC, Cavanagh D (1997) The molecular biology of coronaviruses. Adv Virus Res 48:1-100

Landers TA, Blumenthal T, Weber K (1974) Function and structure in ribonucleic acid phage Q beta ribonucleic acid replicase. The roles of the different subunits in transcription of synthetic templates. J Biol Chem 249:5801-5808

Le H, Tanguay RL, Balasta ML, Wei CC, Browning KS, Metz AM, Goss DJ, Gallie DR (1997) Translation initiation factors eIF-iso4G and eIF-4B interact with the poly(A)-binding protein and increase its RNA binding activity. J Biol Chem 272:16247-16255

Lee HJ, Shieh CK, Gorbalenya AE, Koonin EV, La Monica N, Tuler J, Bagdzhadzhyan A, Lai MM (1991) The complete sequence (22 kilobases) of murine coronavirus gene 1 encoding the putative proteases and RNA polymerase. Virology 180:567-582

Leibowitz JL, DeVries JR (1988) Synthesis of virus-specific RNA in permeabilized murine coronavirus-infected cells. Virology 166:66-75

Leibowitz JL, DeVries JR, Haspel MV (1982) Genetic analysis of murine hepatitis virus strain JHM. J Virol 42:1080-1087

Li HP, Huang P, Park S, Lai MMC (1999) Polypyrimidine tract-binding protein binds to the leader RNA of mouse hepatitis virus and serves as a regulator of viral transcription. J Virol 73:772-777

Li HP, Zhang X, Duncan R, Comai L, Lai MMC (1997) Heterogeneous nuclear ribonucleoprotein A1 binds to the transcription-regulatory region of mouse hepatitis virus RNA. Proc Natl Acad Sci USA 94:9544-9549

Liao CL, Lai MMC (1994) Requirement of the 5' end genomic sequence as an upstream *cis*-acting element for coronavirus subgenomic mRNA transcription. J Virol 68:4727-4737

Lim KP, Liu DX (1998) Characterization of the two overlapping papain-like proteinase domains encoded in gene 1 of the coronavirus infectious bronchitis virus and determination of the C-terminal cleavage site of an 87-kDa protein. Virology 245:303-312

Lim KP, Ng LF, Liu DX (2000) Identification of a novel cleavage activity of the first papain-like proteinase domain encoded by open reading frame 1a of the corona-

virus avian infectious bronchitis virus and characterization of the cleavage products. J Virol 74:1674–1685
Lin YJ, Lai MM (1993) Deletion mapping of a mouse hepatitis virus defective interfering RNA reveals the requirement of an internal and discontiguous sequence for replication. J Virol 67:6110–6118
Lin YJ, Liao CL, Lai MM (1994) Identification of the *cis*-acting signal for minus-strand RNA synthesis of a murine coronavirus: implications for the role of minus-strand RNA in RNA replication and transcription. J Virol 68:8131–8140
Lin YJ, Zhang X, Wu RC, Lai MMC (1996) The 3' untranslated region of coronavirus RNA is required for subgenomic mRNA transcription from a defective interfering RNA. J Virol 70:7236–7240
Linder P, Daugeron MC (2000) Are DEAD-box proteins becoming respectable helicases? Nat Struct Biol 7:97–99
Liu DX, Brierley I, Tibbles KW, Brown TD (1994) A 100-kilodalton polypeptide encoded by open reading frame (ORF) 1b of the coronavirus infectious bronchitis virus is processed by ORF 1a products. J Virol 68:5772–5780
Liu DX, Brown TD (1995) Characterisation and mutational analysis of an ORF 1a-encoding proteinase domain responsible for proteolytic processing of the infectious bronchitis virus 1a/1b polyprotein. Virology 209:420–427
Liu DX, Shen S, Xu HY, Wang SF (1998) Proteolytic mapping of the coronavirus infectious bronchitis virus 1b polyprotein: evidence for the presence of four cleavage sites of the 3C-like proteinase and identification of two novel cleavage products. Virology 246:288–297
Liu Q, Johnson RF, Leibowitz JL (2001) Secondary structural elements within the 3' untranslated region of mouse hepatitis virus strain JHM genomic RNA. J Virol 75:12105–12113
Liu Q, Yu W, Leibowitz JL (1997) A specific host cellular protein binding element near the 3' end of mouse hepatitis virus genomic RNA. Virology 232:74–85
Lohman TM, Bjornson KP (1996) Mechanisms of helicase-catalyzed DNA unwinding. Annu Rev Biochem 65:169–214
Lohmann V, Korner F, Herian U, Bartenschlager R (1997) Biochemical properties of hepatitis C virus NS5B RNA-dependent RNA polymerase and identification of amino acid sequence motifs essential for enzymatic activity. J Virol 71:8416–8428
Lu X, Lu Y, Denison MR (1996) Intracellular and in vitro-translated 27-kDa proteins contain the 3C-like proteinase activity of the coronavirus MHV-A59. Virology 222:375–382
Lu XT, Sims AC, Denison MR (1998) Mouse hepatitis virus 3C-like protease cleaves a 22-kilodalton protein from the open reading frame 1a polyprotein in virus-infected cells and in vitro. J Virol 72:2265–2271
Lu Y, Denison MR (1997) Determinants of mouse hepatitis virus 3C-like proteinase activity. Virology 230:335–342
Lu Y, Lu X, Denison MR (1995) Identification and characterization of a serine-like proteinase of the murine coronavirus MHV-A59. J Virol 69:3554–3559
Luytjes W, Bredenbeek PJ, Noten AF, Horzinek MC, Spaan WJ (1988) Sequence of mouse hepatitis virus A59 mRNA 2: indications for RNA recombination between coronaviruses and influenza C virus. Virology 166:415–422

Ma AS, Moran-Jones K, Shan J, Munro TP, Snee MJ, Hoek KS, Smith R (2002) hnRNP A3, a novel RNA trafficking response element binding protein. J Biol Chem 8:8

Mahy BW, Siddell S, Wege H, ter Meulen V (1983) RNA-dependent RNA polymerase activity in murine coronavirus-infected cells. J Gen Virol 64:103–111

Makino S, Joo M, Makino JK (1991) A system for study of coronavirus mRNA synthesis: a regulated, expressed subgenomic defective interfering RNA results from intergenic site insertion. J Virol 65:6031–6041

Martin JP, Koehren F, Rannou JJ, Kirn A (1988) Temperature-sensitive mutants of mouse hepatitis virus type 3 (MHV-3): isolation, biochemical and genetic characterization. Arch Virol 100:147–160

Mayeda A, Munroe SH, Caceres JF, Krainer AR (1994) Function of conserved domains of hnRNP A1 and other hnRNP A/B proteins. EMBO J 13:5483–5495

Meerovitch K, Svitkin YV, Lee HS, Lejbkowicz F, Kenan DJ, Chan EK, Agol VI, Keene JD, Sonenberg N (1993) La autoantigen enhances and corrects aberrant translation of poliovirus RNA in reticulocyte lysate. J Virol 67:3798–3807

Michael WM, Siomi H, Choi M, Pinol-Roma S, Nakielny S, Liu Q, Dreyfuss G (1995) Signal sequences that target nuclear import and nuclear export of pre-mRNA-binding proteins. Cold Spring Harb Symp Quant Biol 60:663–668

Miller DJ, Schwartz MD, Ahlquist P (2001) Flock house virus RNA replicates on outer mitochondrial membranes in *Drosophila* cells. J Virol 75:11664–11676

Molenkamp R, van Tol H, Rozier BC, van der Meer Y, Spaan WJ, Snijder EJ (2000) The arterivirus replicase is the only viral protein required for genome replication and subgenomic mRNA transcription. J Gen Virol 81:2491–2496

Moyer SA, Baker SC, Horikami SM (1990) Host cell proteins required for measles virus reproduction. J Gen Virol 71:775–783

Moyer SA, Baker SC, Lessard JL (1986) Tubulin: a factor necessary for the synthesis of both Sendai virus and vesicular stomatitis virus RNAs. Proc Natl Acad Sci USA 83:5405–5409

Nanda SK, Leibowitz JL (2001) Mitochondrial aconitase binds to the 3' untranslated region of the mouse hepatitis virus genome. J Virol 75:3352–3362

Nelson GW, Stohlman SA, Tahara SM (2000) High affinity interaction between nucleocapsid protein and leader/intergenic sequence of mouse hepatitis virus RNA. J Gen Virol 81:181–188

Neufeld KL, Richards OC, Ehrenfeld E (1991) Purification, characterization, and comparison of poliovirus RNA polymerase from native and recombinant sources. J Biol Chem 266:24212–24219

Niepmann M (1996) Porcine polypyrimidine tract-binding protein stimulates translation initiation at the internal ribosome entry site of foot-and-mouth-disease virus. FEBS Lett 388:39–42

Niepmann M, Petersen A, Meyer K, Beck E (1997) Functional involvement of polypyrimidine tract-binding protein in translation initiation complexes with the internal ribosome entry site of foot-and-mouth disease virus. J Virol 71:8330–8339

Oglesbee MJ, Liu Z, Kenney H, Brooks CL (1996) The highly inducible member of the 70 kDa family of heat shock proteins increases canine distemper virus polymerase activity. J Gen Virol 77:2125–2135

Osman TA, Buck KW (1997) The tobacco mosaic virus RNA polymerase complex contains a plant protein related to the RNA-binding subunit of yeast eIF-3. J Virol 71:6075–6082

Pardigon N, Strauss JH (1996) Mosquito homolog of the La autoantigen binds to Sindbis virus RNA. J Virol 70:1173–1181

Parsley TB, Towner JS, Blyn LB, Ehrenfeld E, Semler BL (1997) Poly (rC) binding protein 2 forms a ternary complex with the 5′-terminal sequences of poliovirus RNA and the viral 3CD proteinase. RNA 3:1124–1134

Perlman S, Ries D, Bolger E, Chang LJ, Stoltzfus CM (1986) MHV nucleocapsid synthesis in the presence of cycloheximide and accumulation of negative strand MHV RNA. Virus Res 6:261–272

Pinol-Roma S, Dreyfuss G (1992) Shuttling of pre-mRNA binding proteins between nucleus and cytoplasm. Nature 355:730–732

Pinon JD, Mayreddy RR, Turner JD, Khan FS, Bonilla PJ, Weiss SR (1997) Efficient autoproteolytic processing of the MHV-A59 3C-like proteinase from the flanking hydrophobic domains requires membranes. Virology 230:309–322

Pinon JD, Teng H, Weiss SR (1999) Further requirements for cleavage by the murine coronavirus 3C-like proteinase: identification of a cleavage site within ORF1b. Virology 263:471–484

Pogue GP, Huntley CC, Hall TC (1994) Common replication strategies emerging from the study of diverse groups of positive-strand RNA viruses. Arch Virol Suppl 9:181–194

Quadt R, Kao CC, Browning KS, Hershberger RP, Ahlquist P (1993) Characterization of a host protein associated with brome mosaic virus RNA-dependent RNA polymerase. Proc Natl Acad Sci USA 90:1498–1502

Robb JA, Bond CW (1979) Pathogenic murine coronaviruses. I. Characterization of biological behavior in vitro and virus-specific intracellular RNA of strongly neurotropic JHMV and weakly neurotropic A59V viruses. Virology 94:352–370

Rothstein MA, Richards OC, Amin C, Ehrenfeld E (1988) Enzymatic activity of poliovirus RNA polymerase synthesized in *Escherichia coli* from viral cDNA. Virology 164:301–308

Sachs AB, Sarnow P, Hentze MW (1997) Starting at the beginning, middle, and end: translation initiation in eukaryotes. Cell 89:831–838

Sawicki SG, Sawicki DL (1986) Coronavirus minus-strand RNA synthesis and effect of cycloheximide on coronavirus RNA synthesis. J Virol 57:328–334

Schaad MC, Stohlman SA, Egbert J, Lum K, Fu K, Wei T, Jr., Baric RS (1990) Genetics of mouse hepatitis virus transcription: identification of cistrons which may function in positive and negative strand RNA synthesis. Virology 177:634–645

Schalinske KL, Chen OS, Eisenstein RS (1998) Iron differentially stimulates translation of mitochondrial aconitase and ferritin mRNAs in mammalian cells. Implications for iron regulatory proteins as regulators of mitochondrial citrate utilization. J Biol Chem 273:3740–3746

Schiller JJ, Kanjanahaluethai A, Baker SC (1998) Processing of the coronavirus MHV-JHM polymerase polyprotein: identification of precursors and proteolytic products spanning 400 kilodaltons of ORF1a. Virology 242:288–302

Schmid SR, Linder P (1992) D-E-A-D protein family of putative RNA helicases. Mol Microbiol 6:283–291

Schwartz M, Chen J, Janda M, Sullivan M, den Boon J, Ahlquist P (2002) A positive-strand RNA virus replication complex parallels form and function of retrovirus capsids. Mol Cell 9:505–514

Schwarz B, Routledge E, Siddell SG (1990) Murine coronavirus nonstructural protein ns2 is not essential for virus replication in transformed cells. J Virol 64:4784–4791

Seybert A, Hegyi A, Siddell SG, Ziebuhr J (2000) The human coronavirus 229E superfamily 1 helicase has RNA and DNA duplex-unwinding activities with 5′-to-3′ polarity. RNA 6:1056–1068

Shen X, Masters PS (2001) Evaluation of the role of heterogeneous nuclear ribonucleoprotein A1 as a host factor in murine coronavirus discontinuous transcription and genome replication. Proc Natl Acad Sci USA 98:2717–2722

Shi ST, Huang P, Li HP, Lai MMC (2000) Heterogeneous nuclear ribonucleoprotein A1 regulates RNA synthesis of a cytoplasmic virus. EMBO J 19:4701–4711

Shi ST, Schiller JJ, Kanjanahaluethai A, Baker SC, Oh JW, Lai MM (1999) Colocalization and membrane association of murine hepatitis virus gene 1 products and de novo-synthesized viral RNA in infected cells. J Virol 73:5957–5969

Shi ST, Yu GY, Lai MMC (2003) Multiple type A/B heterogeneous nuclear ribonucleoproteins (hnRNPs) can replace hnRNP A1 in mouse hepatitis virus RNA synthesis. J Virol 11:10584–10593

Sims AC, Ostermann J, Denison MR (2000) Mouse hepatitis virus replicase proteins associate with two distinct populations of intracellular membranes. J Virol 74:5647–5654

Siomi H, Dreyfuss G (1995) A nuclear localization domain in the hnRNP A1 protein. J Cell Biol 129:551–560

Snijder EJ, van Tol H, Roos N, Pedersen KW (2001) Non-structural proteins 2 and 3 interact to modify host cell membranes during the formation of the arterivirus replication complex. J Gen Virol 82:985–994

Sokolowski M, Schwartz S (2001) Heterogeneous nuclear ribonucleoprotein C binds exclusively to the functionally important UUUUU-motifs in the human papillomavirus type-1 AU-rich inhibitory element. Virus Res 73:163–175

Spagnolo JF, Hogue BG (2000) Host protein interactions with the 3′ end of bovine coronavirus RNA and the requirement of the poly(A) tail for coronavirus defective genome replication. J Virol 74:5053–5065

Spangberg K, Wiklund L, Schwartz S (2000) HuR, a protein implicated in oncogene and growth factor mRNA decay, binds to the 3′ ends of hepatitis C virus RNA of both polarities. Virology 274:378–390

Stalcup RP, Baric RS, Leibowitz JL (1998) Genetic complementation among three panels of mouse hepatitis virus gene 1 mutants. Virology 241:112–121

Stohlman SA, Baric RS, Nelson GN, Soe LH, Welter LM, Deans RJ (1988) Specific interaction between coronavirus leader RNA and nucleocapsid protein. J Virol 62:4288–4295

Svitkin YV, Ovchinnikov LP, Dreyfuss G, Sonenberg N (1996) General RNA binding proteins render translation cap dependent. EMBO J 15:7147–7155

Tahara S, Bergmann C, Nelson G, Anthony R, Dietlin T, Kyuwa S, Stohlman S (1993) Effects of mouse hepatitis virus infection on host cell metabolism. Adv Exp Med Biol 342:111–116

Tahara SM, Dietlin TA, Bergmann CC, Nelson GW, Kyuwa S, Anthony RP, Stohlman SA (1994) Coronavirus translational regulation: leader affects mRNA efficiency. Virology 202:621–630

Tahara SM, Dietlin TA, Nelson GW, Stohlman SA, Manno DJ (1998) Mouse hepatitis virus nucleocapsid protein as a translational effector of viral mRNAs. Adv Exp Med Biol 440:313–318

Tan BH, Fu J, Sugrue RJ, Yap EH, Chan YC, Tan YH (1996) Recombinant dengue type 1 virus NS5 protein expressed in *Escherichia coli* exhibits RNA-dependent RNA polymerase activity. Virology 216:317–325

Tarun SZ, Jr., Sachs AB (1996) Association of the yeast poly(A) tail binding protein with translation initiation factor eIF-4G. EMBO J 15:7168–7177

Tarun SZ, Jr., Wells SE, Deardorff JA, Sachs AB (1997) Translation initiation factor eIF4G mediates in vitro poly(A) tail-dependent translation. Proc Natl Acad Sci USA 94:9046–9051

Thiel V, Herold J, Schelle B, Siddell SG (2001) Viral replicase gene products suffice for coronavirus discontinuous transcription. J Virol 75:6676–6681

Thiel V, Siddell SG (1994) Internal ribosome entry in the coding region of murine hepatitis virus mRNA 5. J Gen Virol 75:3041–3046

Tijms MA, van Dinten LC, Gorbalenya AE, Snijder EJ (2001) A zinc finger-containing papain-like protease couples subgenomic mRNA synthesis to genome translation in a positive-stranded RNA virus. Proc Natl Acad Sci USA 98:1889–1894

Valcarcel J, Gebauer F (1997) Post-transcriptional regulation: the dawn of PTB. Curr Biol 7:R705–708

van der Meer Y, Snijder EJ, Dobbe JC, Schleich S, Denison MR, Spaan WJ, Locker JK (1999) Localization of mouse hepatitis virus nonstructural proteins and RNA synthesis indicates a role for late endosomes in viral replication. J Virol 73:7641–7657

van Dinten LC, den Boon JA, Wassenaar AL, Spaan WJ, Snijder EJ (1997) An infectious arterivirus cDNA clone: identification of a replicase point mutation that abolishes discontinuous mRNA transcription. Proc Natl Acad Sci USA 94:991–996

van Dinten LC, Rensen S, Gorbalenya AE, Snijder EJ (1999) Proteolytic processing of the open reading frame 1b-encoded part of arterivirus replicase is mediated by nsp4 serine protease and Is essential for virus replication. J Virol 73:2027–2037

van Dinten LC, Wassenaar AL, Gorbalenya AE, Spaan WJ, Snijder EJ (1996) Processing of the equine arteritis virus replicase ORF1b protein: identification of cleavage products containing the putative viral polymerase and helicase domains. J Virol 70:6625–6633

Van Dyke TA, Flanegan JB (1980) Identification of poliovirus polypeptide P63 as a soluble RNA-dependent RNA polymerase. J Virol 35:732–740

Wang Y, Zhang X (1999) The nucleocapsid protein of coronavirus mouse hepatitis virus interacts with the cellular heterogeneous nuclear ribonucleoprotein A1 in vitro and in vivo. Virology 265:96–109

Wang YF, Chen SC, Wu FY, Wu CW (1997) The interaction between human cytomegalovirus immediate-early gene 2 (IE2) protein and heterogeneous ribonucleoprotein A1. Biochem Biophys Res Commun 232:590–594

Weighardt F, Biamonti G, Riva S (1995) Nucleo-cytoplasmic distribution of human hnRNP proteins: a search for the targeting domains in hnRNP A1. J Cell Sci 108:545–555

Williams GD, Chang RY, Brian DA (1995) Evidence for a pseudoknot in the 3' untranslated region of the bovine coronavirus genome. Adv Exp Med Biol 380:511–514

Wu-Baer F, Lane WS, Gaynor RB (1996) Identification of a group of cellular cofactors that stimulate the binding of RNA polymerase II and TRP-185 to human immunodeficiency virus 1 TAR RNA. J Biol Chem 271:4201–4208

Yokomori K, Banner LR, Lai MM (1991) Heterogeneity of gene expression of the hemagglutinin-esterase (HE) protein of murine coronaviruses. Virology 183:647–657

Yokomori K, Lai MM (1991) Mouse hepatitis virus S RNA sequence reveals that nonstructural proteins ns4 and ns5a are not essential for murine coronavirus replication. J Virol 65:5605–5608

Yu W, Leibowitz JL (1995a) A conserved motif at the 3' end of mouse hepatitis virus genomic RNA required for host protein binding and viral RNA replication. Virology 214:128–138

Yu W, Leibowitz JL (1995b) Specific binding of host cellular proteins to multiple sites within the 3' end of mouse hepatitis virus genomic RNA. J Virol 69:2016–2023

Yuan ZH, Kumar U, Thomas HC, Wen YM, Monjardino J (1997) Expression, purification, and partial characterization of HCV RNA polymerase. Biochem Biophys Res Commun 232:231–235

Zhang X, Lai MMC (1995a) Interactions between the cytoplasmic proteins and the intergenic (promoter) sequence of mouse hepatitis virus RNA: correlation with the amounts of subgenomic mRNA transcribed. J Virol 69:1637–1644

Zhang X, Li HP, Xue W, Lai MMC (1999) Formation of a ribonucleoprotein complex of mouse hepatitis virus involving heterogeneous nuclear ribonucleoprotein A1 and transcription-regulatory elements of viral RNA. Virology 264:115–124

Zhang X, Liao CL, Lai MMC (1994) Coronavirus leader RNA regulates and initiates subgenomic mRNA transcription both *in trans* and *in cis*. J Virol 68:4738–4746

Zhang XM, Lai MMC (1995b) Regulation of coronavirus RNA transcription is likely mediated by protein-RNA interactions. Adv Exp Med Biol 380:515–521

Ziebuhr J, Siddell SG (1999) Processing of the human coronavirus 229E replicase polyproteins by the virus-encoded 3C-like proteinase: identification of proteolytic products and cleavage sites common to pp1a and pp1ab. J Virol 73:177–185

Ziebuhr J, Snijder EJ, Gorbalenya AE (2000) Virus-encoded proteinases and proteolytic processing in the Nidovirales. J Gen Virol 81:853–879

Coronavirus Reverse Genetics by Targeted RNA Recombination

P. S. Masters[1] (✉) · P. J. M. Rottier[2]

[1] Laboratory of Viral Disease, Division of Infectious Disease, Wadsworth Center, New York State Department of Health, Albany, NY, USA
masters@wadsworth.org
[2] Virology Division, Department of Infectious Diseases and Immunology, Faculty of Veterinary Medicine and Institute of Biomembranes, Utrecht University, 3584 CL, Utrecht, The Netherlands
p.rottier@vet.uu.nl

1	Introduction	134
2	Coronavirus RNA Recombination	134
3	Targeted RNA Recombination: Methodology and Technical Issues	137
3.1	Original Development of the System	137
3.2	Improving the Donor RNA: DI and Pseudo-DI RNAs	140
3.3	Improving the Recipient Virus: Host Range-Based Selection	142
4	Targeted RNA Recombination: Spectrum of Applications	145
4.1	Virion Structure and Assembly	146
4.2	RNA Replication and Transcription	148
4.3	Pathogenesis	149
4.4	Coronavirus Vaccines and Vectors	151
5	Conclusions and Future Prospects	152
	References	154

Abstract Targeted RNA recombination was the first reverse genetics system devised for coronaviruses at a time when it was not clear whether the construction of full-length infectious cDNA clones would become possible. In its current state targeted RNA recombination offers a versatile and powerful method for the site-directed mutagenesis of the downstream third of the coronavirus genome, which encodes all the viral structural proteins. The development of this system is described, with an emphasis on recent improvements, and multiple applications of this technique to the study of coronavirus molecular biology and pathogenesis are reviewed. Additionally, the relative strengths and limitations of targeted RNA recombination and infectious cDNA systems are contrasted.

1
Introduction

Targeted RNA recombination was developed to address the need for a reverse genetic system for coronaviruses at a time when it was uncertain whether the construction of full-length infectious cDNA clones was technically feasible or, indeed, even possible. As detailed elsewhere in this volume, this goal has now been realized, largely through the tenacity and ingenuity of a handful of investigators. Concurrently, the ensuing decade since its origination has allowed targeted recombination to evolve into a productive methodology that, across the boundaries of multiple laboratories and viral species, has enabled coronavirus studies to take advantage of the opportunities offered by reverse genetics.

In this chapter we begin with a brief background on the prominence of recombination in coronavirus RNA synthesis and then detail how this property has been exploited for the purposes of site-directed mutagenesis of the coronavirus genome. We describe the scientific problems to which targeted recombination has been successfully applied, and finally we comment on the future prospects for this technique. Throughout our discussion emphasis is placed on new developments in the field since the last time this subject was reviewed (Masters 1999).

2
Coronavirus RNA Recombination

RNA recombination is a well-established phenomenon among animal, plant, and bacterial RNA viruses (reviewed in Lai 1992; Nagy and Simon 1997). As a mechanism of genetic exchange, it provides these viruses with a powerful evolutionary attribute. Recombination is concomitant with viral RNA replication. The consensus model for its occurrence is that the viral RNA polymerase, with a nascent RNA strand attached to it, dissociates from its template and resumes RNA synthesis after it has bound elsewhere to the same or to another template. This "copy-choice" or template-switching mechanism was originally established for polioviruses (Kirkegaard and Baltimore 1986), the viral species for which RNA recombination was first demonstrated (Ledinko 1963), but it seems to be generally applicable.

Homologous RNA recombination takes place when there is a switch of templates between regions of high sequence similarity. This particular form of recombination had only been observed for—and was thus believed to be restricted to—positive-strand RNA viruses, but it has recently also been demonstrated for a minus-strand RNA virus (Plyusnin et al. 2002). Homologous RNA recombination occurs at a remarkably high rate among coronaviruses (Lai 1992, 1996). Their huge genome size and particular mode of replication, employing a discontinuous mode of transcription, may favor polymerase template switching (Brian and Spaan 1996). Accordingly, the phenomenon also has been observed for other nidoviruses, particularly the arteriviruses (Li et al. 1999; Yuan et al. 1999; van Vugt et al. 2001).

Experimental evidence for RNA recombination in coronaviruses has rapidly accumulated, ever since its first description in the mid-1980s (Lai et al. 1985). Essentially all of the early work was done with mouse hepatitis virus (MHV) by taking skillful advantage of the availability of distinctive natural viral strains and classic mutants generated in the laboratory. Initially, through the analysis of progeny obtained from coinfection of culture cells or mouse brains with different MHV variants and application of different selection principles (e.g., temperature sensitivity, cell fusion ability, sensitivity to neutralization by specific antibodies), many of the fundamental features of coronavirus recombination were elucidated (Lai et al. 1985; Keck et al. 1987, 1988a,b; Makino et al. 1987). Sequence analyses revealed that recombination can happen virtually anywhere along the genome but that particular virus combinations show preferred crossover regions, probably owing to selective pressure (Banner et al. 1990). Many MHV recombinants were found to have multiple crossovers, consistent with an exceptionally high frequency of recombination. The overall frequency per passage was estimated at approximately 1% per 1,300 nucleotides (or 25% over the entire genome) by long-range mapping using temperature-sensitive mutants (Baric et al. 1990). Similar studies subsequently demonstrated that, within a relatively short interval, the recombination frequency is uniform (Banner and Lai 1991) but it increases progressively from the 5' to the 3' end of the MHV genome, presumably because of participation of subgenomic (sg) RNAs (Fu and Baric 1994). Although homologous RNA recombination has been less extensively studied in other viral species, the experimental demonstration of this phenomenon has not been limited to the group 2 coronavirus MHV. It has been shown as well for the group 3 coronavirus infectious bronchitis virus (IBV) (Kottier et al. 1995) and the group 1 coronavirus transmissible gastroenteritis virus (TGEV) (Sánchez et al. 1999), for the former by coinfection of viruses into embryonated eggs and for the latter by electroporation of defective RNA into infected cells in tissue culture.

Recombination of coronaviruses appears to be a process of significant importance in the wild. Its occurrence has been shown to contribute to the natural evolution of IBV. This highly contagious virus comprises many different serotypes, and new ones emerge regularly, with the result that these viruses escape from host immunity and cause new outbreaks. Although many of the new variants arise by genetic drift as a result of subtle mutations in the spike protein (S) gene, similar to the changes that lead to antigenic drift in influenza viruses, new serotypes apparently also originate from genetic exchange of S gene sequences between different viruses through homologous RNA recombination (Kusters et al. 1990; Cavanagh et al. 1990; Wang et al. 1993; Jia et al. 1995). Of considerable impact on these evolutionary processes is the veterinary practice of vaccination. Large-scale application of (combinations of) live attenuated vaccine viruses drastically enhances the opportunities for recombination. The identification of vaccine-derived sequences in field isolates is therefore not surprising (Kusters et al. 1990; Wang et al. 1993; Lee and Jackwood 2001). Rather, these events actually seem to occur at high frequency and are not restricted to the S gene region, as attested by the complex genetic makeup of IBV strains that carry the footprints of multiple independent recombinations (Jia et al. 1995; Lee and Jackwood 2000).

Homologous RNA recombination also plays an important role in the evolution of feline coronaviruses. These viruses fall into two serotypes, with type I viruses being the most prevalent. Unlike type I viruses, the type II viruses cross-react with canine coronavirus (CCoV) in virus neutralization assays, and sequence analysis of their S genes indeed confirms this relatedness: Serotype II viruses appear to be derived from recombination between type I feline coronaviruses and CCoV (Motokawa et al. 1995; Vennema et al. 1995; Herrewegh et al. 1995). Detailed analyses of two type II strains revealed that each actually resulted from double recombination, with crossover points located both upstream and downstream of the S gene (Herrewegh et al. 1998). Importantly, all of the crossover points were unique, and subsequent sequencing of the 3' genomic region of two additional type II strains showed that the template switches in this region had occurred at different sites in all four viruses: two each in the envelope protein (E) and the membrane protein (M) genes (Vennema 1999). Obviously, these viruses must have arisen from independent recombination events. Although it is not known in which host species the coinfection of feline and canine coronaviruses takes place, these observations suggest that such occurrences are not overly rare.

More generally, RNA recombination is also believed to have been instrumental in the emergence of the three coronavirus groups. Viruses from these groups characteristically differ in the identities and genomic locations of their nonessential genes. These group-specific genes are presumed to have been acquired by recombination, in this case nonhomologous, with cellular or heterologous viral RNAs. A case in point is the hemagglutinin-esterase (HE) gene found in several group 2 coronaviruses as well as in toroviruses. This gene was presumably derived from recombination between an ancestral coronavirus and influenza C virus, as is suggested by its remarkable sequence similarity to the corresponding orthomyxoviral HE gene (Luytjes et al. 1988). Apart from still-undefined roles in interactions with their respective hosts (de Haan et al. 2002a), the functions and possible origins of the other group-specific genes remain elusive.

3
Targeted RNA Recombination: Methodology and Technical Issues

3.1
Original Development of the System

Targeted RNA recombination was devised as a means of introducing specified changes into the coronavirus genome through recombination between a donor synthetic RNA and a recipient parent virus possessing some characteristic that allows it to be counterselected. The genomic changes to be introduced are first generated in a cDNA transcription vector, and donor RNA is transcribed in vitro from this plasmid. After RNA recombination in infected cells, viral progeny bearing the desired alterations are selected on the basis of their possession of a phenotypic property not found in the original recipient virus.

The earliest scheme for targeted RNA recombination came about by the fortunate confluence of a number of separate discoveries. First, as outlined in the previous section, an abundance of experimental work, primarily with MHV, had demonstrated that RNA recombination is a frequent event in the coronavirus infectious cycle. Second, it had recently been shown that each coronavirus sgRNA possesses a negative-strand counterpart (Sethna et al. 1989). Although the original proposal that sgRNAs function as replicons has not proved correct, this key finding made clear that the positive-strand sgRNAs serve as substrates for the viral polymerase, thus rendering them likely participants in polymerase-mediated recombination. Finally, an MHV mutant was found that had the ideal properties for the recipient parent virus. This mutant, Alb4, was among a collection of classic, random mutants isolated on the basis of production of an atypical cytopathic effect at the nonpermissive temperature (39°C) (Sturman et al. 1987). Alb4 is temperature sensitive, but it is not an absolute conditional-lethal mutant, in that it produces plaques at the nonpermissive temperature that are tiny by comparison with the wild type. Additionally, virions of Alb4 are thermolabile, exhibiting a drop in infectious titer of two to three orders of magnitude when held at the nonpermissive temperature for 24 h, a treatment that only minimally affects the viability of the wild type. The lesion in Alb4 was found to reside in the nucleocapsid (N) gene, the gene closest to the 3′ untranslated region (3′ UTR) of the genome, and consists of an 87-nt (in frame) deletion (Koetzner et al. 1992) that removes a 29-amino acid linker connecting two functional domains of the N protein (Parker and Masters 1990).

The experiment establishing the principle of targeted RNA recombination, then, was carried out by cotransfection of mouse cells with the purified genome of Alb4 and a synthetic copy of sgRNA7, which is the smallest of the MHV sgRNAs and serves as the mRNA for N protein (Fig. 1) (Koetzner et al. 1992). The synthetic donor RNA contained the wild-type (undeleted) version of the N gene and was tagged with a presumed nondeleterious 5-nt insertion in the 3′ UTR. The precarious nature of this latter assumption was only revealed years later, when it was found that the insertion had been made in a mutable loop of an RNA secondary structure that is absolutely essential for MHV replication (Hsue and Masters 1997). The viral progeny resulting from the cotransfection were subjected to a heat-killing step, so as to greatly reduce the background of Alb4 parent virus, and candidate recombinants were identified as viruses forming large (i.e., wild-type sized) plaques at the nonpermissive temperature. The presence in the putative recombinants of both the region that is deleted in Alb4 and the 5-nt tag was verified by size or restriction fragment polymorphisms in RT-PCR products from genomic RNA that had been isolated from purified virions. Additionally, the 5-nt tag, which is present in neither wild-type MHV nor the Alb4 mutant, was demonstrated by direct sequencing of genomic RNA of the recombinants. These viruses were thus the first engineered site-specific mutants of a member of the coronavirus family.

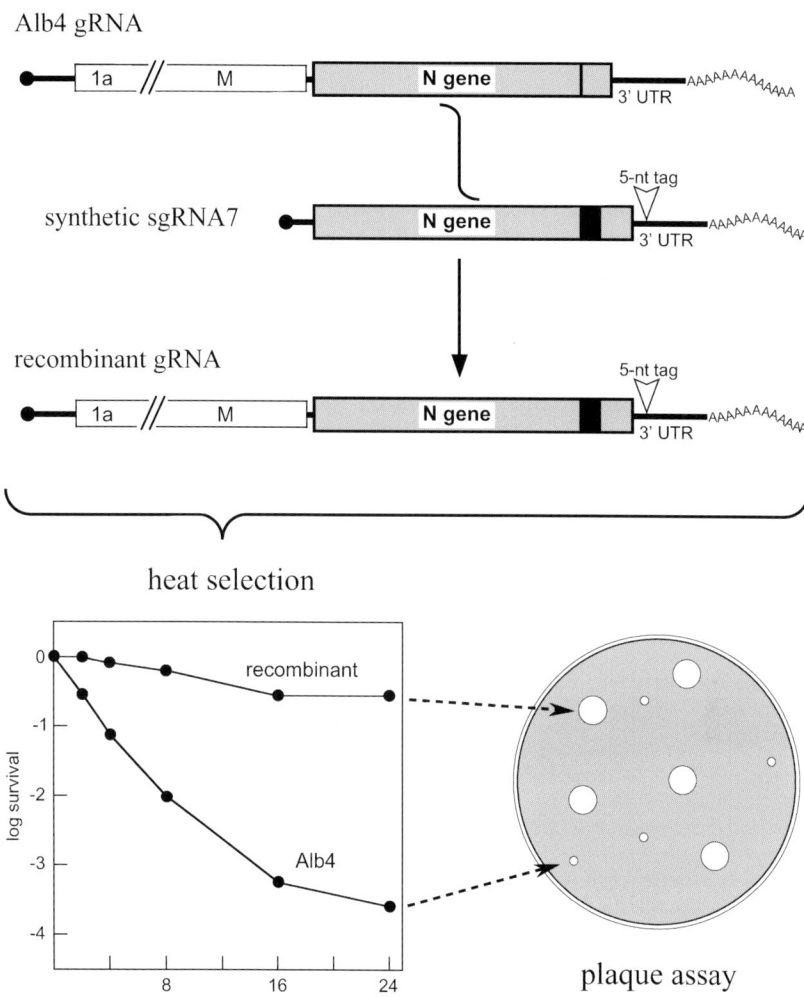

Fig. 1. Earliest implementation of targeted RNA recombination. Genomic RNA (*gRNA*) from the Alb4 mutant and synthetic donor RNA were cotransfected into cells. The donor RNA contained the wild-type N gene, including the 87-nt region that is deleted in Alb4 (*black rectangle*) and was also tagged with a 5-nt insertion in the 3' UTR. After harvest of progeny virus, the Alb4 parent was selectively killed by heat treatment, and recombinants were identified as viruses forming large (wild-type size) plaques at the nonpermissive temperature

3.2
Improving the Donor RNA: DI and Pseudo-DI RNAs

The initial demonstration of targeted RNA recombination was soon followed by a report of the incorporation of genetic markers into the MHV genome by using a defective interfering (DI) RNA, MIDI-C, as the donor RNA (Fig. 2) (van der Most et al. 1992). In this case, a coding-silent marker tagging the region of the Alb4 N gene deletion was successfully transferred from the DI RNA to Alb4 recipient virus. Additionally, it was shown that recombinants bearing MIDI-C-derived markers that had been transduced into gene 1 of wild-type MHV could be isolated by screening, without any prior selection. Most importantly, the efficiency of obtaining recombinants with this DI RNA as the donor appeared to

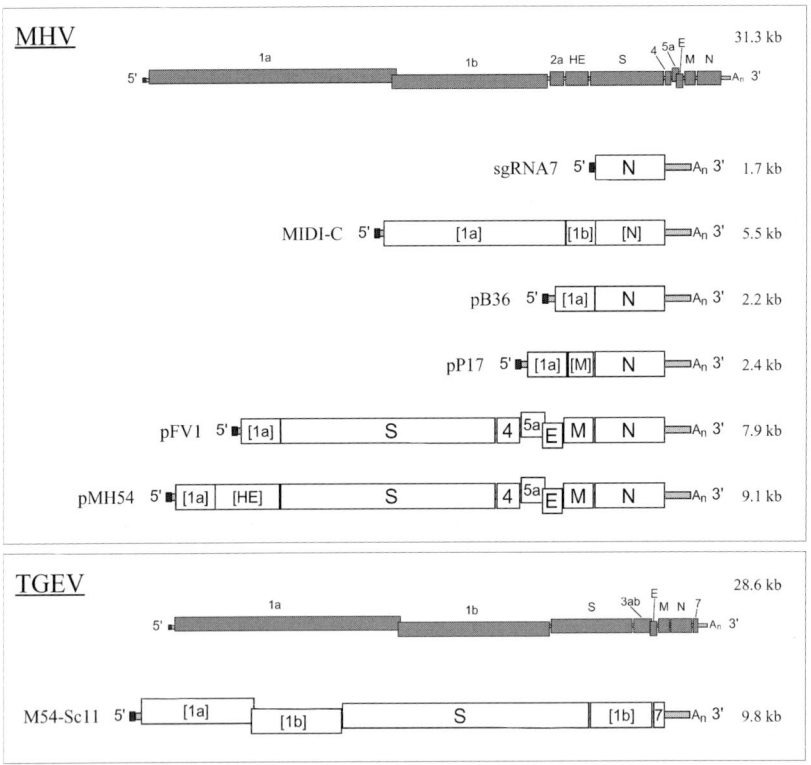

Fig. 2. DI and pseudo-DI donor RNAs used in targeted recombination studies. Shown at the *top* of each panel is the genome from which the donor RNAs are derived. *Brackets* indicate fragments of full-length genes

be significantly higher than had been achieved with sgRNA7 as the donor. Coronavirus DI RNAs are parasitic elements that arise through the accumulation of extensive deletions, which eliminate most of the coding capacity of the genome while retaining *cis*-acting elements essential for viral polymerase recognition. They are therefore not independently viable, but they replicate by feeding into the RNA synthesis machinery provided by a helper virus.

In an attempt to systematically optimize the performance of targeted recombination, the efficiencies of the two types of donor RNA were directly compared (Masters et al. 1994). For this purpose, a nonnatural MHV DI RNA, designated pB36 RNA, was constructed to contain the entire N gene, mimicking the composition of a well-characterized, naturally arising DI RNA of bovine coronavirus (BCoV) (Chang et al. 1994). It was found that this donor DI RNA replicated abundantly and consistently yielded targeted recombinants with Alb4 at an efficiency on the order of 10^{-2}, some two to three orders of magnitude greater than that obtained with sgRNA7 donor RNA. This meant that candidate (large plaque) recombinants could generally be identified directly against the background of small plaques formed by the Alb4 parent, without the need for a heat-killing counterselection step. Because the mechanistic details of RNA recombination remain to be unraveled, it has not yet been resolved whether the increased donor efficiency of DI RNAs results from their replicative competence or from some other intrinsic property. It is possible that the critical feature of DI RNAs is not their ability to replicate per se, but, rather, that they possess some sequence or structural element that brings about their localization to the RNA synthesis compartment, or that facilitates their alignment with homologous regions of the acceptor genome template.

Despite our not understanding precisely why DI RNAs work so well, it was nevertheless straightforward to design additional donor RNAs based on the relatively simple composition of pB36 RNA, which comprises only the 5'- and 3'-terminal segments from the MHV genome, connected by a short heterologous linker (Masters et al. 1994). The inclusion of more material from the 3' end of the genome resulted in progressively larger plasmid vectors for donor RNAs—pP17 (Fischer et al. 1997a), pFV1 (Fischer et al. 1997b), and pMH54 (Kuo et al. 2000) (Fig. 2)—which were collectively capable of transducing mutations into any of the genes downstream of gene 1, the viral replicase gene. The availability of these larger donor RNAs, termed pseudo-DI RNAs because it has never been directly determined whether they are replication competent, consequently places all of the MHV structural genes within the reach of the targeted recombination method. Separately, a similar principle was applied to a different coronavirus species, TGEV, by the insertion of the S protein gene into a naturally occurring DI RNA of that virus (Méndez et al. 1996). However, in this case the donor RNA, M54-SC11 (Fig. 2), was not completely colinear with the 3' end of the recipient genome, and thus the formation of the recombinants that were isolated was dependent upon two crossover events, one upstream and one downstream of the targeted region (Sánchez et al. 1999). Work has also been done toward using a modified naturally occurring DI RNA of IBV in targeted RNA recombination, but the recovery of viable recombinants from this system has not yet been reported (Neuman et al. 2001).

3.3
Improving the Recipient Virus: Host Range-Based Selection

Although the Alb4 mutant was invaluable in moving coronavirus genetics from classic to molecular capabilities, a fundamental limitation of the scheme described above is that selection against Alb4 makes sense only if the mutant being sought is more fit than Alb4 at the nonpermissive temperature. This precondition still allows the selection of a wide variety of mutants, but it places a restriction on the range of problems to which a genetic system could potentially be applied. Two studies made use of alternative strategies to circumvent this fitness precondition. In one study, an RT-PCR-based screen of large pools of candidate recombinant plaques was employed to identify clustered charged-to-alanine mutations made in the E gene (Fischer et al. 1998). The fact that the resulting E protein mutants were temperature sensitive and thermolabile explained why they could not be isolated by a heat-killing selection. A second means around the fitness requirement was to carry out a different type of selection. In this case, neutralization with monoclonal antibodies specific for the S protein of MHV strain A59 (the strain to which Alb4 belongs) was used to obtain recombinants that had incorporated the S gene of MHV strain 4 (Phillips et al. 1999). However, both of these alternatives had disadvantages. Mutant identification by screening is extremely labor intensive and of uncertain efficiency, and strain-specific monoclonal antibody selection is applicable only under special circumstances.

Superseding these two particular exceptions, a very powerful positive selection strategy was enabled by the creation of an interspecies chimeric mutant of MHV in which the ectodomain of the S protein was replaced with its counterpart from feline infectious peritonitis virus

(FIPV) (Kuo et al. 2000). This substitution had its foundation in work done with viruslike particles (VLPs), which had suggested that the determinants for functional S protein incorporation into virions reside solely in the transmembrane domain and the endodomain of the molecule (Godeke et al. 2000). Because both MHV and FIPV are stringently species specific in tissue culture, the interspecies chimeric mutant, named fMHV, was readily obtained by a targeted RNA recombination experiment that selected for a virus that had acquired the ability to grow in feline cells. It soon became apparent that the inverse of this selection would provide significantly greater flexibility in the construction of MHV mutants than the Alb4-based targeted recombination scheme. The use of fMHV as the recipient virus with donor RNA transcribed from a pMH54-derived vector, which would restore the region encoding the MHV S gene ectodomain, should, in principle, allow the selection of recombinants harboring any nonlethal MHV mutation (Fig. 3, top panel). No matter how fragile its phenotype, the constructed mutant should be identifiable on the basis of its having regained the ability to grow in murine cells, in contrast to the fMHV parent, which can only grow in feline cells. The feasibility and utility of this strategy have now been proven repeatedly in multiple laboratories.

The strength of the host range-based selection has been most dramatically demonstrated by its ability to recover a mutant with a two-residue truncation of the carboxy terminus of the M protein (Kuo and Masters 2002) and a mutant with the critical E gene entirely deleted (Kuo and Masters 2003). Both of these mutants are severely impaired, forming tiny plaques at all temperatures and yielding infectious titers that are, at most, orders of magnitude lower than those of the wild type. In the initial selection of these and a number of other highly defective mutants, the recombinants being sought were identified as tiny plaques among a mixture of tiny and wild-type-sized plaques (Kuo and Masters 2002, 2003). Analysis of the latter showed that they were reconstructed wild-type viruses, which had arisen via a second crossover event occurring downstream of the restored MHV S gene, but upstream of the mutation of interest (Fig. 3, top panel). Although double crossovers occur with lower frequency than a single crossover, they can constitute a significant fraction of the initial recombinants in cases in which the wild type has a marked growth advantage over the constructed mutant. To preclude the possibility of the second crossover event, a variant of fMHV, designated fMHV.v2, has been constructed in which the gene order downstream of the S gene has been rearranged (Fig. 3, middle panel). The use of fMHV.v2 eliminates the background of progeny recombinants generated

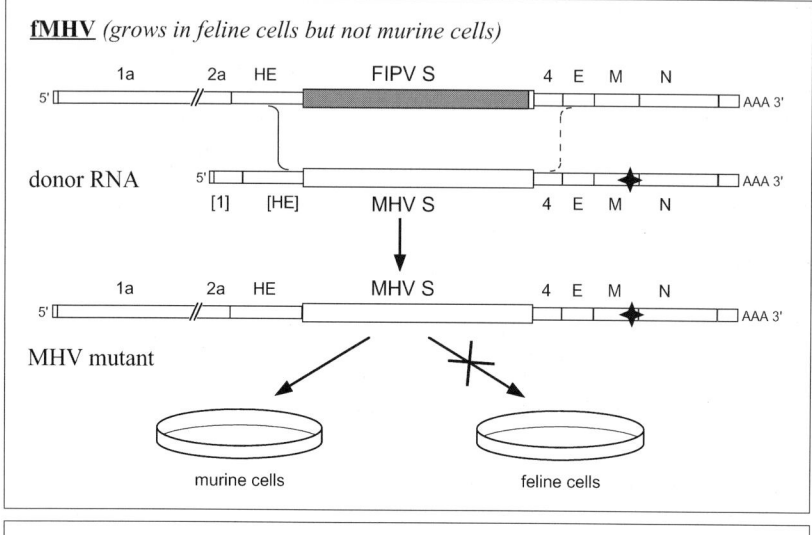
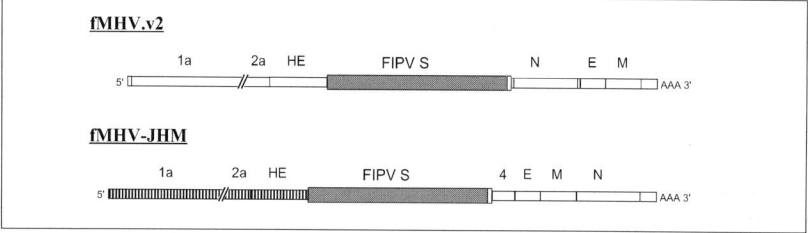
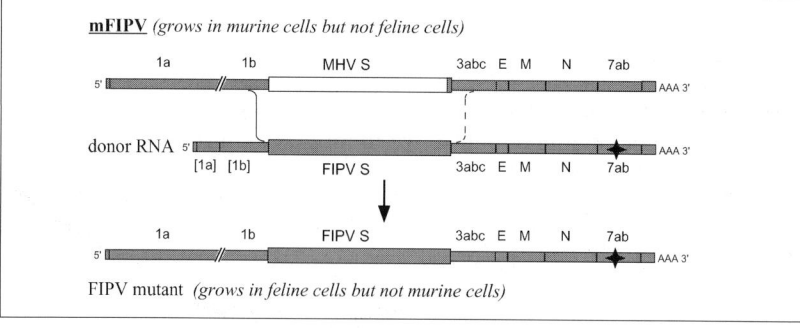

Fig. 3. Host range-based selection. *Top panel*: Selection strategy with the interspecies chimeric coronavirus fMHV, which contains the portion of the S gene encoding the ectodomain of the spike protein of FIPV (*shaded rectangle*) but is otherwise identical to MHV. fMHV is able to grow in feline cells but cannot grow in murine cells. In targeted recombination with donor RNA that restores the MHV S ectodomain, a single crossover (*solid line*), within the HE gene, can generate a recombinant that has reacquired the ability to grow in murine cells and has also incorporated an engi-

by second crossovers, and it is of particular utility in the recovery of unselected markers that are debilitating or that are located far downstream of the S gene (Goebel et al. 2004).

In addition to depth, the host range-based selection system has been shown to have breadth. The fMHV structural genes have been incorporated into the JHM strain of MHV, resulting in fMHV-JHM (Fig. 3, middle panel) (Ontiveros et al. 2001). This chimeric virus, in conjunction with the appropriate JHM strain counterpart of the pMH54 vector, has been used to construct site-directed mutants in MHV-JHM, thereby providing proof of principle for the applicability of this system to MHV strains other than strain A59. A more far-reaching extension of the method has been achieved by the construction of mFIPV, an interspecies chimeric mutant of FIPV in which the ectodomain of the S protein has been replaced with that of MHV (Haijema et al. 2003). This virus, which, as expected, has a host cell species permissivity exactly the converse of that of fMHV, provides the starting point for construction of site-directed mutations in the structural and nonstructural genes of FIPV (Fig. 3, bottom panel). These results establish host range-based selection as a general blueprint for the carrying out of reverse genetics in all coronaviruses, or at least in those that exhibit some level of host range restriction in tissue culture.

4
Targeted RNA Recombination: Spectrum of Applications

The impact of reverse genetic systems on progress in virology has been overwhelmingly demonstrated for most families of viruses over the last two decades, much to the frustration of many a coronavirologist. The new availability of multiple systems for engineering coronaviral genomes suddenly provides these investigators with unexpected opportunities, requiring choices to be made. These choices will be guided by the

neered mutation (*star*). A potential second crossover (*broken line*) would regenerate a wild-type recombinant lacking the mutation. *Middle panel*: Variant interspecies recipient viruses fMHV.v2, which greatly reduces the probability of the undesired downstream second crossover, and fMHV-JHM, which can be used to construct mutants of the JHM strain. *Bottom panel*: Selection strategy with mFIPV, entirely analogous to the fMHV scheme (*top panel*)

particular research question and by the practical and theoretical limitations of the various engineering systems. Because of the restrictions inherent in its selection principle, targeted RNA recombination in its present format will retain its greatest value in the study and manipulation of functions specified by the genomic regions downstream of the polymerase gene. The vast potential of this technology for coronavirus research can perhaps best be envisaged just by looking back at the first ten years of its existence. What follows is a brief survey of important contributions that the targeted recombination approach has made in the different areas of its application. Unless otherwise specified, the work discussed refers to MHV.

4.1
Virion Structure and Assembly

By their nature, the earliest versions of the targeted recombination method revolved around the N gene. They allowed the mapping of the extreme thermolability of virions of MHV-A59 mutants Alb4 and Alb1 to a deletion (Koetzner et al. 1992) and to a point mutation (Masters et al. 1994), respectively, in this gene. By an analysis of a panel of independently isolated revertant viruses this thermolability could, for Alb4, subsequently be attributed to a disturbed RNA binding capacity of the N protein. By the use of targeted recombination, critical evidence was obtained linking the restoration of the wild-type phenotype to a single reverting amino acid mutation, different for each revertant, in a domain of the N protein to which RNA binding had been previously mapped (Peng et al. 1995a). It was therefore somewhat surprising that major parts of this domain, as well as the segment that is deleted in Alb4, could be exchanged without penalty by the corresponding domain of the BCoV N protein (Peng et al. 1995b). The resulting MHV-BCoV chimeric viruses were viable and thermally stable. In contrast, for other regions of the N protein, such as the terminal domains, interspecies exchange was not tolerated, presumably because these regions are involved in protein-protein interactions that are specific for each virus.

Interactions between M molecules are thought to provide the major force for the assembly of the coronavirus envelope (Rottier 1995; Vennema et al. 1996). In a mutational study investigating the primary structural requirements of the M protein for assembly of VLPs from coexpressed M and E proteins, it was found that stringent structural conditions must be satisfied for envelope formation. In particular, the extreme carboxy terminus of M was shown to be crucial in this system (de Haan et al. 1998). The mere deletion of the terminal residue (MΔ1) almost completely abolished assembly, whereas an M protein mutant additionally lacking the penultimate residue (MΔ2) was entirely assembly incompetent. By contrast, when these deletions and other mutations in the carboxy-terminal domain were transferred to the MHV genome by targeted recombination, the resulting effects were generally much less severe, or were even absent. The MΔ1 viral mutant, for instance, had no detectable defect. Apparently, in the context of the complete virion, changes that are devastating in the VLP system can be accommodated by other stabilizing interactions, most likely between the envelope and the nucleocapsid. Although the MΔ2 viral mutant could not be identified in this study and was thus considered nonviable (de Haan et al. 1998), the power of the host range-based selection system later enabled its isolation (Kuo and Masters 2002). The MΔ2 virus formed tiny plaques in tissue culture and grew extremely poorly, and on passage, revertants with strongly improved growth properties rapidly emerged. Genetic analysis of a large number of second-site revertants, combined with the targeted reintroduction of some of the reverting mutations back into the MΔ2 mutant genome, identified residues both in the M protein and in the N protein that could compensate for the two-residue deletion. This provided compelling evidence for a structural interaction between the carboxy termini of these two proteins in MHV.

Despite its minute abundance in virions, the E protein is a critical factor in the assembly of coronaviruses. Its function, however, is still unresolved. To study the role of E, clustered charged-to-alanine mutations were introduced into the protein through targeted recombination (Fischer et al. 1998). Three viable mutant viruses were obtained, two of which were temperature sensitive whereas the third had a wild-type phenotype. Both temperature-sensitive mutants were markedly thermolabile when grown at the permissive temperature. When virions of one of these E mutants were viewed by electron microscopy, particles with strikingly aberrant shapes were observed. These data indicated an important role for the E protein in virion morphogenesis and stability. Remarkably, however, it has recently become clear that this role is not essential. Again, because of the power of the host range-based selection system of targeted recombination, it has been possible to isolate a mutant of MHV from which the E gene is entirely deleted (Kuo and Masters 2003; de Haan and Rottier, unpublished results). Although the ΔE mutant produces tiny plaques with an unusual morphology, has a slow growth rate, and grows to low infectious titer, it is, nevertheless, completely viable. Curiously, the E protein appears to be an absolute requirement for the group 1 coronavirus TGEV; growth of TGEV from which the E gene has been deleted is essentially dependent on *in trans* complementation by expressed E protein (Curtis et al. 2002; Ortego et al. 2002).

4.2
RNA Replication and Transcription

Because all intergenic regions, including their associated transcription-regulating sequences (TRSs), as well as the 3' UTR, are accessible for manipulation by targeted RNA recombination, this methodology allows the study of many questions related to viral replication and transcription. An initial foray in this direction sought to define functionally equivalent segments of the 3' UTRs of MHV and BCoV. This led to the identification of a conserved bulged stem-loop secondary structure at the upstream end of the 3' UTR, adjacent to the stop codon of the N gene (Hsue and Masters 1997). The stem-loop was shown to be essential for virus viability as well as for DI RNA replication. More recent work (Hsue et al. 2000; Goebel et al. 2004), using mutational analysis combined with chemical and enzymatic probing, has refined the picture of this structure and has delineated its relationship with a downstream, and partially overlapping, RNA pseudoknot that was first discovered in BCoV (Williams et al. 1999). The mutually exclusive nature of the stem-loop and the pseudoknot suggests that they are components of a molecular switch, functioning to mediate some event during RNA synthesis.

Coronaviruses have a genome organization in which the order of the essential genes (5'-polymerase-S-E-M-N-3') is strictly conserved, despite the high frequency of RNA recombination of these viruses. To find out whether this fixed gene order is in some way a vital property, deliberate rearrangements were introduced into the viral genome through targeted recombination. All attempted gene rearrangements were found to be tolerated, generally with surprisingly little effect on the growth characteristics of the recombinant viruses in cell culture or, for one virus tested, in the mouse host (de Haan et al. 2002b).

The factors that determine the relative efficiencies of synthesis of coronavirus sgRNAs are as yet poorly understood. Evidence indicates that the identity of the TRS, its sequence context, and its genomic position can all contribute to the process. The fortuitous effect of one or more of three nucleotide changes introduced into a donor vector for targeted recombination (pMH54, Fig. 2), to create a convenient restriction site upstream of the gene 4 TRS, illustrates the importance of the TRS sequence context (Ontiveros et al. 2001; de Haan et al. 2002a). For unknown reasons this modification led to a dramatic (at least sevenfold) upregulation of sgRNA4 synthesis. Other examples of unexpected context effects were observed in some genomically rearranged viruses (de Haan et al. 2002b). For instance, relocation of the gene 4–5a/E-M cluster to a location between the polymerase and S genes (mutant MHV-EMSmN) resulted in a strong increase in the level of synthesis of the (now) largest sgRNA, by comparison with its wild-type counterpart. The opposite was observed after relocation of the M gene to a position immediately downstream of the S gene (mutant MHV-SMEN): The sgRNA specifying the M protein was hardly detectable. Obviously, much more systematic work will need to be done to provide clear insights into these complex issues. To explore other questions related to coronavirus RNA synthesis, targeted RNA recombination has also been employed for the insertion of a new transcription unit into the MHV genome (Hsue and Masters 1999), as well as for the embedding of a high-affinity binding site for a putative host transcriptional factor (Shen and Masters 2001).

4.3
Pathogenesis

The ability to study the effect of targeted mutations in the viral genome on the course of a natural infection represents an essential tool with which to rigorously address the interplay between host and virus. This is illustrated most impressively by a series of studies on the role of the MHV S protein in viral pathogenesis. Modifications ranging from single amino acid changes to complete spike replacements were applied. In the first category the simple substitution Q159L in the receptor binding domain of MHV-A59 S protein significantly reduced viral virulence; replication in the liver and, consequently, the extent of viral hepatitis were strongly decreased (Leparc-Goffart et al. 1998). Similarly, amino acid substitutions at the S1/S2 cleavage site indicated that efficient cleavage and cell-cell fusion are not necessary for virulence (Hingley et al. 2002). In the category of large-scale modifications, the replacement of the entire S gene of MHV-A59, a moderately neurovirulent virus, by that of MHV-4, which is highly neurovirulent, resulted in a chimeric virus with dramatically increased neurovirulence. Although replication in the brain was not elevated, viral antigen staining and inflammation in the central nervous system were increased (Phillips et al. 1999). The acquired spike apparently conveys to the chimeric virus most of the pathogenic properties of its cognate virus. This interpretation was confirmed in an analogous exchange involving the MHV-2 S gene. In this case, the non-demyelinating phenotype of the latter virus was passed on to MHV-A59, a demyelinating strain (Das Sarma et al. 2000). Still further support for the role of S as the primary determinant of pathogenicity was provided by an analysis of the chimeric viruses for their ability to induce hepatitis after intrahepatic inoculation (Navas et al. 2001). The level of replication in the liver and the extent of hepatocellular damage paralleled those of the virus from which the spike had been obtained, that is, MHV-A59 carrying an MHV-4, MHV-A59, or MHV-2 spike exhibited low, moderate, or high replication and pathology, respectively. Finally, a series of chimeric viruses containing intramolecularly recombed MHV-4/MHV-A59 S genes in the MHV-A59 background was tested to further explore the determinants of neurovirulence within the MHV-4 spike (Phillips et al. 2001, 2002). Reciprocal exchanges of the S1 and S2 subunits, and of parts of the hypervariable region of S1, yielded viruses that replicated well in vitro but were generally severely attenuated in mice. These results suggest that such modifications disturb interactions within the S protein that are important for efficient infection in the mouse brain.

A critical role of the S protein in pathogenesis was also demonstrated for TGEV. By replacement, through targeted recombination, of most of the S gene of a respiratory TGEV isolate by that of a virus with enteric tropism, recombinants were obtained that had acquired the latter property (Sánchez et al. 1999). These recombinant viruses thereby also gained the ability to replicate to high titers in the porcine enteric tract, as well as the marked virulence that is the distinguishing trait of the enteropathogenic parent virus.

In addition to the genes encoding the polymerase and canonical structural proteins coronaviruses have a number of other genes, forming characteristic sets in each coronavirus group, the functions of which are as yet unknown. None of these genes is essential for replication, as was demonstrated by targeted recombination for MHV (Fischer et al. 1997a; Ontiveros et al. 2001; de Haan et al. 2002a) and for FIPV (Haijema et al. 2003). Targeted inactivation of gene 4 in MHV-JHM did not affect the virulence of this virus, whether it was inoculated intracranially or intranasally, nor were the pathological effects in the central nervous system any different from those of the wild type (Ontiveros et al. 2001). More drastic genetic changes of group-specific genes in the MHV-A59 background, however, were clearly attenuating. In this situation, viruses were constructed deleting genes 4 and 5a, genes 2a and HE, or all four genes, the latter deletions creating a "minimal" coronavirus. Removal of genes 4 and 5a, but not that of genes 2a and HE, reduced viral growth in cell culture slightly yet significantly. In intracranially inoculated mice, however, the virulence of all three deletion mutants was clearly reduced (de Haan et al. 2002a). For FIPV, the deletion of the group-specific genes 3a, 3b, and 3c or genes 7a and 7b did not substantially influence in vitro growth properties; in contrast, the "minimal" virus lacking all five of these genes was strongly impaired. The oronasal inoculation of cats with these deletion viruses, at a dose confirmed to be lethal for wild-type FIPV, remained without clinical consequences. That the animals had actually been infected was clear from their development of virus-neutralizing antibodies (Haijema et al. 2003). These deletion studies suggest that the nonessential genes encode functions important for host-virus interactions.

4.4
Coronavirus Vaccines and Vectors

The technology of targeted recombination has already displayed a number of features that will be essential for the development of coronaviruses as vectors for vaccination and therapy. One is the capability of rendering these viruses avirulent by the deletion of particular genes, as demonstrated for MHV and FIPV. In the latter case, viruses lacking either the 3abc or the 7ab gene cluster were indeed shown to serve as live-attenuated vaccine candidates, because cats infected with these mutants were protected against subsequent challenge with an otherwise lethal dose of virulent FIPV (Haijema et al. 2003). Second, the ability to genetically rearrange coronavirus genomes provides a critical safety asset, because it will allow the construction of vaccine or vector viruses that, because of judiciously modified gene orders, should have vanishingly small probabilities of generating viable progeny through recombination with coronaviruses in the wild. Third, the potential to retarget coronaviruses by modification of their S proteins, on which the current host range selection system for recombinant viruses is also based, constitutes another important feature that might be further developed to enable the directing of vectors to predefined cellular surface antigens.

Finally, for their use both as vectors and as carrier vaccines, the demonstrated ability of coronaviruses to incorporate and express foreign genes is obviously essential. Green fluorescent protein (GFP) was the first nonviral protein to be expressed by a coronavirus (Fischer et al. 1997b). The recombinant MHV containing the GFP gene inserted in place of gene 4 grew as well as the wild type did, but its level of GFP expression was poor. A slightly different construct containing the "enhanced" GFP gene, again replacing gene 4 but also in the context of the upregulated TRS4 of pMH54, yielded a virus that replicated as well as wild-type virus both in vitro and in the mouse central nervous system (Das Sarma et al. 2002). This virus produced fluorescence during infection in vitro and in mouse brain, and GFP expression was stably maintained through at least six passages in tissue culture. In another study luciferase gene expression cassettes were inserted at various positions in the MHV genome. Whereas the *Renilla* luciferase gene remained stable over eight passages, irrespective of its location, the firefly luciferase gene was lost quite rapidly as a result of the acquisition of deletions. Luciferase expression levels appeared to increase when the gene was positioned closer to the 3' end of the genome (de Haan et al. 2003). Moreover, the simultaneous synthesis of both luciferase activities from a single engineered virus demonstrated the potential for the use of coronaviruses as multivalent expression vectors.

5
Conclusions and Future Prospects

Within the span of nearly a decade, targeted recombination has established itself as a powerful and versatile technique for the reverse genetics of the 3' third of the coronavirus genome, which encompasses the region encoding all of the structural genes. The past two years, however, have seen the opening of a new frontier in coronavirus reverse genetics, with reports of the assembly of infectious cDNAs for TGEV (Almazán et al. 2000; Yount et al. 2000), HCoV-229E (Thiel et al. 2001), IBV (Casais et al. 2001), MHV (Yount et al. 2002), and SARS-CoV (Yount et al. 2003). These recent developments raise the question of whether targeted RNA recombination will retain interest only as an historic relic. We think that this is unlikely to be the case. It is more probable that each reverse genetic system will have its own specific advantages under a particular set of experimental circumstances. At this moment, one can only tentatively comment on the relative strengths and limitations of targeted RNA recombination and infectious cDNAs for coronavirus reverse genetics. The targeted recombination system is at a fairly mature stage of development. By contrast, work with the infectious clone systems is sufficiently early in exploring their potentiality that it is not clear how hardy or manipulable these systems may become.

The capability of paramount value that is provided by the infectious clones, no matter what the burden in experimental labor, is access to gene 1. The capacity to site-specifically mutagenize the exceedingly large viral RNA polymerase gene will undoubtedly play a major role in the acquisition of an understanding of the workings of this complex machinery. Except for its periphery, gene 1 is effectively out of the range of targeted RNA recombination, because the construction of donor RNA vectors entering this region is hindered by precisely the same technical problems that made the assembly of infectious cDNAs so formidable a task. A second unique characteristic of the infectious clones is their potential to provide the means by which the "passage zero" situation can be examined for intentionally lethal mutant constructs. This property has been elegantly and forcefully exploited with the infectious clone of equine arterivirus in the study of nidovirus RNA synthesis (van Dinten et al. 1997; Tijms et al. 2001; Pasternak et al. 2001). However, in order for similar studies to be executed with coronavirus infectious clones, platforms need to be devised that can produce ample amounts of viral genome (and its resulting gene 1 translation product) in the initial round of launch. This must be done without the generation of significant levels of other RNA species that have the propensity to confound analysis or interfere with RNA synthesis. As of this writing, the reported infectious clone systems are not yet sufficiently robust to enable these types of experiments.

For work involving coronavirus structural genes, targeted RNA recombination is likely to remain the method of choice for many studies. One reason for this is its relative ease of manipulation. The largest of the donor RNA vectors are still threefold smaller than an entire genome. Thus mutagenesis at the DNA level can generally be carried out without subcloning steps. A second strong asset of targeted recombination is that the host range-based selection system has demonstrated both its efficiency, in straightforward isolation of desired mutants, and its power, in recovery of extremely defective mutants such as the M protein truncation and the E gene deletion (Kuo and Masters 2002, 2003). Finally, targeted recombination lends itself well to studies involving domain exchange between different proteins (Peng et al. 1995b), the exchange of genomic elements (Hsue and Masters 1997), or the creation of mutants containing multiple mutations. In these cases the system, through its own selection of allowable crossover sites, can reveal which substitutions retain functionality and which are lethal. Related to this, the targeted recombination system establishes a stringent criterion for the lethality of a given mutation. If markers, silent or otherwise, upstream and downstream of the mutation in question can be transferred from a single donor RNA to progeny recombinants, while the mutation itself is excluded by multiple crossover events, then this argues strongly that the mutation produces a lethal phenotype (de Haan et al. 1998; Hsue et al. 2000). In this situation the donor RNA provides its own internal control. As mentioned above, a similarly convincing standard of lethality for the infectious clones will require a more vigorous RNA production at passage zero. The sum of these considerations makes it likely that targeted recombination will serve a useful role in coronavirus genetics for some time to come.

References

Almazán F, González JM, Pénzes Z, Izeta A, Calvo E, Plana-Durán J, Enjuanes L (2000) Engineering the largest RNA virus genome as an infectious bacterial artificial chromosome. Proc Natl Acad Sci USA 97:5516–5521

Banner LR, Keck JG, Lai MMC (1990) A clustering of RNA recombination sites adjacent to a hypervariable region of the peplomer gene of murine coronavirus. Virology 175:548–555

Banner LR, Lai MMC (1991) Random nature of coronavirus RNA recombination in the absence of selective pressure. Virology 185:441–445

Baric RS, Fu K, Schaad MC, Stohlman SA (1990) Establishing a genetic recombination map for murine coronavirus strain A59 complementation groups. Virology 177:646–656

Brian DA, Spaan WJM (1997) Recombination and coronavirus defective interfering RNAs. Semin Virol 8:101–111

Casais R, Thiel V, Siddell SG, Cavanagh D, Britton P (2001) Reverse genetics system for the avian coronavirus infectious bronchitis virus. J Virol 75:12359–12369

Cavanagh D, Davis P, Cook J, Li D (1990) Molecular basis of the variation exhibited by avian infectious bronchitis coronavirus (IBV). Adv Exp Med Biol 276:369–372

Chang R-Y, Hofmann MA, Sethna PB, Brian DA (1994) A *cis*-acting function for the coronavirus leader in defective interfering RNA replication. J Virol 68:8223–8231

Curtis KM, Yount B, Baric RS (2002) Heterologous gene expression from transmissible gastroenteritis virus replicon particles. J Virol 76:1422–1434.

Das Sarma J, Fu L, Tsai JC, Weiss SR, Lavi E (2000) Demyelination determinants map to the spike glycoprotein gene of coronavirus mouse hepatitis virus. J Virol 74:9206–9213

Das Sarma J, Scheen E, Seo SH, Koval M, Weiss SR (2002) Enhanced green fluorescent protein expression may be used to monitor murine coronavirus spread in vitro and in the mouse central nervous system. J Neurovirol 8:381–391

deHaan CAM, Kuo L, Masters PS, Vennema H, Rottier PJM (1998) Coronavirus particle assembly: primary structure requirements of the membrane protein. J Virol 72:6838–6850

de Haan CAM, Masters PS, Shen X, Weiss S, Rottier PJM (2002a) The group-specific murine coronavirus genes are not essential, but their deletion, by reverse genetics, is attenuating in the natural host. Virology 296:177–189 de Haan CAM, Volders H, Koetzner CA, Masters PS, Rottier PJM (2002b) Coronaviruses maintain viability despite dramatic rearrangements of the strictly conserved genome organization. J Virol 76:12491–12502

de Haan CAM, van Genne L, Stoop JN, Volders H, Rottier PJM (2003) Coronaviruses as vectors: position dependence of foreign gene expression. J Virol 77:11312–11323

Fischer F, Peng D, Hingley ST, Weiss SR, Masters PS (1997a) The internal open reading frame within the nucleocapsid gene of mouse hepatitis virus encodes a structural protein that is not essential for viral replication. J Virol 71:996–1003

Fischer F, Stegen CF, Koetzner CA, Masters PS (1997b) Analysis of a recombinant mouse hepatitis virus expressing a foreign gene reveals a novel aspect of coronavirus transcription. J Virol 71:5148–5160

Fischer F, Stegen CF, Masters PS, Samsonoff WA (1998) Analysis of constructed E gene mutants of mouse hepatitis virus confirms a pivotal role for E protein in coronavirus assembly. J Virol 72:7885–7894

Fu K, Baric RS (1994) Map locations of mouse hepatitis virus temperature-sensitive mutants: confirmation of variable rates of recombination. J Virol 68:7458–7466

Godeke GJ, de Haan CA, Rossen JW, Vennema H, Rottier PJM (2000) Assembly of spikes into coronavirus particles is mediated by the carboxy-terminal domain of the spike protein. J Virol 74:1566–1571

Goebel SJ, Hsue B, Dombrowski TF, Masters PS (2004) Characterization of the RNA components of a putative molecular switch in the 3' untranslated region of the murine coronavirus genome. J Virol 78:669–682

Haijema BJ, Volders H, Rottier PJM (2003) Switching species tropism: an effective way to manipulate the feline coronavirus genome. J Virol 77:4528–4538

Herrewegh AA, Vennema H, Horzinek MC, Rottier PJM, de Groot RJ (1995) The molecular genetics of feline coronaviruses: comparative sequence analysis of the ORF7a/7b transcription unit of different biotypes. Virology 212:622–631

Herrewegh AA, Smeenk I, Horzinek MC, Rottier PJM, de Groot RJ (1998) Feline coronavirus type II strains 79-1683 and 79-1146 originate from a double recombination between feline coronavirus type I and canine coronavirus. J Virol 72:4508–4514

Hingley ST, Leparc-Goffart I, Seo SH, Tsai JC, Weiss SR (2002) The virulence of mouse hepatitis virus strain A59 is not dependent on efficient spike protein cleavage and cell-to-cell fusion. J Neurovirol 8:400–410

Hsue B, Masters PS (1997) A bulged stem-loop structure in the 3' untranslated region of the genome of the coronavirus mouse hepatitis virus is essential for replication. J Virol 71:7567–7578

Hsue B, Masters PS (1999) Insertion of a new transcriptional unit into the genome of mouse hepatitis virus. J Virol 73:6128–6135

Hsue B, Hartshorne T, Masters PS (2000) Characterization of an essential RNA secondary structure in the 3' untranslated region of the murine coronavirus genome. J Virol 74:6911–6921

Jia W, Karaca K, Parrish CR, Naqi SA (1995) A novel variant of avian infectious bronchitis virus resulting from recombination among three different strains. Arch Virol 140:259–271

Keck JG, Stohlman SA, Soe LH, Makino S, Lai MMC (1987) Multiple recombination sites at the 5'-end of murine coronavirus RNA. Virology 156:331–341

Keck JG, Matsushima GK, Makino S, Fleming JO, Vannier DM, Stohlman SA, Lai MMC (1988a) In vivo RNA-RNA recombination of coronavirus in mouse brain. J Virol 62:1810–1813

Keck JG, Soe LH, Makino S, Stohlman SA, Lai MMC (1988b) RNA recombination of murine coronaviruses: recombination between fusion-positive mouse hepatitis virus A59 and fusion-negative mouse hepatitis virus 2. J Virol 62:1989–1998

Kirkegaard K, Baltimore D (1986) The mechanism of RNA recombination in poliovirus. Cell 47:433–443

Koetzner CA, Parker MM, Ricard CS, Sturman LS, Masters PS (1992) Repair and mutagenesis of the genome of a deletion mutant of the coronavirus mouse hepatitis virus by targeted RNA recombination. J Virol 66:1841–1848

Kottier SA, Cavanagh D, Britton P (1995) Experimental evidence of recombination in coronavirus infectious bronchitis virus. Virology 213:569–580

Kuo L, Godeke G-J, Raamsman MJB, Masters PS, Rottier PJM (2000) Retargeting of coronavirus by substitution of the spike glycoprotein ectodomain: crossing the host cell species barrier. J Virol 74:1393–1406

Kuo L, Masters PS (2002) Genetic evidence for a structural interaction between the carboxy termini of the membrane and nucleocapsid proteins of mouse hepatitis virus. J Virol 76:4987–4999

Kuo L, Masters PS (2003) The small envelope protein E is not essential for murine coronavirus replication. J Virol 77:4597–4608

Kusters JG, Jager EJ, Niesters HGM, van der Zeijst BAM (1990) Sequence evidence for RNA recombination in field isolates of avian coronavirus infectious bronchitis virus. Vaccine 8:605–608

Lai MMC, Baric RS, Makino S, Keck JG, Egbert J, Leibowitz JL, Stohlman SA (1985) Recombination between nonsegmented RNA genomes of murine coronaviruses. J Virol 56:449–456

Lai MMC (1992) RNA recombination in animal and plant viruses. Microbiol Rev 56:61–79

Lai MMC (1996) Recombination in large RNA viruses: coronaviruses. Semin Virol 7:381–388

Ledinko N (1963) Genetic recombination with poliovirus type 1: studies of crosses between a normal horse serum-resistant mutant and several guanidine-resistant mutants of the same strain. Virology 20:107–119

Lee CW, Jackwood MW (2000) Evidence of genetic diversity generated by recombination among avian coronavirus IBV. Arch Virol 145:2135–2148

Lee CW, Jackwood MW (2001) Spike gene analysis of the DE072 strain of infectious bronchitis virus: origin and evolution. Virus Genes 22:85–91

Leparc-Goffart I, Hingley ST, Chua MM, Phillips J, Lavi E, Weiss SR (1998) Targeted recombination within the spike gene of murine coronavirus mouse hepatitis virus-A59: Q159 is a determinant of hepatotropism. J Virol 72:9628–9636

Li K, Chen Z, Plagemann P (1999) High-frequency homologous genetic recombination of an arterivirus, lactate dehydrogenase-elevating virus, in mice and evolution of neuropathogenic variants. Virology 258:73–83

Luytjes W, Bredenbeek PJ, Noten AFH, Horzinek MC, Spaan WJM (1988) Sequence of mouse hepatitis virus A59 mRNA2: indications for RNA recombination between coronaviruses and influenza C virus. Virology 166:415–422

Makino S, Fleming JO, Keck JG, Stohlman SA, Lai MMC (1987) RNA recombination of coronaviruses: localization of neutralizing epitopes and neuropathogenic determinants on the carboxyl terminus of peplomers. Proc Natl Acad Sci USA 84:6567–6571

Masters PS, Koetzner CA, Kerr CA, Heo Y (1994) Optimization of targeted RNA recombination and mapping of a novel nucleocapsid gene mutation in the coronavirus mouse hepatitis virus. J Virol 68:328–337

Masters PS (1999) Reverse genetics of the largest RNA viruses. Adv Virus Res 53:245–264

Méndez A, Smerdou C, Izeta A, Gebauer F, Enjuanes L (1996) Molecular characterization of transmissible gastroenteritis coronavirus defective interfering genomes: packaging and heterogeneity. Virology 217:495–507

Motokawa K, Hohdatsu T, Aizawa C, Koyama H, Hashimoto H (1995) Molecular cloning and sequence determination of the peplomer protein gene of feline infectious peritonitis virus type I. Arch Virol 140:469–480

Nagy PD, Simon A (1997) New insights into the mechanisms of RNA recombination. Virology 235:1–9

Navas S, Seo S-H, Chua MM, Das Sarma J, Lavi E, Hingley ST, Weiss SR (2001) Murine coronavirus spike protein determines the ability of the virus to replicate in the liver and cause hepatitis. J Virol 75:2452–2457

Neuman B, Cavanagh D, Britton P (2001) Use of defective RNAs containing reporter genes to investigate targeted recombination for avian infectious bronchitis virus. Adv Exp Med Biol 494:513–518

Ontiveros E, Kuo L, Masters PS, Perlman S (2001) Inactivation of expression of gene 4 of mouse hepatitis virus strain JHM does not affect virulence in the murine CNS. Virology 289:230–238

Ortego J, Escors D, Laude H, Enjuanes L (2002) Generation of a replication-competent, propagation-deficient virus vector based on the transmissible gastroenteritis coronavirus genome. J Virol 76:11518–11529

Parker MM, Masters PS (1990) Sequence comparison of the N genes of five strains of the coronavirus mouse hepatitis virus suggests a three domain structure for the nucleocapsid protein. Virology 179:463–468

Pasternak AO, van den Born E, Spaan WJM, Snijder EJ (2001) Sequence requirements for RNA strand transfer during nidovirus discontinuous subgenomic RNA synthesis. EMBO J 20:7220–7228

Peng D, Koetzner CA, Masters PS (1995a) Analysis of second-site revertants of a murine coronavirus nucleocapsid protein deletion mutant and construction of nucleocapsid protein mutants by targeted RNA recombination. J Virol 69:3449–3457

Peng D, Koetzner CA, McMahon T, Zhu Y, Masters PS (1995b) Construction of murine coronavirus mutants containing interspecies chimeric nucleocapsid proteins. J Virol 69:5475–5484

Phillips JJ, Chua MM, Lavi E, Weiss SR (1999) Pathogenesis of chimeric MHV4/MHV-A59 recombinant viruses: the murine coronavirus spike protein is a major determinant of neurovirulence. J Virol 73:7752–7760

Phillips JJ, Chua M, Seo SH, Weiss SR (2001) Multiple regions of the murine coronavirus spike glycoprotein influence neurovirulence. J Neurovirol 7:421–431

Phillips JJ, Chua MM, Rall GF, Weiss SR (2002) Murine coronavirus spike glycoprotein mediates degree of viral spread, inflammation, and virus-induced immunopathology in the central nervous system. Virology 301:109–120

Plyusnin A, Kukkonen SK, Plyusnina A, Vapalahti O, Vaheri A (2002) Transfection-mediated generation of functionally competent Tula hantavirus with recombinant S RNA segment. EMBO J 21:1497–1503

Rottier PJM (1995) The coronavirus membrane glycoprotein. In: Siddell SG (ed) The Coronaviridae. Plenum Press, New York, pp 115–139

Sánchez CM, Izeta A, Sánchez-Morgado JM, Alonso S, Sola I, Balasch M, Plana-Durán J, Enjuanes L (1999) Targeted recombination demonstrates that the spike gene of transmissible gastroenteritis coronavirus is a determinant of its enteric tropism and virulence. J Virol 73:7607–7618

Sethna PB, Hung S-L, Brian DA (1989) Coronavirus subgenomic minus-strand RNAs and the potential for mRNA replicons. Proc Natl Acad Sci USA 86:5626–5630

Shen X, Masters PS (2001) Evaluation of the role of heterogeneous nuclear ribonucleoprotein A1 as a host factor in murine coronavirus discontinuous transcription and genome replication. Proc Natl Acad Sci USA 98:2717–2722

Sturman LS, Eastwood C, Frana MF, Duchala C, Baker F, Ricard CS, Sawicki SG, Holmes KV (1987) Temperature-sensitive mutants of MHV-A59. Adv Exp Med Biol 218:159–168

Thiel V, Herold J, Schelle B, Siddell SG (2001) Infectious RNA transcribed in vitro from a cDNA copy of the human coronavirus genome cloned in vaccinia virus. J Gen Virol 82:1273–1281

Tijms MA, van Dinten LC, Gorbalenya AE, Snijder EJ (2001) A zinc finger-containing papain-like protease couples subgenomic mRNA synthesis to genome translation in a positive-stranded RNA virus. Proc Natl Acad Sci USA 98:1889–1894

van der Most RG, Heijnen L, Spaan WJM, de Groot RJ (1992) Homologous RNA recombination allows efficient introduction of site-specific mutations into the genome of coronavirus MHV-A59 via synthetic co-replicating RNAs. Nucl Acids Res 20:3375–3381

van Dinten LC, den Boon JA, Wassenaar ALM, Spaan WJM, Snijder EJ (1997) An infectious arterivirus cDNA clone: identification of a replicase point mutation that abolishes discontinuous mRNA transcription. Proc Natl Acad Sci USA 94:991–996

van Vugt JJ, Storgaard T, Oleksiewicz MB, Botner A (2001) High frequency RNA recombination in porcine reproductive and respiratory syndrome virus occurs preferentially between parental sequences with high similarity. J Gen Virol 82:2615–2620

Vennema H, Poland A, Floyd-Hawkins K, Pedersen NC (1995) A comparison of the genomes of FECVs and FIPVs and what they tell us about the relationships between feline coronaviruses and their evolution. Feline Pract 23:40–44

Vennema H, Godeke G-J, Rossen JWA, Voorhout WF, Horzinek MC, Opstelten D-J E, Rottier PJM (1996) Nucleocapsid-independent assembly of coronavirus-like particles by co-expression of viral envelope protein genes. EMBO J 15:2020–2028

Vennema H (1999) Genetic drift and genetic shift during feline coronavirus evolution. Vet Microbiol 69:139–141

Wang L, Junker D, Collisson EW (1993) Evidence of natural recombination within the S1 gene of infectious bronchitis virus. Virology 192:710–716

Williams GD, Chang RY, Brian DA (1999) A phylogenetically conserved hairpin-type 3' untranslated region pseudoknot functions in coronavirus RNA replication. J Virol 73:8349–8355

Yount B, Curtis KM, Baric RS (2000) Strategy for systematic assembly of large RNA and DNA genomes: transmissible gastroenteritis virus model. J Virol 74:10600–10611

Yount B, Denison MR, Weiss SR, Baric RS (2002) Systematic assembly of a full-length infectious cDNA of mouse hepatitis virus strain A59. J Virol 76:11065–11078

Yount B, Curtis KM, Fritz EA, Hensley LE, Jahrling PB, Prentice E, Denison MR, Geisbert TW, Baric RS (2003) Reverse genetics with a full-length infectious cDNA of severe acute respiratory syndrome coronavirus. Proc Natl Acad Sci USA 100:12995–13000

Yuan S, Nelsen CJ, Murtaugh MP, Schmitt BJ, Faaberg KS (1999) Recombination between North American strains of porcine reproductive and respiratory syndrome virus. Virus Res 61:87–98

Coronavirus Reverse Genetics and Development of Vectors for Gene Expression

L. Enjuanes (✉) · I. Sola · S. Alonso · D. Escors · S. Zúñiga

Department of Molecular and Cell Biology, Centro Nacional de Biotecnología, CSIC, Campus Universidad Autónoma, 28049 Cantoblanco, Madrid, Spain
L.Enjuanes @ cnb.uam.es

1	Introduction	162
2	Pathogenesis Induced by Group 1 Coronaviruses	164
3	Engineering Coronavirus Genome	165
4	Essential Genes Required for TGEV Replication	168
5	Transcription-Regulating Sequences	170
5.1	Control of Transcription in TGEV	173
5.2	Effect of TRS Copy Number on Transcription	177
6	Expression Systems Based on Group 1 Coronaviruses	177
6.1	Helper-Dependent Expression Systems	178
6.2	Single Genome Coronavirus Vectors	180
6.3	Replication-Competent, Propagation-Deficient Coronavirus-Derived Expression Systems	182
7	Coronavirus Vector Cloning Capacity	183
8	Insertion Site, Stability, and Expression Levels	184
9	Molecular Basis of Group 1 Coronavirus Tropism	186
10	Modulation of Coronavirus Vector Virulence	187
11	Biosafety in Coronavirus-Derived Vectors	188
12	Conclusions	189
	References	190

Abstract Knowledge of coronavirus replication, transcription, and virus–host interaction has been recently improved by engineering of coronavirus infectious cDNAs. With the transmissible gastroenteritis virus (TGEV) genome the efficient (>40 μg per 10^6 cells) and stable (>20 passages) expression of the foreign genes has been shown. Knowledge of the transcription mechanism in coronaviruses has been significantly increased, making possible the fine regulation of foreign gene expression. A new family of vectors based on single coronavirus genomes, in which essential genes have been deleted, has emerged including replication-competent, propagation-deficient vectors. Vector biosafety is being increased by relocating the RNA packaging signal to the position previously occupied by deleted essential genes, to prevent the rescue of fully competent viruses that might arise from recombination events with wild-type field coronaviruses. The large cloning capacity of coronaviruses (>5 kb) and the possibility of engineering the tissue and species tropism to target expression to different organs and animal species, including humans, has increased the potential of coronaviruses as vectors for vaccine development and, possibly, gene therapy.

1
Introduction

Reverse genetics for coronaviruses has been initially achieved by targeted recombination (Masters 1999) (see the chapter by Masters and Rottier, this volume). Recently, the first coronavirus infectious cDNA clones have been constructed for transmissible gastroenteritis coronavirus (TGEV) (Almazán et al. 2000; Yount et al. 2000), human coronavirus (HCoV) 229E (Thiel et al. 2001a) (see the chapter by Thiel and Siddell, this volume), severe and acute respiratory syndrome coronavirus (SARS-CoV) (Yount et al. 2003), mouse hepatitis virus (MHV) (Yount et al. 2002) (see the chapter by Baric and Sims, this volume), and avian coronavirus (Casais et al. 2001).

The construction of virus vectors derived from RNA viruses is a comprehensive process that for optimum performance requires at least (1) the availability of an infectious cDNA clone; (2) knowledge of the virus transcription mechanism to optimize mRNA levels; (3) determination of the essential and nonessential genes to create room for heterologous genes; (4) understanding of the molecular basis of virus tropism, in order to control the species- and tissue specificity of the vector; (5) control of virus virulence, in order to generate attenuated vectors; and (6) design of a strategy for vector safety. In this chapter the progress on these aspects will be reviewed. The chapter will focus on the advances in the generation of virus vectors based on coronavirus genomes by reverse genetics, frequently using the TGEV genome as a model.

TGEV is an enveloped virus containing an internal core (Fig. 1A), formed by the genomic RNA, the N protein, and the M protein carboxy-terminus. Dissociation of the core by chaotropic agents leads to the release of a helical nucleoprotein composed of the genomic RNA and the N protein. All M protein molecules are embedded within the membrane. TGEV M protein presents two topologies. In one-third of the molecules both the amino and the carboxy terminus face the outside of the virion,

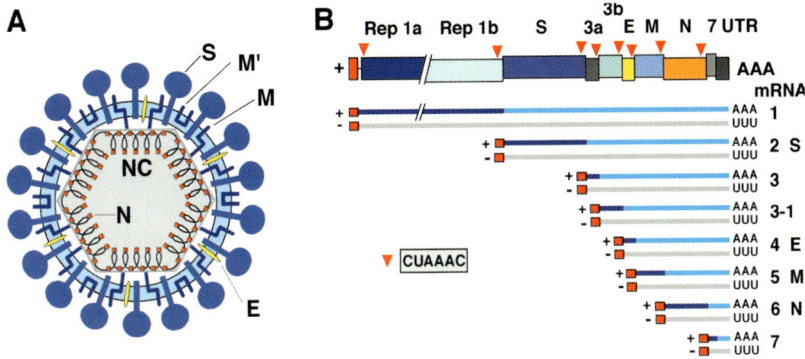

Fig. 1A, B. Coronavirus structure and genome organization. **A** Diagram of coronavirus structure using TGEV as a prototype. The scheme shows the envelope, the core, and the nucleocapsid structure. *S*, spike protein; *M* and *M'*, large membrane proteins with the amino terminus facing the external surface of the virion and the carboxy terminus toward the inside or the outside face of the virion, respectively; *E*, small envelope protein; *N*, nucleoprotein; *NC*, nucleocapsid. **B** Representation of a prototype TGEV genome and subgenomic RNAs. *Beneath the top bar* a set of positive- and negative-sense mRNA species synthesized in infected cells is shown. *Dark and semidark thin lines* (+), mRNA sequences translated and nontranslated into viral proteins, respectively. *Light lines* (−), RNAs complementary to the different mRNAs. Poly(A) and Poly(U) tails are indicated by *AAA* or *UUU*. *Rep 1a* and *Rep 1b*, replicase genes; other acronyms as in **A**

whereas in the other two-thirds the carboxy terminus is inside and is integrated within the core, being essential to maintain its structure (Escors et al. 2001a, b). In addition, the virus envelope contains two other proteins, S and E (Enjuanes et al. 2000a). The S protein is responsible for attachment and entry into cells and is the major inducer of TGEV-neutralizing antibodies (Suñé et al. 1990).

The TGEV genome is a single molecule of positive-sense, single-stranded RNA of 28.5 kb, which is infectious and contains eight functional genes, four of which, the spike (S), envelope (E), membrane (M), and nucleoprotein (N), encode structural proteins (Fig. 1B). The genes are arranged in the order 5'-Rep1a-1b-S-3a-3b-E-M-N-7-3'. TGEV mRNAs consist of seven to eight types of varying sizes, depending on the strain (Penzes et al. 2001).

2
Pathogenesis Induced by Group 1 Coronaviruses

Coronaviruses are classified in three groups according to genetic analysis (González et al. 2003; Siddell 1995). Group 1 includes coronaviruses infecting human, porcine, canine, and feline species, closely related in sequence and, in some cases, also antigenically (Sánchez et al. 1990). Coronaviruses are associated mainly with respiratory, enteric, hepatic, and central nervous system diseases. In humans and fowl, coronaviruses primarily cause upper respiratory tract infections, whereas porcine and bovine coronaviruses establish enteric infections that result in severe economic loss (USDA 2002). Human CoV are responsible for 10%–20% of all common colds and have been implicated in gastroenteritis, high- and low-respiratory tract infections, and rare cases of encephalitis (Denison 1999). HCoV have also been associated with infant necrotizing enterocolitis (Resta et al. 1985) and are tentative candidates for multiple sclerosis (Denison 1999). Recently, a new SARS-CoV has emerged, infecting more than 8,000 people and causing more than 800 deaths in 5 months (Drosten et al. 2003; Holmes and Enjuanes 2003; Marra et al. 2003; Snijder et al. 2003; Thiel et al. 2003a).

Epithelial cells are the main targets of porcine coronaviruses. Widely distributed cells such as macrophages are also infected. TGEV is an enteropathogenic coronavirus that replicates in both villous epithelial cells of the small intestine and in lung cells. A nonenteropathogenic virus related to TGEV, the porcine respiratory coronavirus (PRCV), appeared in Europe in the 1980s (Callebaut et al. 1988; Pensaert et al. 1986) and later on in North America (Vaughn and Paul 1993; Vaughn et al. 1995; Wesley et al. 1990b). This virus replicates to high titers in the respiratory tract and undergoes limited replication in submucosal cells of the small intestine (Cox et al. 1990a, b). In contrast to TGEV, PRCV infection of swine resulted no clinical signs of disease (Duret et al. 1988; Pensaert et al. 1986; Wesley et al. 1990b).

A TGEV-like disease was associated with porcine epidemic diarrhea coronavirus (PEDV) (Pensaert and De Bouck 1978). This virus, closely related to TGEV in sequence but antigenically distinct (Sánchez et al. 1990), also infects the enteric tract of swine. Nevertheless, in contrast to TGEV, PEDV does not infect the lungs. Probably, TGEV and PEDV use different receptors, because TGEV easily grows in porcine cells whereas PEDV does not and, in contrast, PEDV replicates in monkey (Vero) cells only permissive to certain strains of TGEV (J.M. Sanchez and L. Enjuanes, unpublished data).

Canine coronavirus (CCoV) usually produces a mild gastroenteritis, although some virus strains cause a more severe and sometimes fatal diarrhea. Feline coronavirus (FCoV) causes a disease involving an antibody-dependent enhancement of infection and immunocomplex-induced lesions. Two serotypes of feline enteric coronavirus have been identified that cause feline infectious peritonitis by infecting macrophages (Olsen 1993).

3
Engineering Coronavirus Genome

Two types of expression vectors have been developed for coronaviruses (Fig. 2). One requires two components (helper-dependent expression system) (Fig. 2A) and the other a single genome that is modified either by targeted recombination (Masters 1999) (Fig. 2B.1) or by engineering a cDNA encoding an infectious RNA. Infectious cDNA clones are available for porcine (Fig. 2B.2 and B.3), human (Fig. 2B.4), avian, and murine coronaviruses as indicated above. Infectious cDNA clones have also been constructed for the *Arteriviridae* family closely related to coronaviruses (de Vries et al. 2000; Meulenberg et al. 1998; van Dinten et al. 1997).

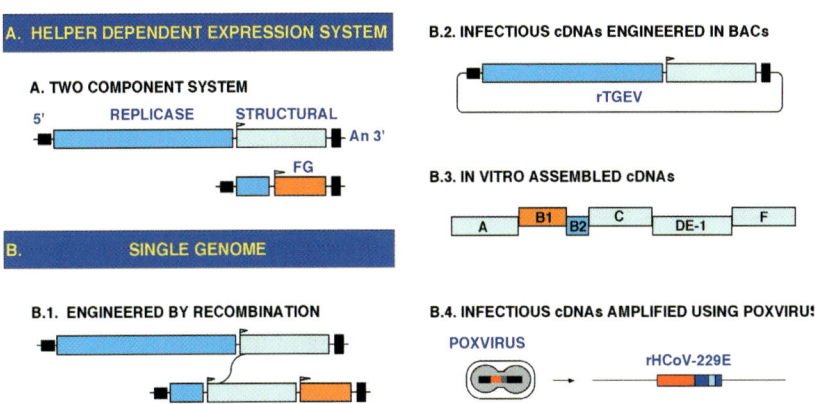

Fig. 2A, B. Coronavirus derived expression systems. A Helper-dependent expression system based on two components, the helper virus and a minigenome carrying the foreign gene (*FG*). *An*, poly A. B Single genome engineered by targeted recombination (*B.1*), by assembling an infectious cDNA clone derived from TGEV genome in BACs (*B.2*), by the in vitro ligation of six cDNA fragments (*B.3*), or by using poxviruses as the cloning vehicle (*B.4*)

An infectious coronavirus cDNA clone was first obtained for TGEV (Almazán et al. 2000; González et al. 2002). The strategy used to clone this cDNA was based on three points: (1) The construction was started from a defective minigenome (DI) that was stably and efficiently replicated by helper viruses (Izeta et al. 1999). During the filling in of minigenome deletions, a cDNA fragment that was toxic to the bacterial host was identified. This fragment was reintroduced into the cDNA in the last cloning step. (2) Transcription of the long coronavirus RNA genome, including a 5′ cap, in the nucleus is essential for its infectivity. The RNA was expressed in a process mediated by the recognition of the cytomegalovirus (CMV) promoter by the cellular polymerase II. This process was followed by a second amplification in the cytoplasm driven by the viral polymerase and (3) increase of viral cDNA stability within bacteria by cloning the cDNA as a bacterial artificial chromosome (BAC) producing a maximum of two plasmid copies per cell. Following this procedure, an infectious TGEV cDNA clone that produces a virulent virus infecting both the enteric and the respiratory tract of swine was engineered (Almazán et al. 2000). The stable propagation of a TGEV full-length cDNA in bacteria as a BAC has been improved by the insertion of an intron to disrupt the toxic region identified in the viral genome (Fig. 3) (González et al. 2002). The viral RNA was transcribed in the cell nucleus under the control of the CMV promoter, and the intron was efficiently removed during translocation of this RNA to the cytoplasm. Intron insertion in two different positions allowed stable plasmid amplification for at least 200 generations. Infectious TGEV was efficiently recovered from cells transfected with the modified cDNAs. The great advantage of this system is that coronavirus reverse genetics only involves standard recombinant DNA technologies performed within bacteria. The experience in our laboratory over more than 5 years has proven that this approach leads to the efficient rescue of mutants in all viral genes.

A second procedure to assemble a full-length infectious construct of TGEV was based on the in vitro ligation of six adjoining cDNA subclones that spanned the entire TGEV genome. Each clone was engineered with unique flanking interconnecting junctions that determine a precise assembly with only the adjacent cDNA subclones, resulting in a full-length TGEV cDNA. In vitro transcripts derived from the full-length TGEV construct were infectious (Yount et al. 2000).

An infectious cDNA clone has also been constructed for HCoV-229E (Thiel et al. 2001a), another member of the Group 1 coronaviruses, MHV (Yount et al. 2002) (see the chapter by Baric and Sims, this volume) and SARS-CoV (Yount et al. 2003), which are Group 2 coronavirus

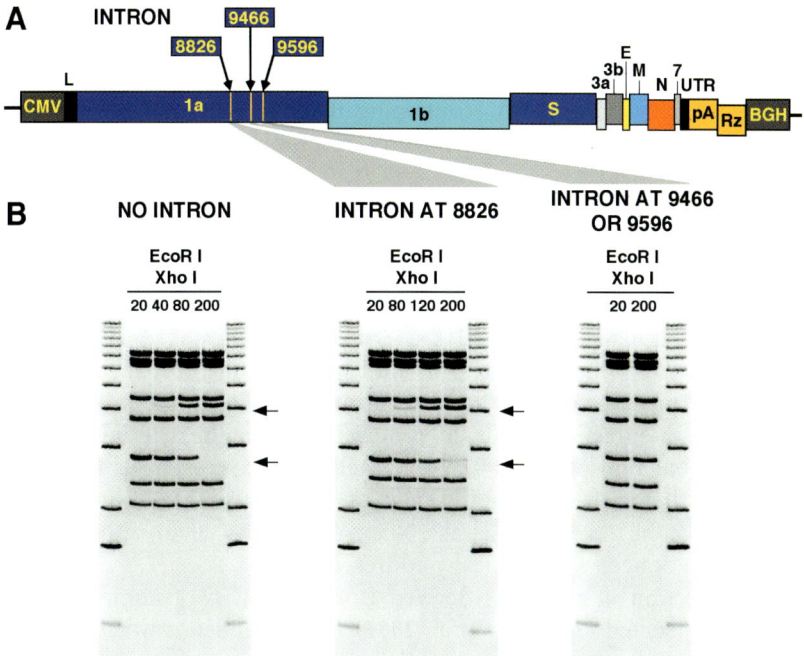

Fig. 3A, B. Intron insertion to stabilize TGEV full-length cDNA. **A** Strategy for the insertion of the 133-nt intron at the indicated positions of the TGEV sequence. **B** Analysis of the three intron-containing TGEV full-length cDNAs in *E. coli* cells. The *Eco*RI-*Xho*I restriction patterns of the three plasmids extracted from *E. coli* cells grown for the indicated number of generations are shown. *Arrows* indicate disappearance or appearance of a band. *M*, molecular mass markers

members, and IBV (Casais et al. 2001), a member of coronavirus Group 3. cDNA copies of the HCoV-229E and the MHV genomes have been cloned and propagated in vaccinia virus. Briefly, the full-length genomic cDNA clone of HCoV-229E is assembled by in vitro ligation and then cloned into the vaccinia virus DNA under the control of the T7 promoter. Recombinant vaccinia viruses containing the HCoV-229E genome are recovered after transfection of the vaccinia virus DNA into cells infected with fowlpox virus. In a second phase, the vaccinia virus DNA is purified and used as a template for in vitro transcription of HCoV-229E genomic RNA, which is transfected into susceptible cells for the recovery of infectious recombinant coronavirus. An IBV cDNA clone was assembled using the same strategy reported for HCoV-229E with some modifications. Similarly to HCoV-229E, the IBV genomic cDNA is assembled downstream of the T7 promoter by in vitro ligation and cloned into the vaccinia virus DNA. However, recovery of recombinant IBV is done after in situ synthesis of infectious IBV RNA by transfection of restricted recombinant vaccinia virus DNA (containing the IBV genome) into primary chicken cells previously infected with a recombinant fowlpox expressing T7 RNA polymerase.

A replicon has been constructed with the HCoV-229E genome (Thiel et al. 2001b). This replicon included the 5′ and 3′ ends of the HCoV-229E genome, the replicase gene of this virus, and a reporter gene coding for green fluorescent protein (GFP). RNA transcribed from this cDNA and transfected into BHK-21 cells led to only 0.1% of the cells showing strong fluorescence. This indicated that the coronavirus replicase gene products suffice for discontinuous subgenomic mRNA transcription, in agreement with the requirements for the arterivirus replicase (Molenkamp et al. 2000). Nevertheless, coexpression of N protein seems to increase rescue efficiency of infectious virus from cDNAs (Almazán et al. 2004; Thiel et al. 2003b; Yount et al. 2000), although this issue has not been systematically addressed.

A collection of replicons derived from the TGEV genome has also been constructed (Almazán et al. 2004). These replicons were launched from the cell nucleus with the CMV promoter, were efficiently rescued in the presence of N protein, expressed a heterologous gene in more than 83% of transfected cells, and had low or no cytopathogenicity for human cells.

4
Essential Genes Required for TGEV Replication

One of the most distinguishing features of the nidovirus genome is the conservation of the domain organization in the polyproteins pp1a and pp1ab, involved in genome replication, which are expressed by ribosomal frameshifting and polyprotein cleavage by viral proteases. A comparative analysis of replicative polyproteins of coronaviruses and arteriviruses identified the most variable regions in the N-terminal half of pp1a (Bonilla et al. 1994; Lee et al. 1991; Nelsen et al. 1999). Further insight into the *Coronaviridae* family showed some conserved domains in the pp1a/pp1b polyproteins between coronaviruses and toroviruses. The observation that some domains are not conserved in all coronaviruses and toroviruses indicated that these might be nonessential for the viral life cycle. In addition, all conserved arterivirus domains were found to be smaller than their coronavirus counterparts, indicating that some sequences in the coronavirus genome could be dispensable for virus replicative functions (Gorbalenya 2001). Nevertheless, attempts to delete the replicase nonconserved domains to determine whether they are essential have not yet been made.

ORFs 3a, 3b, and 7 of TGEV encode nonstructural proteins (Enjuanes and Van der Zeijst 1995; Ortego et al. 2003; Sola et al. 2003) that are nonessential for virus replication in cell culture. Although in the enteric and virulent Miller strain of TGEV expression of mRNAs from ORFs 3a and 3b has been observed, another virulent TGEV isolate (McGoldrick et al. 1999) includes a large deletion in ORF 3a, suggesting that this ORF is not required for replication in the enteric tract or to lead to virulent isolates. In other attenuated TGEV strains with a growth essentially limited to the respiratory tract, including the Purdue strain, the subgenomic mRNA corresponding to ORF 3b is not transcribed because this gene is preceded by a noncanonical transcription-regulating sequence (TRS) (O'Connor and Brian 1999, 2000; Wesley et al. 1989). These PRCV variants have deletions of varying sizes within ORF 3a and 3b (Vaughn et al. 1995; Wesley et al. 1991). The lack of enteric tropism and attenuation of these TGEV strains has been associated with a deletion of around 670 nt located at the 5' end of the S gene and not with deletion of genes 3a and 3b (Sánchez et al. 1992, 1999).

The engineering of a TGEV genome with all the genes separated by unique restriction endonuclease sites (Ortego et al. 2003) allowed the systematic deletion of each ORF to analyze whether they are essential or dispensable for virus growth (Fig. 4). TGEV with deleted 3a and 3b genes (rTGEV-Δ3) showed growth kinetics and mRNAs levels similar to those of the parental virus in cell cultures, demonstrating that the deletion of ORFs 3a and 3b did not affect either viral replication or transcription. In in vivo experiments, rTGEV-Δ3 virus kept the replication efficiency and tropism of the wild-type virus with a very small reduction in virulence, confirming that these properties were not significantly influenced by genes 3a and 3b (Sola et al. 2003).

TGEV ORF 7 is located downstream of the essential N gene. With the TGEV infectious cDNA clone including unique engineered restriction endonuclease sites, a deletion mutant was generated (rTGEV-Δ7) in which gene 7 expression has been abrogated (Ortego et al. 2003). The rTGEV-Δ7 contained a deletion spanning 21 nt upstream of the ORF 7 AUG and the first nucleotides of this ORF. Recombinant rTGEV-Δ7 showed standard kinetics in cell culture, indicating that the protein encoded by gene 7 was not essential for TGEV replication in tissue culture

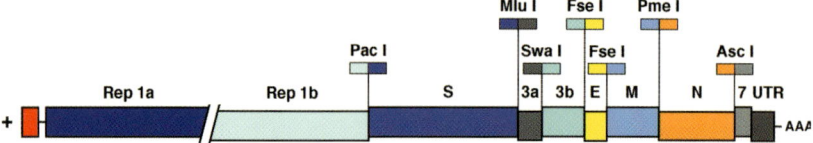

Fig. 4. Essential genes required for TGEV replication. The diagram shows a scheme of the TGEV genome in which the partial overlapping between genes has been resolved by duplicating sequences at the 5′ end of each gene, starting at gene S, in order to be able to delete a gene without affecting the expression of the flanking ones. The duplicated sequences include termination codons of the preceding gene and TRSs including the CS (5′-CUAAAC-3′). In addition, the indicated unique restriction endonuclease sites (*top of the bars*) were introduced to create insertion sites for heterologous genes. The name of the gene is indicated *on top of each bar*

(Ortego et al. 2003; Sola et al. 2003). Interestingly, in vivo infection with rTGEV-Δ7 showed an additional reduction in virus replication in the lung and gut compared with the parental virus and in virulence, indicating that TGEV gene 7 influences virus pathogenesis (Ortego et al. 2003).

5
Transcription-Regulating Sequences

To optimize virus vector expression levels it is essential to improve (1) replication levels without increasing virulence, (2) the accumulation levels of total mRNA, and (3) translation from mRNA. These objectives can only be achieved by knowing the mechanisms involved in these processes. To accomplish this goal, a brief review of mRNA transcription in coronavirus (particularly TGEV) is provided.

Coronavirus RNA synthesis occurs in the cytoplasm via a negative-strand RNA intermediate that contains short stretches of oligo(U) at the 5′ end. Both genome-size and subgenomic negative-strand RNAs, which correspond in number of species and size to those of virus-specific mRNAs, have been detected (Brian 2001; Sawicki et al. 2001). Coronavirus mRNAs have a leader sequence at their 5′ ends. Preceding every transcription unit on the viral genomic RNA, there is a conserved *core sequence* (CS) that is identical to a sequence located at the 3′ end of the leader sequence (CS-L). This sequence motif constitutes part of the signal for subgenomic mRNA transcription. The common 5′ leader sequence is only found at the very 5′ terminus of the genome, which implies that the synthesis of subgenomic mRNAs involves fusion of noncontiguous sequences. The mechanism involved in this process is under debate; nevertheless, the *discontinuous transcription during negative-strand RNA synthesis* model is compatible with most of the experimental evidence (Pasternak et al. 2001; Sawicki et al. 2001; van Marle et al. 1999; Zúñiga et al. 2004) (see chapter by Sawicki and Sawicki, this volume). As the leader-mRNA junction occurs during synthesis of the negative strand, within the sequence complementary to the CS (cCS), the nature of the CS is considered crucial for mRNA synthesis.

The CS motif includes six nucleotides that are highly conserved in all the genes of the same coronavirus (Fig. 5A). In addition, the 5' and 3' flanking sequences are partially conserved in the different genes of related viruses and these flanking sequences influence the activity of the CS (Fig. 5A) (Alonso et al. 2002a). Therefore, we consider that the TRSs could be divided into three sequence blocks, the CS, and the 5' (5'-TRS) and 3' (3'-TRS) flanking sequences (Fig. 5B). The most frequently used CS of coronaviruses belonging to Group 1 (hexamer 5'-CUAAAC-3') and Group 2 (heptamer 5'-UCUAAAC-3') share homology (Fig. 5C), whereas the CS of coronaviruses belonging to Group 3, like that of IBV, has the most divergent sequence (5'-CUUAACAA-3') (Fig. 5B). Also, arterivirus CSs have a sequence (5'-UCAACU-3') that partially resembles that of IBV.

Transcription levels may be influenced by many factors. Three of these, probably the most relevant, are as follows. (1) Potential base-pairing between the leader 3' end and sequences complementary to the TRS preceding the "body" (coding sequence) (cTRS-B), which guide the fusion between the nascent negative strand and the leader TRS. A minimum complementarity is required between the TRS located at the end of the leader (TRS-L) and the cTRS-B of each gene. Although several studies on the effect of the extension of this complementarity on mRNA synthesis have been performed (Alonso et al. 2002a; Joo and Makino 1992; La Monica et al. 1992; Makino and Joo 1993; Makino et al. 1991; Shieh et al. 1987; van der Most et al. 1994) the length of optimum TRSs has not been accurately defined. (2) Proximity of a gene to the 3' end. Because the TRSs are considered slow down or stop signals for the replicase complex, the smaller mRNAs should be the most abundant. Although this has been shown to be the case in the *Mononegavirales* (Wertz et al. 1998), and in coronaviruses shorter mRNAs are generally more abundant, the relative abundance of coronavirus mRNAs is not strictly related to their proximity to the 3' end (Alonso et al. 2002a; Penzes et al. 2001). (3) Potential interaction of proteins with the TRS RNA and protein–protein interactions that could regulate transcription levels. The as-

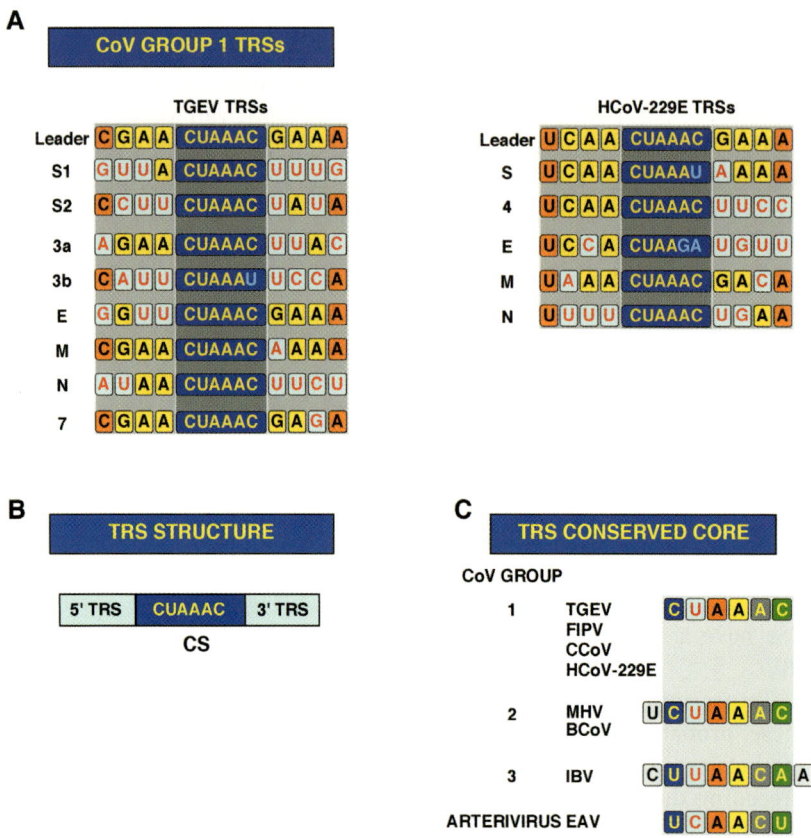

Fig. 5A–C. Coronavirus transcription-regulating sequences. **A** TRSs from two Group 1 CoVs (TGEV and HCoV-229E). CSs sequences are represented in *white letters* inside *dark boxes*. *Yellow boxes* highlight the identity among the sequences immediately flanking the CS at both 5′ and 3′ ends. **B** Group 1 coronavirus TRS sequence, including the highly conserved CS sequence (5′-CUAAAC-3′) and the flanking 5′ and 3′ TRS. **C** Sequence of the most frequently used CS of coronaviruses and arteriviruses

sociation of the nascent RNA chain with the leader TRS is probably mediated by the approximation of the TRS-L to the TRS-B of each gene through RNA-protein and protein–protein interactions.

The influence of the CS in transcription has been analyzed in detail in the arteriviruses (Pasternak et al. 2001, 2003; van Marle et al. 1999) and in coronaviruses (Alonso et al. 2002a; Zúñiga et al. 2004). With infectious cDNA clones of EAV and TGEV it has been shown that mRNA synthesis requires base-pairing interaction between the leader TRS and a cTRS in the nascent viral negative strand. The construction of double mutants in which a mutant leader CS was combined with the corresponding mutant of the body CS resulted in the restoration of the specific mRNA synthesis, initially suggesting that the sequence of the CS per se is not crucial, as long as the possibility for CS base-pairing is maintained. Nevertheless, it has been shown that other factors, besides leader-body base-pairing, also play a role in mRNA synthesis and that the primary sequence (or secondary structure) of TRSs may dictate strong base preferences at certain positions (Pasternak et al. 2001). Detailed analyses of the TRSs used in the arteriviruses (van Marle et al. 1999), MHV (Zhang and Liu 2000), BCoV (Ozdarendeli et al. 2001), and TGEV (Sola et al. 2003; Zúñiga et al. 2004) indicate that noncanonical CS sequences may also be used for strand transfer in the discontinuous mRNA synthesis in the *Nidovirales*.

5.1
Control of Transcription in TGEV

Template switching during synthesis of the negative RNA to join the leader is probably mediated by RNA–protein and protein–protein interactions (Fig. 6). In the RNA–protein interaction, primary or secondary RNA structure of sequences flanking the CS are probably involved. Therefore, identification of the nature and size of these sequences should help to understand the transcription mechanism. In TGEV, the TRSs have been characterized with a helper-dependent expression system based on coronavirus minigenomes, by studying the synthesis of subgenomic RNAs (sgmRNAs). TGEV TRSs include the CS (5'-CUAAAC-3'), which promotes from 100- to 1,000-fold increase of mRNA synthesis when this CS is located in the appropriate context (i.e., flanked by the appropriate 5' and 3' sequences) (Alonso et al. 2002a). The relevant sequences contributing to TRS activity have been studied by extending the CS 5' upstream and 3' downstream. Sequences from virus genes flanking the CS influenced transcription levels from moderate (10- to 20-fold variation) to complete mRNA synthesis silencing. For example, a canonical CS at nt 120 downstream of the initiation codon of the S gene did not lead to production of the corresponding mRNA because of the nature of the flanking sequences (Alonso et al. 2002a). The effect of 5' flanking sequences on transcription has also been studied using full-length genomes by inserting between genes N and 7 an expression cassette identical to those studied in the minigenome. The highest

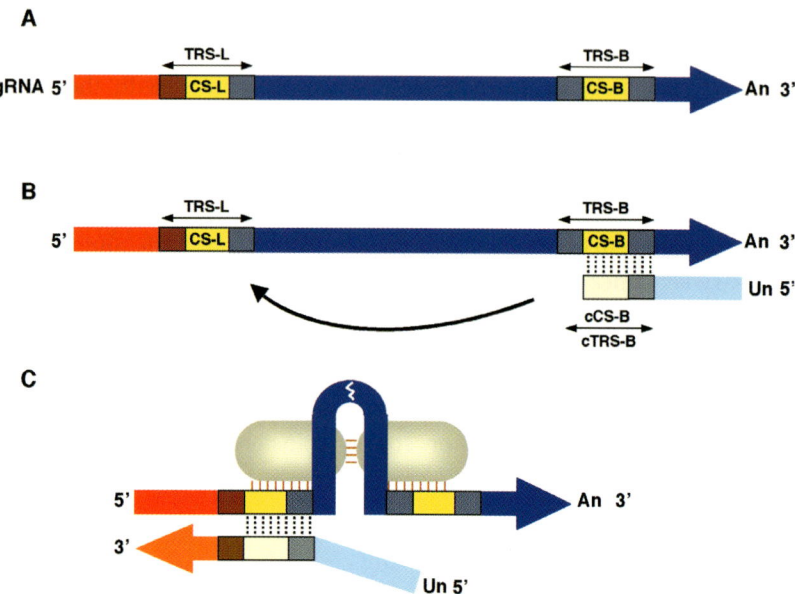

Fig. 6A–C. Diagram of the elements involved in coronavirus transcription. **A** Sequence elements probably involved in the discontinuous synthesis of the negative RNA strand. *CS-L* and *CS-B*, leader and body CSs; *TRS-L* and *TRS-B*, transcription-regulating sequences from leader and body; *An*, Poly(A). **B** Scheme of the discontinuous transcription during negative-strand synthesis and of the sequence elements involved. *cCS-B* and *cTRS-B* represent the CS-B and TRS-B complementary sequences, respectively; *Un*, Poly(U). **C** Leader and body sequences are probably located in close proximity in higher-order structures maintained by RNA-protein and protein-protein interactions

ratio of mRNA to genomic RNA was reached with 5' TRSs of about 88 nt, in agreement with the results obtained with the minigenomes and confirming the importance of sequences flanking the CS 5' end (Alonso et al. 2002a).

The influence of 3' TRSs on sgmRNA synthesis has also been studied with helper-dependent expression systems (minigenomes) encoding GUS in which the CS 5' flanking sequences of the expression cassette were maintained constant. In seven constructs the CS was flanked at the 3' end by each of the sequences flanking the CS in the virus genes S, 3a, 3b, E, M, N, and 7. In addition, another construct was designed to extend in 12 nucleotides the potential base-pairing between the 3' end of the leader and cTRS. In this construct, the added nucleotides were identical to those present in the genomic RNA downstream of the CS present at the 3′ end of the leader. GUS expression by minigenomes with the CS 3′ flanking sequences derived from the different virus genes differed by 10-fold, being maximum for minigenomes with a 3′ TRS derived from M gene or with an extended complementarity to TRS-L. The results indicated that CS 3′ flanking sequences have a large influence on sgmRNA accumulation and on protein synthesis (Alonso et al. 2002a). With the helper-dependent expression system, an optimized TRS has been designed comprising 88 nt from the N gene 5′ TRS, the CS, and 3 nt from the M gene 3′ TRS.

With full-length TGEV genomes, the insertion of a cassette for expressing heterologous genes between the N and 7 viral genes led to a new genetic organization of the 3′ end of recombinant viruses. As a consequence, a major species of sgmRNAs were generated from the non-canonical CS 5′-CUAAAA-3′. It was shown that extension of complementarity between CS flanking sequences and leader RNA was associated to transcription activation from a noncanonical CS. This observation suggests that additional base-pairing between the leader RNA and sequences flanking the CS motif could compensate for the absence of complete complementarity between leader and noncanonical CSs at the junction site (Sola et al. 2003; Zúñiga et al. 2004). Interestingly, other noncanonical CSs identical to the one described above (5′-CUAAAA-3′), present throughout the TGEV genome, did not promote transcription. These data confirm that in full-length coronavirus genomes the sequences flanking the CS also play a major role regulating sgmRNA expression levels.

The role of base-pairing between the nascent minus-strand RNA synthesized during transcription and the TRS-L was studied with full-length TGEV genomes (Zúñiga et al. 2004). Each nucleotide of the leader CS and the CS located 5′ upstream of the nonessential gene 3a were mutated to prevent base-pairing between the TRS-L and cTRS-B. In a parallel set of experiments, complementary mutations or mutations to preserve non-Watson-Crick base-pairing were also introduced. Interestingly, the relative amounts of mRNA transcribed were related to the free energy (ΔG) of TRS-L to cTRS-B duplex formation (Zúñiga et al. 2004).

Overall, these data indicated that complementarity between the TRS-L and the nascent negative RNA strand plays a major role in coronavirus transcription regulation. The experimental data are compatible with a working model for coronavirus transcription that includes three steps (Fig. 7) (Zúñiga et al. 2004): (1) interaction between the 5′ and 3′ ends of coronavirus genome to bring the TRS-L in close proximity to nascent minus RNA; (2) continuous base-pairing scanning between the

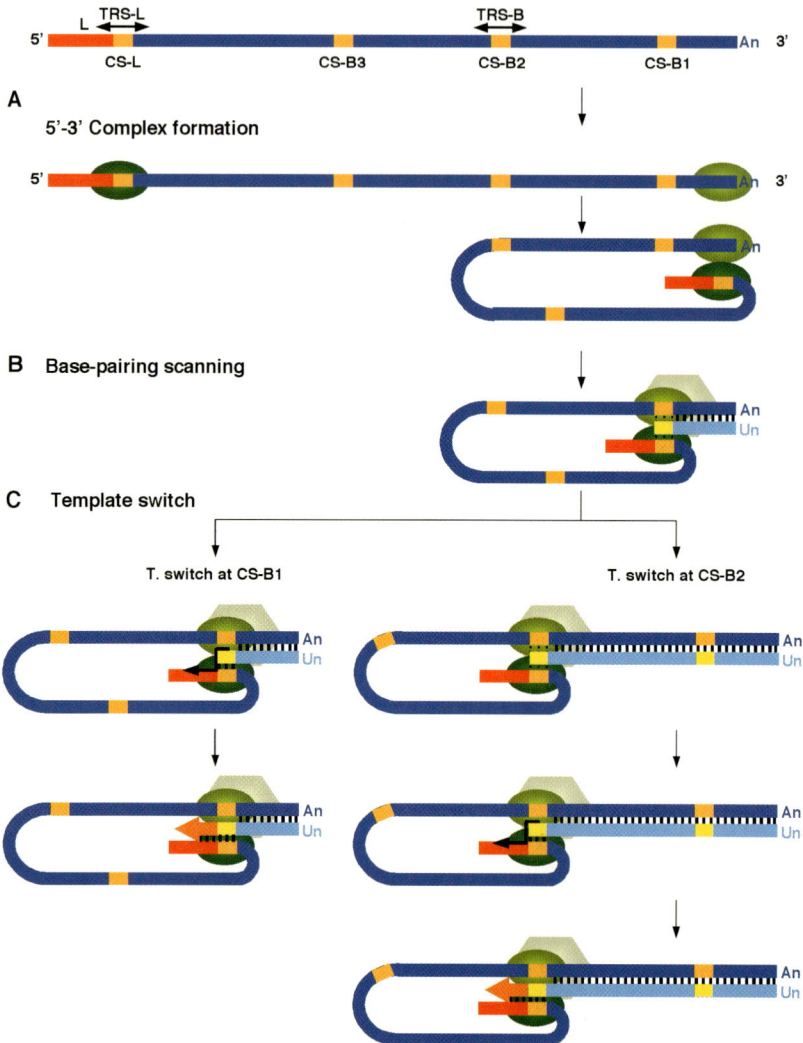

Fig. 7A–C. Three-step working model of coronavirus transcription. **A** 5′-3′ complex formation step. Proteins binding the 5′ and 3′ end TGEV sequences are represented by the *green balls*. Leader sequence is colored in *red*, CS sequences are colored in *yellow*. *An*, poly(A) tail. **B** Base-pairing scanning step. Minus-strand RNA is in a lighter color compared with positive-strand RNA. The transcription complex is represented by the *hexagon*. *Vertical dotted bars* represent the base-pairing scanning by the TRS-L sequence in the transcription process. *Vertical solid bars* indicate complementarity between gRNA and the nascent minus strand. *Un*, poly(U) tail. **C** Template switch step. The *thin arrow* indicates the switch in the template made by the transcription complex to complete the synthesis of (−) sgRNA

TRS-L and the nascent minus RNA strand; and (3) decision between template switch to copy the leader or continuation of RNA synthesis using as template contiguous genome sequences. This model explains the formation of sgmRNAs of different length and the synthesis of alternative sgmRNAs in sequence domains with high sequence identity with the TRS-L, observed by sequencing more than 90 leader-to-body junctions of sgmRNA generated by introducing point mutations within CS-L and the CS-B preceding the gene 3a (Zúñiga et al. 2004).

5.2
Effect of TRS Copy Number on Transcription

Studies on coronavirus transcription were performed with more than one contiguous TRS in order to express the same mRNA. The accumulated sgmRNA remained nearly the same for constructs with one, two, or three TRSs, and transcription preferentially occurred in all cases at the 3′-most TRS (Joo and Makino 1995; Krishnan et al. 1996; Stirrups et al. 2000; van Marle et al. 1995). A similar result was obtained by using arterivirus recombinant viruses in which three TRSs were introduced (Pasternak et al. 2004).

6
Expression Systems Based on Group 1 Coronaviruses

Coronaviruses have several advantages as vectors over other viral expression systems: (1) Coronaviruses are single-stranded RNA viruses that replicate in the cytoplasm without a DNA intermediary, making integration of the virus genome into the host cell chromosome unlikely (Lai and Cavanagh 1997). (2) These viruses have the largest RNA virus genome and, in principle, have room for the insertion of large foreign genes (Enjuanes et al. 2000a; Masters 1999). (3) A pleiotropic secretory immune response is best induced by the stimulation of gut-associated lymphoid tissues. Because coronaviruses in general infect mucosal surfaces, both respiratory and enteric, they may be used to target the antigen to the enteric and respiratory areas to induce a strong secretory immune response. (4) The tropism of coronaviruses may be engineered by modifying the S gene (Ballesteros et al. 1997; Kuo et al. 2000; Leparc-Goffart et al. 1998; Sánchez et al. 1999). (5) Nonpathogenic coronavirus strains infecting most species of interest (human, porcine, bovine, canine, feline, and avian) are available and therefore suitable to develop safe virus vectors. (6) Infectious coronavirus cDNA clones are available to design expression systems.

6.1
Helper-Dependent Expression Systems

Helper-dependent expression systems have been developed for coronaviruses from Groups 1, 2, and 3 (Enjuanes et al. 2001). Expression systems based on Group 1 coronaviruses have been developed for the porcine and human coronaviruses, because minigenomes are only available for these two coronavirus species. With TGEV-derived minigenomes an expression system has been assembled (Izeta et al. 1999). TGEV-derived RNA minigenomes were successfully transcribed in vitro with T7 polymerase and amplified after transfection into susceptible cells infected with a helper virus. TGEV-derived minigenomes of 3.3 (M33), 3.9 (M39), and 5.4 (M54) kb were efficiently used for the expression of heterologous genes (Alonso et al. 2002a, b). The smallest minigenome efficiently replicated and encapsidated by the helper virus was 3.3 kb in length (Izeta et al. 1999), although it has been shown that for replication and packaging of minigenomes including the β-glucuronidase (GUS) gene the most 5′ 649 nt and the most 3′ 278 nt may be sufficient (Escors et al. 2003).

With the M39 minigenome, a two-step amplification system was developed based on the cloning of a cDNA copy of the minigenome behind the CMV promoter (Izeta et al. 1999). Minigenome RNA is first transcribed in the nucleus by the cellular RNA *pol II*, and the RNA is then translocated into the cytoplasm, where it is amplified by the replicase of the helper virus. GUS and a surface glycoprotein (ORF5) that is the major protective antigen of the porcine respiratory and reproductive syndrome virus (PRSSV) have been expressed using this vector (Alonso et al. 2002b). GUS expression levels with an optimized TRS ranged between 2 and 8 μg of protein per 10^6 cells. Protein levels were dependent on the extent of transcription and also on translation regulation because the presence of an appropriate Kozak context led to higher protein expression levels (Alonso et al. 2002a). Expression of GUS gene and PRRSV ORF5 with these minigenomes has been demonstrated in the epithelial cells of alveoli and in scattered pneumocytes of swine lungs, which led to the induction of a strong immune response to these antigens (Alonso et al. 2002b).

HCoV-229E, another member of coronavirus Group 1, has also been used to express new sgmRNAs (Thiel et al. 1998). A synthetic RNA including 646 nt from the 5' end plus 1,465 nt from the 3' end was amplified by the helper virus. Nevertheless, little information has been provided on the expression of heterologous genes with this system.

Most of the work in coronavirus Group 2 helper-dependent systems has been done with MHV defective RNAs (Liao et al. 1995; Lin and Lai 1993; Zhang et al. 1997). Three heterologous genes have been expressed with the MHV system, chloramphenicol acetyltransferase (CAT), hemagglutinin-esterase (HE), and interferon-γ (IFN-γ). Expression of CAT or HE was detected only in the first two passages because the minigenome used lacks the packaging signal (Liao and Lai 1995). When virus vectors expressing CAT and HE were inoculated intracerebrally into mice, HE- or CAT-specific subgenomic mRNAs were only detected in the brains at days 1 and 2, indicating that the genes in the DI vector were expressed only in the early stage of viral infection (Zhang et al. 1998). An MHV DI RNA was also developed as a vector for expressing IFN-γ. The murine IFN-γ was secreted into culture medium as early as 6 h after transfection and reached a peak level at 12 h. No inhibition of virus replication was detected with the IFN-γ produced by the DI RNA in cell cultures, but infection of susceptible mice with DI RNA producing IFN-γ caused significantly milder disease, accompanied by less virus replication than that caused by virus containing a control DI vector (Lai et al. 1997; Zhang et al. 1997).

Group 3 coronavirus-derived helper-dependent expression systems are based on IBV. A defective RNA (CD-61) derived from the Beaudette strain of the IBV virus was used as an RNA vector for the expression of two reporter genes, luciferase and CAT (Penzes et al. 1996; Stirrups et al. 2000). With IBV minigenomes, CAT expression levels between 1 and 2 μg per 10^6 cells have been described. Therefore, the highest expression levels have been obtained with a two-step amplification system based on TGEV-derived minigenomes with optimized TRSs (Alonso et al. 2002a; Izeta et al. 1999).

Expression systems based on minigenomes have the advantage of a large cloning capacity, but with the limitation of a reduced stability. For instance, TGEV-derived helper-dependent expression systems expressing GUS or PRRSV ORF5 synthesized the heterologous gene for more than five passages, but at this time smaller minigenomes were detected that finally displaced the larger ones (Alonso et al. 2002b).

6.2
Single Genome Coronavirus Vectors

Coronavirus expression systems based on Group 1 viruses have been derived from TGEV and HCoV-229E. With a TGEV infectious cDNA, the GFP gene of 0.72 kb was cloned in two positions of the RNA genome, either by replacing the nonessential 3a and 3b genes or between genes N and 7. The engineered genome with the GFP gene at the position of ORFs 3a and 3b was very stable (>30 passages in cultured cells) and led to the production of high protein levels (50 μg/10^6cells) (Fig. 8) (Sola et al. 2003). Therefore, expression levels with coronavirus-based vectors are similar to those described for vectors derived from other positive-strand RNA viruses such as Sindbis virus (Agapov et al. 1998; Frolov et al. 1996). The stability of viruses with expression cassettes at different positions was variable. For instance, GFP or GUS expression units inserted between genes N and 7 led to unstable viruses (Alonso et al. 2004; Sola et al. 2003). With TGEV-derived vectors expressing GFP, the acquisition of immunity by newborn piglets breast-fed by immunized sows (i.e., lactogenic immunity) was demonstrated (Sola et al. 2003).

Recombinant TGEVs have also been assembled by in vitro junction of six cDNA fragments encoding a full-length genome, in which GFP gene has replaced ORF3a, leading to the production of a TGEV that grew to titers of 10^8 pfu/ml and expressed GFP in a high proportion of cells (Curtis et al. 2002).

HCoV-229E has also been used as the base for expression systems using either the full-length infectious cDNA clone (Thiel et al. 2001a) or an autonomous replicating subgenomic RNA (replicon) (Thiel et al. 2003b) (see chapter by Thiel and Siddell, this volume). In each case, it has been shown that it is possible to insert transcriptional cassettes.

An infectious cDNA clone for Group 2 coronaviruses has been constructed for MHV (Yount et al. 2002). This system will provide significant advantages in the analysis of coronavirus replication and transcription by complementing a large collection of temperature-sensitive mutants of MHV. Reverse genetics in coronaviruses of Group 2 has also

Fig. 8A–C. GFP expression levels with coronavirus based vectors. A Schemes of BACs containing the TGEV genome in which an expression cassette with the GFP gene has been inserted replacing genes 3a and 3b (*top bar*) or between genes N and 7 (*lower bar*), respectively. B The relative proportion of cells expressing GFP was evaluated by

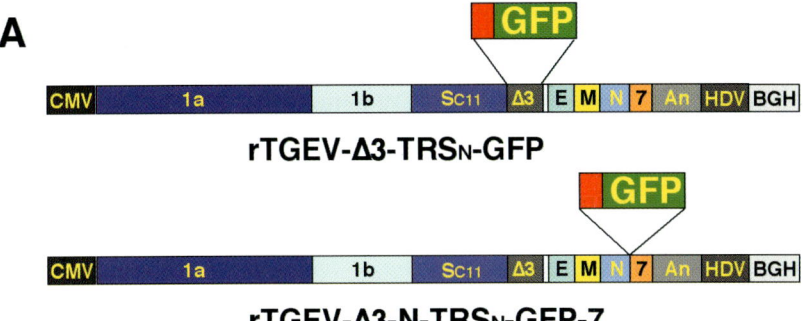

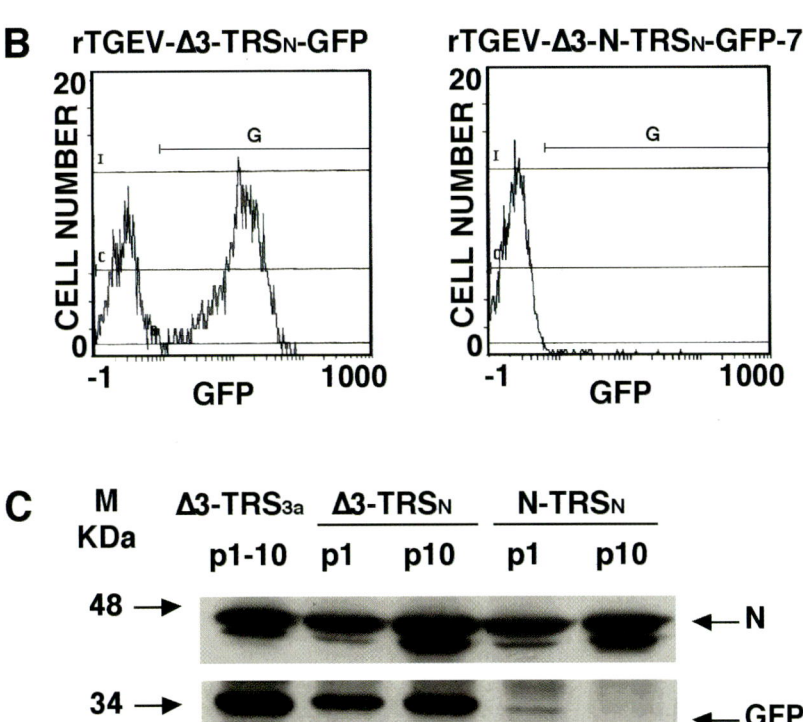

cytofluorometry. C The amount of protein synthesized by the virus vectors with expression modules inserted at genes 3a and 3b ($\Delta 3$) or between genes N and 7 (N), using TRSs from gene 3a (*TRS3a*) or from gene N (*TRSN*) was analyzed by Western blot with GFP-specific monoclonal antibodies (αGFP). The amount of viral nucleoprotein was evaluated in the same samples with N-specific monoclonal antibodies (αN) as an internal control

been efficiently performed by targeted recombination between a helper virus and either nonreplicative or replicative coronavirus-derived RNAs and was also used to express heterologous genes. For instance, the gene encoding GFP was inserted into MHV between the S and E genes, resulting in the creation of the largest known RNA viral genome (Fischer et al. 1997). Coronavirus expression systems derived from Group 3 coronavirus are based on infectious IBV cDNA clones assembled with a strategy similar to that reported for HCoV-229E (Casais et al. 2001) (see chapter by Baric and Sims, this volume).

6.3
Replication-Competent, Propagation-Deficient Coronavirus-Derived Expression Systems

Replication-competent, propagation-deficient virus vectors, based on TGEV genomes deficient in the essential gene E obtained with E^+ packaging cell lines have been developed (Ortego et al. 2002). Two types of cell lines expressing TGEV E protein have been selected (Fig. 9). One cell line transiently expresses E protein using the noncytopathic Sindbis virus replicon pSINrep21 (Frolov et al. 1999). Another cell line was ob-

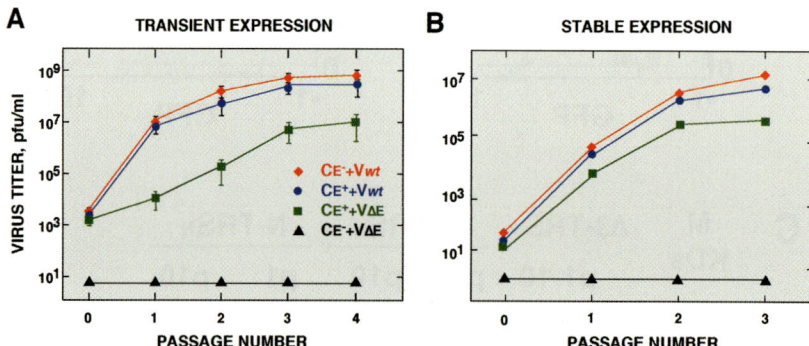

Fig. 9A, B. Rescue of recombinant TGEV-ΔE from cDNA in cells transiently or stably expressing E protein. **A** Titer of recombinant TGEVs rescued from cDNA in BHK cells expressing porcine aminopeptidase N (pAPN) (CE^-) or the same cells transiently expressing TGEV E protein (CE^+) with a Sindbis virus replicon. Cells were transfected either with rTGEV-*wt* (*Vwt*) or with rTGEV-ΔE virus (*V*ΔE). **B** Titer of recombinant TGEVs rescued from cDNA in BHK cells expressing pAPN (CE^-) or the same cells stably expressing TGEV E protein (CE^+) under the control of the CMV promoter. Cells were transfected either with rTGEV-*wt* (*Vwt*) or with rTGEV-ΔE virus (*V*ΔE). Error bars represent standard deviations of the mean from four experiments

tained in which the E gene is stably expressed under the control of CMV promoter. Recombinant TGEVs, deficient in the essential E gene, reached high titers (1×10^7 pfu/ml) in cells transiently expressing the TGEV E protein, whereas this titer was up to 5×10^5 pfu/ml in packaging cell lines stably expressing E protein. Virus titers were directly related to E protein expression levels (Ortego et al. 2002). Recovered virions showed the same morphology and stability at different pH and temperatures from the wild-type virus. The titers of the rescued viruses will most probably be increased by transforming new cell types with higher TGEV replication levels, leading to efficacious expression systems, or by increasing E protein accumulation.

A second strategy for the construction of replication-competent, propagation-deficient TGEV genomes expressing heterologous genes involves the assembly of an infectious cDNA from six cDNA fragments ligated in vitro (Curtis et al. 2002). The defective virus with the essential E gene deleted was complemented by the expression of E gene using the Venezuelan equine encephalitis (VEE) replicon. However, titers of recombinant virus expressing the GFP were at least 10- to 100-fold lower (around 10^4 pfu/ml) than with the system that used stably transformed cells or the Sindbis virus vector to complement E gene deletion (Ortego et al. 2002).

A multigene RNA replicon based on HCoV-229E has been developed (Thiel et al. 2003b) containing the 5' and 3' ends of this virus, the entire human coronavirus replicase gene, and three reporter genes [i.e., CAT gene, luciferase (LUC) gene, and the GFP gene]. Each reporter gene is located downstream of a human coronavirus TRS, which is required for the synthesis of individual mRNAs. The transfection of the vector and human coronavirus nucleocapsid protein mRNA into BKH-21 cells resulted in the expression of the CAT, LUC, and GFP reporter proteins. In addition, it has been shown that human coronavirus-based vector RNA can be packed into propagation-deficient pseudovirions that, in turn, can be used to transduce immature and mature human dendritic cells.

7
Coronavirus Vector Cloning Capacity

Coronavirus helper-dependent expression systems based on minigenomes have a theoretical cloning capacity close to 27 kb, because an RNA with a size of about 3 kb is efficiently amplified and packaged by the helper virus and the virus genome has about 30 kb. In contrast, the theoretical cloning capacity for an expression system based on a single coronavirus genome, like TGEV, with the current available knowledge, is between 3 and 3.5 kb, taking into account that (1) the nonessential genes 3a (0.2 kb), 3b (0.73 kb), and 7 (0.24 kb) have been deleted, leading to a viable virus (Sola et al. 2003); (2) the standard S gene can be replaced by the S gene of a PRCV mutant with a deletion of 0.67 kb; and (3) both DNA and RNA viruses may accept genomes with sizes up to 105% of the wild-type genome.

In propagation-deficient coronavirus vectors, in which one or more of the essential genes S, E, M, and N would have been deleted, the cloning capacity could be increased over the former 3.5 kb by an additional 0.25–4.5 kb. This cloning capacity will probably be enlarged in the near future when nonessential domains of the replicase gene are deleted (see above).

8
Insertion Site, Stability, and Expression Levels

TGEV-derived helper-dependent expression systems have the advantage of a high cloning capacity. In contrast, they have a limited stability mainly derived from the presence of the foreign gene, because TGEV minigenomes of 9.7, 5.4, 3.9, and 3.3 kb, in the absence of the heterologous gene, were amplified, packaged, and efficiently propagated for at least 30 passages, without generating new dominant minigenome RNAs (Izeta et al. 1999; Méndez et al. 1996). In contrast, the insertion of genes such as GUS, TGEV S protein, and PRRSV ORF5 in the M39 minigenome led to the appearance of new smaller minigenomes that could easily be detected at passage 5. Nevertheless, with this minigenome heterologous gene expression was observed for about 10 passages. Minigenome stability is highly dependent on the nature of the foreign gene, as TGEV or IBV minigenomes expressing luciferase gene were lost in the first 2–3 passages (Stirrups et al. 2000).

The stability of the expression systems is also conditioned by the type of polymerases involved in minigenome amplification and mRNA transcription (Agapov et al. 1998). The TGEV-derived vector is based on expression of the minigenome using the CMV promoter. In this case, eukaryotic RNA polymerase II expresses the minigenome with an estimated error frequency of 5×10^{-6} (de Mercoyrol et al. 1992), which is lower than the error accumulation frequency of 10^{-4} to 10^{-5} during in vitro transcription of minigenome RNAs with T7 DNA-dependent RNA-polymerase (Boyer et al. 1992; Sooknanan et al. 1994). In addition, the eukaryotic RNA polymerase II has additional mechanisms to ensure even more accurate transcription (Thomas et al. 1998). After transfection of in vitro produced RNA, synthesis of DI-RNA and mRNAs by the viral RNA-dependent RNA-polymerase should have an accumulation of mutations with a relatively higher frequency of 10^{-3} to 10^{-4} (de Mercoyrol et al. 1992; Ward et al. 1988). Therefore, an improvement in expression stability should be observed by using expression systems initiated by DNA transfection.

To study the effect of TRS position within the minigenome on expression levels, a cassette encoding the GUS gene was inserted at different nucleotide distances (0.9, 1.6, 2.8, and 3.3 kb) from the 5' end of the TGEV minigenome M39 (Alonso et al. 2002a). The mRNA levels were high in the two insertion sites closer to the 3' end of the minigenome and slightly increased in the most 3' site. In contrast, in a minigenome derived from the HCoV-229E, an expression cassette was cloned into three different positions (at 1.1, 1.3, and 1.8 kb from the 5' end of a minigenome of about 2.1 kb) and the mRNA levels decreased for inserts located closer to the 3' end (Thiel et al. 1998). The experiments performed with TGEV and HCoV-229E apparently led to different results. With TGEV, and possibly with HCoV-229, some insertion sites were too close to the ends of the minigenomes and may have affected essential primary or secondary structures required for their replication. It is possible that in coronavirus the variation of expression levels with insertion site is mainly influenced by the sequences flanking the TRS in each position and that the relative position itself plays a less prominent role. In fact, in a systematic study using MHV, a 0.4-kb region including a TRS of 12 nt flanked upstream and downstream by 0.2-kb fragments was inserted at seven different positions of a 2.2-kb minigenome (Jeong et al. 1996). The 12-nt TRS core was flanked by upstream and downstream sequences in order to prevent the influence of variable flanking sequences within the different insertion sites. In all insertion sites, the level both of the minigenome and of the mRNA produced were similar, suggesting that the position of the insert along the minigenome had little influence on the mRNA expression level.

In full-length genomes, a correlation between the proximity to the 3' end of the genome and the relative efficiency of mRNA synthesis from a given TRS has been observed in several viral systems, including coronaviruses such as MHV and TGEV (Alonso et al. 2002a; de Haan et al. 2003; Hiscox et al. 1995; Sola et al. 2003; van Marle et al. 1995), arteriviruses such as EAV (Pasternak et al. 2003), and the *Mononegavirales*

(Iverson and Rose 1981; Wertz et al. 1998). In the TGEV single genome vector, in order to increase heterologous gene expression levels, an expression cassette encoding the GFP gene was inserted at the 3' end of the genome. Insertion of the expression cassette between TGEV genes N and 7 resulted in an unstable virus, leading to the complete deletion of the additional transcriptional unit. In contrast, insertion of similar expression cassettes replacing ORFs 3a and 3b led to stable expression of GFP (Sola et al. 2003). Therefore, the location of the insertion, and not the nature of the gene, was most likely responsible for the instability. The origin of TGEV recombinant virus instability was mediated either by homologous recombination promoted by the presence of duplicated viral sequences or by nonhomologous recombination yielding a virus that had lost the GFP gene and also the 5' end of gene 7. Therefore, in addition to similarity-essential recombination, similarity-nonessential recombination (Nagy and Simon 1997) may also lead to the instability of these viruses. The instability of expression cassettes inserted at the 3' end of the genome seems a general phenomenon because, in addition, it has been shown that several expression modules encoding the GUS gene were also unstable at this position of the genome but not at the ORF 3a site (Alonso et al. 2004). Furthermore, insertion of other sequence fragments (i.e., 3' end 141 nt of N gene or 717 nt of GFP) between the N gene and the 3' UTR of MHV also produced genomic instability (Hsue and Masters 1999).

9
Molecular Basis of Group 1 Coronavirus Tropism

Group 1 coronaviruses attach to host cells through the S glycoprotein by recognition of pAPN, the cell receptor (Delmas et al. 1992; Yeager et al. 1992). Group 2 coronaviruses use the carcinoembryonic antigen-related cell adhesion molecules (CEACAM) as receptors, or the angiotensin-converting enzyme 2 in the case of SARS-CoV (Li et al. 2003). Engineering the S gene led to changes both in tissue- and species specificity (Ballesteros et al. 1997; Kuo et al. 2000; Leparc-Goffart et al. 1998; Sánchez et al. 1999). Driving vector expression to different tissues may preferentially induce a specific type of immune response, that is, mucosal immunity by targeting the expression to gut-associated lymph nodes. Both tissue- and species specificity have been modified by engineering coronavirus genomes.

Enteric infection by TGEV requires virus binding to pAPN, mediated by an S protein domain encoded by nt 1,518–2,184, and also to a second factor (possibly binding to a coreceptor) mediated by an S protein domain encoded by nt 217–665. In fact, changes of two amino acids included in this S protein domain suffice to restore the enteric infection by a respiratory virus strain (Sánchez et al. 2004). Therefore, the S protein domain encoded by nt 217–665 dictates TGEV enteric tropism (Sánchez et al. 1999, 2004), whereas binding to APN alone is not sufficient to infect the enteric tract. Interestingly, MHV tropism is also influenced by the binding of a domain located at the N-terminus of murine coronavirus spike protein, in positions equivalent to those required for the potential coreceptor binding in TGEV (Kubo et al. 1994).

TGEV species specificity has been extended to infect canine and human cells by replacing the S gene of TGEV with S genes from canine (Riquelme et al. 2004) and human (Ortego et al. 2004) coronaviruses. In this case, a replication-competent, propagation-deficient TGEV has been used for safety because these viruses can only grow in packaging cell lines; therefore, they cannot be propagated from cell to cell in the host.

Animal model systems based on transgenic mice expressing the human APN (hAPN) are being developed (Lassnig et al. 2004; Wentworth and Holmes 2001; Wentworth et al. 2001) to study the molecular basis of human coronavirus, vector-host interaction (i.e., potential side-effects), and immune responses elicited by the virus vector. The transgenic mice express high amounts of hAPN in lungs, gut, spleen, liver, and brain. Cells derived from these mice replicate HCoV-229E (Lassnig et al. 2004; Wentworth and Holmes 2001; Wentworth et al. 2001), and infection of transgenic mice has been shown in one of the systems (Lassnig et al. 2004). The possibility of engineering coronavirus vectors with defined tissue- or species specificity and the development of laboratory animal model systems increases the potential use of this novel vector family.

10
Modulation of Coronavirus Vector Virulence

Tropism changes in general lead to a change in virulence. Certainly this is the case in porcine coronaviruses with a virulence directly related to their ability to grow in the enteric tract (Sánchez et al. 1999). Porcine coronaviruses exclusively growing in the lungs with titers higher than 1×10^6 pfu/g of tissue lead to no obvious clinical symptoms. Changes in the S gene modify virus tropism and also virulence (das Sarma et al.

2000; Gallagher and Buchmeier 2001; Leparc-Goffart et al. 1998; Navas et al. 2001; Phillips et al. 1999; Sánchez et al. 1999; Taguchi et al. 1995). Similarly, cell-to-cell virus spreading is influenced by S protein-dependent fusion activity, also affecting coronavirus pathogenicity (Gallagher and Buchmeier 2001; Luo et al. 1999).

Most coronavirus genes partially overlap, and TGEV is not an exception. To study the effect of nonessential gene deletion on virulence, unique restriction endonuclease sites and sequence duplications were introduced at the 5' end of each TGEV gene as shown above (Fig. 4). Because the TRS is also located at the 5' end of each gene, the insertion of duplicated sequences and of restriction endonuclease sites could alter virus pathogenicity. In fact, with the use of TGEV it has been shown that the introduction of one restriction endonuclease site between each pair of adjacent genes leads to a decrease in enteric tract virus growth and virulence by more than 10-fold and 5-fold, respectively (Ortego et al. 2003). The simultaneous modification of the 5' end of all essential genes (S, E, M, and N) led to a virus growth and virulence reduction of 100-fold and 12-fold, respectively. Therefore, this approach can be used to control virus vector virulence.

Gene expression among the nonsegmented negative-stranded RNA viruses is controlled by the highly conserved order of genes relative to the single transcriptional promoter. Rearrangement of the genes of vesicular stomatitis virus eliminates clinical disease in the natural host and is considered a new strategy for vaccine development (Flanagan et al. 2001). In coronavirus, a change in gene order led to virus attenuation in FIPV (see chapter by Masters and Rottier, this volume) (de Haan et al. 2002).

Deletion of nonessential genes 3a and 3b led to variable results in Group 1 coronaviruses. Although deletion of these genes reduced TGEV virulence very little (Sola et al. 2003), FIPV was clearly attenuated (Haijema et al. 2003). In contrast, deletion of the nonessential gene 7 attenuated both TGEV and FIPV. Therefore, modification of certain nonessential genes may be used as an efficient approach to reduce the virulence of coronavirus vectors.

11
Biosafety in Coronavirus-Derived Vectors

Application of virus vectors to humans requires a reduction of the risk to levels below those of conventional medical interventions (i.e., administration of a safe vaccine). Coronavirus vectors based on the TGEV genome have been engineered to infect human cells by replacing the porcine coronavirus S gene with that of human coronaviruses (Ortego et al. 2004). To increase the safety of the human-adapted vector, a replication-competent, propagation-deficient virus, in which two essential genes (E and N) were deleted, is being modified by introducing mutations that abrogate the activity of the RNA packaging signal (Ψ) and relocating an active Ψ between the two deleted genes (Escors et al. 2003). A recombination event leading to the recovery of the essential genes will most likely lead to loss of the packaging signal, generating a nonviable virus.

12
Conclusions

Both helper-dependent expression systems, based on two components, and single genome vectors constructed by targeted recombination, or by using infectious cDNAs, have been developed for coronaviruses. The sequences that regulate transcription have been characterized with helper-dependent expression systems and full-length infectious cDNA clones. Minigenome-based expression systems have the advantage of their large cloning capacity, in principle higher than 27 kb, produce reasonable amounts of heterologous antigens (2–8 μg/10^6 cells), show a limited stability (synthesis of heterologous gene is maintained for around 10 passages), and elicit strong immune responses. In contrast, coronavirus vectors based on single genomes have at present a limited cloning capacity (around 5 kb) and expression levels of heterologous genes are 10-fold higher than those of helper-dependent systems (>50 μg/10^6 cells) and are very stable (>30 passages). The possibility of expressing different genes under the control of TRSs with programmable strength and engineering tissue and species tropism indicates that coronavirus vectors are very flexible. High expression levels have been obtained with replication-competent, propagation-deficient vectors based on coronavirus genomes. Thus coronavirus-based vectors are emerging with high potential for vaccine development and, possibly, for gene therapy.

Acknowledgements This work has been supported by grants from the Comisión Interministerial de Ciencia y Tecnología (CICYT), La Consejería de Educación y Cultura de la Comunidad de Madrid, and Fort Dodge Veterinaria from Spain and the by European Communities (Life Sciences Program, Key Action 2: Infectious Diseases). DE, IS, SZ, and SA received contracts from the EU projects QLRT-1999-30739, QLRT-2000-00874, and QLRT-1999-00002, respectively).

References

Agapov EV, Frolov I, Lindenbach BD, Pragai BM, Schlesinger S, Rice CM (1998) Noncytopathic Sindbis virus RNA vectors for heterologous gene expression. Proc Natl Acad Sci USA 95:12989–12994

Almazán F, González JM, Pénzes Z, Izeta A, Calvo E, Plana-Durán J, Enjuanes L (2000) Engineering the largest RNA virus genome as an infectious bacterial artificial chromosome. Proc Natl Acad Sci USA 97:5516–5521

Almazán F, Galán C, Enjuanes L (2004) The nucleoprotein is required for efficient coronavirus genome replication. In press

Alonso S, Izeta A, Sola I, Enjuanes L (2002a) Transcription regulatory sequences and mRNA expression levels in the coronavirus transmissible gastroenteritis virus. J Virol 76:1293–1308

Alonso S, Sola I, Teifke J, Reimann I, Izeta A, Balach M, Plana-Durán J, Moormann RJM, Enjuanes L (2002b) In vitro and in vivo expression of foreign genes by transmissible gastroenteritis coronavirus-derived minigenomes. J Gen Virol 83:567–579

Alonso S, Sola I, Zúñiga S, Plana-Durán J, Enjuanes L (2004) Induction of neutralizing antibodies against porcine respiratory and reproductive syndrome virus antibodies (PRRSV) ORF 5 by coronavirus derived vectors. Submitted for publication

Ballesteros ML, Sánchez CM, Enjuanes L (1997) Two amino acid changes at the N-terminus of transmissible gastroenteritis coronavirus spike protein result in the loss of enteric tropism. Virology 227:378–388

Bonilla PJ, Gorbalenya AE, Weiss SR (1994) Mouse hepatitis virus strain A59 RNA polymerase gene ORF 1a: heterogeneity among MHV strains. Virology 198:736–740

Boyer JC, Bebenek K, Kunkel TA (1992) Unequal human immunodeficiency virus type 1 reverse transcriptase error rates with RNA and DNA templates. Proc Natl Acad Sci USA 89:6919–6923

Brian DA (2001) Nidovirus genome replication and subgenomic mRNA synthesis. Pathways followed and *cis*-acting elements required. In: Lavi E, Weiss S and Hingley ST (eds) Nidoviruses. Plenum Press, New York Adv. Exp. Med. Biol., vol 494, pp 415–428

Callebaut P, Correa I, Pensaert M, Jiménez G, Enjuanes L (1988) Antigenic differentiation between transmissible gastroenteritis virus of swine and a related porcine respiratory coronavirus. J Gen Virol 69:1725–1730

Casais R, Thiel V, Siddell SG, Cavanagh D, Britton P (2001) Reverse genetics system for the avian coronavirus infectious bronchitis virus. J Virol 75:12359–12369

Cox E, Hooyberghs J, Pensaert MB (1990a) Sites of replication of a porcine respiratory coronavirus related to transmissible gastroenteritis virus. Res Vet Sci 48:165–169

Cox E, Pensaert MB, Callebaut P, van Deun K (1990b) Intestinal replication of a porcine respiratory coronavirus closely related antigenically to the enteric transmissible gastroenteritis virus. Vet Microbiol 23:237–243

Curtis KM, Yount B, Baric RS (2002) Heterologous gene expression from transmissible gastroenteritis virus replicon particles. J Virol 76:1422–1434

das Sarma J, Fu L, Tsai JC, Weiss SR, Lavi E (2000) Demyelination determinants map to the spike glycoprotein gene of coronavirus mouse hepatitis virus. J Virol 74:9206–9213

de Haan CAM, Volders H, Koetzner CA, Masters PS, Rottier PJM (2002) Coronavirus maintain viability despite dramatic rearrangements of the strictly conserved genome organization. J Virol 76:12491–12502

de Haan CAM, van Genne L, Stoop JN, Volders H, Rottier JMP (2003) Coronaviruses as vectors: position dependence of foreign gene expression. J Virol 77:11312–11323

Delmas B, Gelfi J, L'Haridon R, Vogel LK, Norén O, Laude H (1992) Aminopeptidase N is a major receptor for the enteropathogenic coronavirus TGEV. Nature 357:417–420

de Mercoyrol L, Corda Y, Job C, Job D (1992) Accuracy of wheat-germ RNA polymerase II. General enzymatic properties and effect of template conformational transition from right-handed B-DNA to left-handed Z-DNA. Eur J Biochem 206:49–58

Denison MR (1999) The common cold. Rhinoviruses and coronaviruses. In: Dolin R and Wringht PF (eds) Viral infections of the respiratory tract. Marcel Dekker, Inc., New York Lung Biology in Health and Disease, vol 127, pp 253–280

de Vries AAF, Glaser AL, Raamsman MJB, de Haan CAM, Sarnataro S, Godeke GJ, Rottier PJM (2000) Genetic manipulation of equine arteritis virus using full-length cDNA clones: separation of overlapping genes and expression of a foreign epitope. Virolog *Coronaviridae*. In: van Regenmortel MHV, Fauquet CM, Bishop DHL, Carsten EB, Estes MK, Lemon SM, McGeoch DJ, Maniloff J, Mayo MA, Pringle CR and Wickner RB (eds) Virus taxonomy. Classification and nomenclature of viruses. Academic Press, San Diego, California, pp 835–849

Enjuanes L, Sola I, Almazán F, Ortego J, Izeta A, González JM, Alonso S, Sánchez-Morgado JM, Escors D, Calvo E, Riquelme C, Sánchez CM (2001) Coronavirus derived expression systems. J Biotech 88:183–204

Escors D, Ortego J, Laude H, Enjuanes L (2001a) The membrane M protein carboxy terminus binds to transmissible gastroenteritis coronavirus core and contributes to core stability. J Virol 75:1312–1324

Escors D, Camafeita E, Ortego J, Laude H, Enjuanes L (2001b) Organization of two transmissible gastroenteritis coronavirus membrane protein topologies within the virion and core. J Virol 75:12228–12240

Escors D, Izeta A, Capiscol MC, Enjuanes L (2003) Transmissible gastroenteritis coronavirus packaging signal is located at the 5' end of the virus genome. J Virol 77:7890–7892

Fischer F, Stegen CF, Koetzner CA, Masters PS (1997) Analysis of a recombinant mouse hepatitis virus expressing a foreign gene reveals a novel aspect of coronavirus transcription. J Virol 71:5148–5160

Flanagan EB, Zamparo JM, Ball LA, Rodriguez L, Wertz GW (2001) Rearrangement of the genes of vesicular stomatitis virus eliminates clinical disease in the natural host: new strategy for vaccine development. J Virol 75:6107–6114

Frolov I, Hoffman TA, Prágai BM, Dryga SA, Huang HV, Schlesinger S, Rice CM (1996) Alphavirus-based expression vectors: strategies and applications. Proc Natl Acad Sci USA 93:11371–11377

Frolov I, Agapov E, Hoffman TA, Prágai BM, Lippa M, Schlesinger S, Rice CM (1999) Selection of RNA replicons capable of persistent noncytopathic replication in mammalian cells. J Virol 73:3854–3865

Gallagher T, Buchmeier MJ (2001) Coronavirus spike proteins in viral entry and pathogenesis. Virology 279:371–374

González JM, Penzes Z, Almazán F, Calvo E, Enjuanes L (2002) Stabilization of a full-length infectious cDNA clone of transmissible gastroenteritis coronavirus by the insertion of an intron. J Virol 76:4655–4661

González JM, Gomez-Puertas P, Cavanagh D, Gorbalenya AE, Enjuanes L (2003) A comparative sequence analysis to revise the current taxonomy of the family *Coronaviridae*. Arch Virol 148:2207–2235

Gorbalenya AE (2001) Big nidovirus genome. When count and order of domains matter. In: Lavi E, Weiss S and Hingley ST (eds) The Nidoviruses (Coronaviruses and Arteriviruses). Kluwer Academic/Plenum Publishers, New York, vol 494, pp 1–17

Haijema BJ, Volders H, Rottier PJM (2003) Switching species tropism: an effective way to manipulate the feline coronavirus genome. J Virol 77:4528–4538

Hiscox JA, Mawditt KL, Cavanagh D, Britton P (1995) Investigation of the control of coronavirus subgenomic mRNA transcription by using T7-generated negative-sense RNA transcripts. J Virol 69:6219–6227

Holmes KV, Enjuanes L (2003) The SARS coronavirus: a postgenomic era. Science 300:1377–1378

Hsue B, Masters PS (1999) Insertion of a new transcriptional unit into the genome of mouse hepatitis virus. J Virol 73:6128–6135

Iverson LE, Rose JK (1981) Localized attenuation and discontinuous synthesis during vesicular stomatitis virus transcription. Cell 23:477–484

Izeta A, Smerdou C, Alonso S, Penzes Z, Méndez A, Plana-Durán J, Enjuanes L (1999) Replication and packaging of transmissible gastroenteritis coronavirus-derived synthetic minigenomes. J Virol 73:1535–1545

Jeong YS, Repass JF, Kim Y-N, Hwang S-M, Makino S (1996) Coronavirus transcription mediated by sequences flanking the transcription consensus sequence. Virology 217:311–322

Joo M, Makino S (1992) Mutagenic analysis of the coronavirus intergenic consensus sequence. J Virol 66:6330–6337

Joo M, Makino S (1995) The effect of two closely inserted transcription consensus sequences on coronavirus transcription. J Virol 69:272–280

Krishnan R, Chang RY, Brian DA (1996) Tandem placement of a coronavirus promoter results in enhanced mRNA synthesis from the downstream-most initiation site. Virology 218:400–405

Kubo H, Yamada YK, Taguchi F (1994) Localization of neutralizing epitopes and the receptor-binding site within the amino-terminal 330 amino acids of the murine coronavirus spike protein. J Virol 68:5403–5410

Kuo L, Godeke G-J, Raamsman MJB, Masters PS, Rottier PJM (2000) Retargeting of coronavirus by substitution of the spike glycoprotein ectodomain: crossing the host cell species barrier. J Virol 74:1393–1406

Lai MMC, Cavanagh D (1997) The molecular biology of coronaviruses. Adv Virus Res 48:1-100

Lai MMC, Zhang X, Hinton D, Stohlman S (1997) Modulation of mouse hepatitis virus infection by defective-interfering RNA-mediated expression of viral proteins and cytokines. J Neurovirol 3: S33-S34

La Monica N, Yokomori K, Lai MMC (1992) Coronavirus mRNA synthesis: identification of novel transcription initiation signals which are differentially regulated by different leader sequences. Virology 188:402–407

Lassnig C, Sánchez CM, Enjuanes L, Muller M (2004) Obtention of transgenic mice subceptible to human coronavirus infection. Submitted for publication

Lee HJ, Shieh CK, Gorbalenya AE, Koonin EV, Lamonica N, Tuler J, Bagdzhadzhyan A, Lai MMC (1991) The complete sequence (22 kilobases) of murine coronavirus gene-1 encoding the putative –327

Lin YJ, Lai MMC (1993) Deletion mapping of a mouse hepatitis virus defective interfering RNA reveals the requirement of an internal and discontinuous sequence for replication. J Virol 67:6110–6118

Luo Z, Matthews AM, Weiss SR (1999) Amino acid substitutions within the leucine zipper domain of the murine coronavirus spike protein cause defects in oligomerization and the ability to induce cell-to-cell fusion. J Virol 73:8152–8159

Makino S, Joo M, Makino JK (1991) A system for study of coronavirus messenger RNA synthesis: a regulated, expressed subgenomic defective interfering RNA results from intergenic site insertion. J Virol 65:6031–6041

Makino S, Joo M (1993) Effect of intergenic consensus sequence flanking sequences on coronavirus transcription. J Virol 67:3304–3311

Marra MA, Jones SJM, Astell CR, Holt RA, Brooks-Wilson A, Butterfield YSN, Khattra J, Asano JK, Barber SA, Chan SY, Cloutier A, Coughlin SM, Freeman D, Girn N, Griffith OL, Leach SR, Mayo M, McDonald H, Montgomery SB, Pandoh PK, Petrescu AS, Robertson AG, Schein JE, Siddiqui A, Smailus DE, Stott JM, Yang GS, Plummer F, Andonov A, Artsob H, Bastien N, Bernard K, Booth TF, Bowness D, Czub M, Drebot M, Fernando L, Flick R, Garbutt M, Gray M, Grolla A, Jones S, Feldmann H, Meyers A, Kabani A, Li Y, Normand S, Stroher U, Tipples GA, Tyler S, Vogrig R, Ward D, Watson B, Brunham RC, Krajden M, Petric M, Skowronski DM, Upton C, Roper RL (2003) The genome sequence of the SARS-associated coronavirus. Science 300:1399–1404

Masters PS (1999) Reverse genetics of the largest RNA viruses. Adv Virus Res 53:245–264

McGoldrick A, Lowings JP, Paton DJ (1999) Characterisation of a recent virulent transmissible gastroenteritis virus from Britain with a deleted ORF 3a. Arch Virol 144:763–770

Méndez A, Smerdou C, Izeta A, Gebauer F, Enjuanes L (1996) Molecular characterization of transmissible gastroenteritis coronavirus defective interfering genomes: packaging and heterogeneity. Virology 217:495–507

Meulenberg JJM, Bos-de Ruijter JNA, van de Graaf R, Wenswoort G, Moormann RJM (1998) Infectious transcripts from cloned genome-length cDNA of porcine reproductive and respiratory syndrome virus. J Virol 72:380–387

Molenkamp R, van Tol H, Rozier BC, van der Meer Y, Spaan WJ, Snijder EJ (2000) The arterivirus replicase is the only viral protein required for genome replication and subgenomic mRNA transcription. J Gen Virol 81:2491–2496

Nagy PD, Simon AE (1997) New insights into the mechanisms of RNA recombination. Virology 235:1-9

Navas S, Seo SH, Chua MM, das Sarma J, Lavi E, Hingley ST, Weiss SR (2001) Murine coronavirus spike protein determines the ability of the virus to replicate in the liver and cause hepatitis. J Virol 75:2452–2457

Nelsen CJ, Murtaugh MP, Faaberg KS (1999) Porcine reproductive and respiratory syndrome virus comparison: divergent evolution on two continents. J Virol 73:270–280

O'Connor BJ, Brian DA (1999) The major product of porcine transmissible gastroenteritis coronavirus gene 3b is an integral membrane glycoprotein of 31 kDa. Virology 256:152–161

O'Connor JB, Brian DA (2000) Downstream ribosomal entry for translation of coronavirus TGEV gene 3b. Virology 269:172–182

Olsen CW (1993) A review of feline infectious peritonitis virus: molecular biology, immunopathogenesis, clinical aspects, and vaccination. Vet Microbiol 36:1-37

Ortego J, Escors D, Laude H, Enjuanes L (2002) Generation of a replication-competent, propagation-deficient virus vector based on the transmissible gastroenteritis coronavirus genome. J Virol 76:11518–11529

Ortego J, Sola I, Almazan F, Ceriani JE, Riquelme C, Balasch M, Plana-Durán J, Enjuanes L (2003) Transmissible gastroenteritis coronavirus gene 7 is not essential but influences in vivo virus replication and virulence. Virology 308:13–22

Ortego J, DeDiego ML, Enjuanes L (2004) Novel human vector based on coronavirus genomes. Submitted for publication

Ozdarendeli A, Ku S, Rochat S, Senanayake SD, Brian DA (2001) Downstream sequences influence the choice between a naturally occurring noncanonical and closely positioned upstream canonical heptameric fusion motif during bovine coronavirus subgenomic mRNA synthesis. J Virol 75:7362–7374

Pasternak AO, van den Born E, Spaan WJM, Snijder EJ (2001) Sequence requirements for RNA strand transfer during nidovirus discontinuous subgenomic RNA synthesis. EMBO J 20:7220–7228

Pasternak AO, van den Born E, Spaan WJM, Snijder EJ (2003) The stability of the duplex between sense and antisense transcription-regulating sequences is a crucial factor in arterivirus subgenomic mRNA synthesis. J Virol 77:1175–1183

Pasternak AO, Spaan WJM, Snijder EJ (2004) Regulation of relative abundance of arterivirus subgenomic mRNAs. J Virol. In press

Pensaert M, Callebaut P, Vergote J (1986) Isolation of a porcine respiratory, non-enteric coronavirus related to transmissible gastroenteritis. Vet Quart 8:257–260

Pensaert MB, De Bouck P (1978) A new coronavirus-like particle associated with diarrhea in swine. Arch Virol 58:243–247

Penzes Z, Wroe C, Brown TDK, Britton P, Cavanagh D (1996) Replication and packaging of coronavirus infectious bronchitis virus defective RNAs lacking a long open reading frame. J Virol 70:8660–8668

Penzes Z, González JM, Calvo E, Izeta A, Smerdou C, Mendez A, Sánchez CM, Sola I, Almazán F, Enjuanes L (2001) Complete genome sequence of transmissible gastroenteritis coronavirus PUR46-MAD clone and evolution of the Purdue virus cluster. Virus Genes 23:105–118

Phillips JJ, Chua MM, Lavi E, Weiss SR (1999) Pathogenesis of chimeric MHV4/MHV-A59 recombinant viruses: the murine coronavirus spike protein is a major determinant of neurovirulence. J Virol 73:7752–7760

Resta S, Luby JP, Rosenfeld CD, Siegel JD (1985) Isolation and propagation of a human enteric coronavirus. Science 229:978–981

Riquelme C, Ortego J, Izeta A, Plana-Durán J, Enjuanes L (2004) Engineering a recombinant canine coronavirus with reduced virulence using an infectious cDNA clone of transmissible gastroenteritis coronavirus. Submitted for publication

Sánchez CM, Jiménez G, Laviada MD, Correa I, Suñé C, Bullido MJ, Gebauer F, Smerdou C, Callebaut P, Escribano JM, Enjuanes L (1990) Antigenic homology among coronaviruses related to transmissible gastroenteritis virus. Virology 174:410–417

Sánchez CM, Gebauer F, Suñé C, Méndez A, Dopazo J, Enjuanes L (1992) Genetic evolution and tropism of transmissible gastroenteritis coronaviruses. Virology 190:92–105

Sánchez CM, Izeta A, Sánchez-Morgado JM, Alonso S, Sola I, Balasch M, Plana-Durán J, Enjuanes L (1999) Targeted recombination demonstrates that the spike gene of transmissible gastroenteritis coronavirus is a determinant of its enteric tropism and virulence. J Virol 73:7607–7618

Sánchez CM, Sola I, Sánchez-Morgado JM, Enjuanes L (2004) The amino terminus of transmissible gastroenteritis coronavirus spike protein dictates the enteric tropism of the virus. In press

Sawicki DL, Wang T, Sawicki SG (2001) The RNA structures engaged in replication and transcription of the A59 strain of mouse hepatitis virus. J Gen Virol 82:386–396

Shieh C-k, Soe LH, Makino S, Chang M-F, Stohlman SA, Lai MMC (1987) The $5'$-end sequence of the murine coronavirus genome: implications for multiple fusion sites in leader-primed transcription. Virology 156:321–330

Siddell SG (1995) The *Coronaviridae*. Plenum Press, New York

Snijder EJ, Bredenbeek PJ, Dobbe JC, Thiel V, Ziebuhr J, Poon LLM, Guan Y, Rozanov M, Spaan WJM, Gorbalenya AE (2003) Unique and conserved features of genome and proteome of SARS-coronavirus, and early split-off from the coronavirus group 2 lineage. J Mol Biol 331:991–1004

Sola I, Alonso S, Zúñiga S, Balach M, Plana-Durán J, Enjuanes L (2003) Engineering transmissible gastroenteritis virus genome as an expression vector inducing lactogenic immunity. J Virol 77:4357–4369

Sooknanan R, Howes M, Read L, Malek LT (1994) Fidelity of nucleic acid amplification with avian myeloblastosis virus reverse transcriptase and T7 RNA polymerase. BioTechniques 17:1077–1085

Stirrups K, Shaw K, Evans S, Dalton K, Casais R, Cavanagh D, Britton P (2000) Expression of reporter genes from the defective RNA CD-61 of the coronavirus infectious bronchitis virus. J Gen Virol 81:1687–1698

Suñé C, Jiménez G, Correa I, Bullido MJ, Gebauer F, Smerdou C, Enjuanes L (1990) Mechanisms of transmissible gastroenteritis coronavirus neutralization. Virology 177:559–569

Taguchi F, Kubo H, Takahashi H, Suzuki H (1995) Localization of neurovirulence determinant for rats on the S1 subunit of murine coronavirus JHMV. Virology 208:67–74

Thiel V, Siddell SG, Herold J (1998) Replication and transcription of HCV 229E replicons. Adv Exp Med Biol 440:109–114

Thiel V, Herold J, Schelle B, Siddell S (2001a) Infectious RNA transcribed in vitro from a cDNA copy of the human coronavirus genome cloned in vaccinia virus. J Gen Virol 82:1273–1281

Thiel V, Herold J, Schelle B, Siddell SG (2001b) Viral replicase gene products suffice for coronavirus discontinuous transcription. J Virol 75:6676–6681

Thiel V, Ivanov KA, Putics A, Hertzig T, Schelle B, Bayer S, Wessbrich B, Snijder EJ, Rabenau H, Doerr HW, Gorbalenya AE, Ziebuhr J (2003a) Mechanisms and enzymes involved in SARS coronavirus genome expression. J Gen Virol 84:2305–2315

Thiel V, Karl N, Schelle B, Disterer P, Klagge I, Siddell SG (2003b) Multigene RNA vector based on coronavirus transcription. J Virol 77:9790–9798

Thomas MJ, Platas AA, Hawley DK (1998) Transcriptional fidelity and proofreading by RNA polymerase II. Cell 93:627–637

USDA (2002) Part II: reference of swine health and health management in the United States, 2000. National Animal Health Monitoring System

van der Most RG, De Groot RJ, Spaan WJM (1994) Subgenomic RNA synthesis directed by a synthetic defective interfering RNA of mouse hepatitis virus: a study of coronavirus transcription initiation. J Virol 68:3656–3666

van Dinten LC, den Boon JA, Wassenaar ALM, Spaan WJM, Snijder EJ (1997) An infectious arterivirus cDNA clone: identification of a replicase point mutation that abolishes discontinuous mRNA transcription. Proc Natl Acad Sci USA 94:991–996

van Marle G, Luytjes W, Van der Most RG, van der Straaten T, Spaan WJM (1995) Regulation of Coronavirus mRNA transcription. J Virol 69:7851–7856

van Marle G, Dobbe JC, Gultyaev AP, Luytjes W, Spaan WJM, Snijder EJ (1999) Arterivirus discontinuous mRNA transcription is guided by base pairing between sense and antisense transcription-regulating sequences. Proc Natl Acad Sci USA 96:12056–12061

Vaughn EM, Paul PS (1993) Antigenic and biological diversity among transmissible gastroenteritis virus isolates of swine. Vet Microbiol 36:333–347

Vaughn RM, Halbur PG, Paul PS (1995) Sequence comparison of porcine respiratory coronaviruses isolates reveals heterogeneity in the S, 3, and 3-1 genes. J Virol 69:3176–3184

Ward CD, Stokes MAM, Flanagan JB (1988) Direct measurement of the poliovirus RNA polymerase error frequency in vitro. J Virol 62:558–562

and, recently, recombinant mouse hepatitis virus, representing group III and group II coronaviruses, respectively. We describe how vaccinia virus-mediated homologous recombination can be used to introduce specific mutations into the coronavirus genomic cDNA during its propagation in vaccinia virus and how recombinant coronaviruses can be isolated. Finally, we describe how the coronavirus reverse genetic system has now been extended to the generation of coronavirus replicon RNAs.

1
Introduction

The development of systems for manipulating the coronavirus genome using traditional reverse genetic approaches has presented a considerable technological challenge because of both the genome size and the instability of specific coronavirus cDNA sequences in bacterial systems. However, recently, reverse genetic systems for a number of coronaviruses have been established using non-traditional approaches which are based on the use of bacterial artificial chromosomes (Almazan et al. 2000), the in vitro ligation of coronavirus cDNA fragments (Yount et al. 2000) and the use of vaccinia virus as a vector for the propagation of coronavirus genomic cDNAs (Thiel et al. 2001a). With the systems now available, it is possible to genetically modify coronavirus genomes at will. Recombinant viruses with gene inactivations, deletions or attenuating modifications can be generated and used to study the role of specific gene products in viral replication or pathogenesis. Genetically attenuated viruses can be produced which are potential vaccine candidates and modified coronavirus genomes have been developed as eukaryotic, multigene expression vectors (Thiel et al. 2003). In this article, we shall describe the reverse genetic system that is based on the use of vaccinia virus cloning vectors. This system represents a generic approach to coronavirus reverse genetics and was first described for the generation of recombinant human coronavirus 229E (HCoV 229E)(Thiel et al. 2001a), representing a group I coronavirus. Subsequently, we have used the same approach to generate recombinant avian infectious bronchitis coronavirus (IBV) (Casais et al. 2001) and, recently, recombinant mouse hepatitis virus (MHV-A59) (Coley et al., manuscript in preparation), representing group III and group II coronaviruses, respectively.

The basic strategy for the generation of recombinant coronaviruses can be divided into three phases.

The Assembly of a Full-Length Coronavirus Genomic cDNA. This normally involves the generation of numerous subgenomic cDNA fragments

Reverse Genetics of Coronaviruses Using Vaccinia Virus Vectors

V. Thiel[1] (✉) · S. G. Siddell[2]

[1] Research Department, Cantonal Hospital St. Gallen, St. Gallen, Switzerland
volker.thiel@kssg.ch
[2] Department of Pathology and Microbiology, School of Medical and Veterinary Sciences, University of Bristol, Bristol, UK

1	Introduction	200
2	The Use of Vaccinia Virus as a Vector for Coronavirus cDNA	202
2.1	Cloning of Full-Length Coronavirus cDNA into the Vaccinia Virus Genome	202
2.2	Mutagenesis of Cloned Coronavirus cDNA	204
2.3	Rescue of Recombinant Coronaviruses	206
2.3.1	Rescue of Recombinant Coronaviruses Using Full-Length In Vitro Transcripts	207
2.3.2	Rescue of Recombinant Coronavirus Using Full-Length cDNA	209
2.3.3	Expression of the Nucleocapsid Protein Facilitates the Rescue of Recombinant Coronaviruses	210
3	Recombinant Coronaviruses	211
3.1	Analysis of HCoV 229E Replicase Polyprotein Processing	212
3.2	Analysis of IBV Spike Chimeras	215
3.3	Recombinant MHV Is Fully Pathogenic in Mice	215
4	Generation of Replicon RNAs	217
4.1	Replicase Gene Products Suffice for Discontinuous Transcription	217
4.2	Generation of Autonomously Replicating RNAs	219
5	Development of Coronavirus-Based Multigene Vectors	221
5.1	Multigene Expression Using Coronavirus-Based Vectors	222
5.2	Coronavirus-Based Vectors as Potential Vaccines	223
6	Discussion	225
References		226

Abstract In this article, we describe the reverse genetic system that is based on the use of vaccinia virus cloning vectors. This system represents a generic approach to coronavirus reverse genetics and was first described for the generation of recombinant human coronavirus 229E representing a group I coronavirus. Subsequently, the same approach has been used to generate recombinant avian infectious bronchitis coronavirus

Wentworth DE, Holmes KV (2001) Molecular determinants of species specificity in the coronavirus receptor aminopeptidase N (CD13): influence of N-linked glycosylation. J Virol 75:9741–9752

Wentworth DE, Tresnan DB, Lerman I, Levis R, Shapiro LH, Holmes KV (2001) Subceptibility of transgenic mice expressing the receptor for human coronavirus-229E. In: ASV 20th Annual Meeting, University of Wisconsin-Madison, Madison, p 157

Wertz GW, Perepelitsa VP, Ball LA (1998) Gene rearrangement attenuates expression and lethality of a nonsegmented negative strand RNA virus. Proc Natl Acad Sci USA 95:3501–3506

Wesley RD, Cheung AK, Michael DM, Woods RD (1989) Nucleotide sequence of coronavirus TGEV genomic RNA: evidence of 3 mRNA species between the peplomer and matrix protein genes. Virus Res 13:87–100

Wesley RD, Woods RD, Hill HT, Biwer JD (1990b) Evidence for a porcine respiratory coronavirus, antigenically similar to transmissible gastroenteritis virus, in the United States. J Vet Diagn Invest 2:312–317

Wesley RD, Woods RD, Cheung AK (1991) Genetic analysis of porcine respiratory coronavirus, an attenuated variant of transmissible gastroenteritis virus. J Virol 65:3369–3373

Yeager CL, Ashmun RA, Williams RK, Cardellichio CB, Shapiro LH, Look AT, Holmes KV (1992) Human aminopeptidase N is a receptor for human coronavirus 229E. Nature 357:420–422

Yount B, Curtis KM, Baric RS (2000) Strategy for systematic assembly of large RNA and DNA genomes: the transmissible gastroenteritis virus model. J Virol 74:10600–10611

Yount B, Denison MR, Weiss SR, Baric RS (2002) Systematic assembly of a full length infectious cDNA of mouse hepatitis virus stain A59. J Virol 76:11065–11078

Yount B, Curtis KM, Fritz EA, Hensley LE, Jahrling PB, Prentice E, Denison MR, Geisbert TW, Baric RS (2003) Reverse genetics with a full-length infectious cDNA of severe acute respiratory syndrome coronavirus. Proc Natl Acad Sci USA 100:12995–13000

Zhang X, Hinton DR, Cua DJ, Stohlman SA, Lai MMC (1997) Expression of interferon-γ by a coronavirus defective-interfering RNA vector and its effect on viral replication, spread, and pathogenicity. Virology 233:327–338

Zhang X, Hinton DR, Park S, Parra B, Liao C-L, Lai MMC (1998) Expression of hemagglutinin/esterase by a mouse hepatitis virus coronavirus defective-interfering RNA alters viral pathogenesis. Virology 242:170–183

Zhang X, Liu R (2000) Identification of a noncanonical signal for transcription of a novel subgenomic mRNA of mouse hepatitis virus: implication for the mechanism of coronavirus RNA transcription. Virology 278:75–85

Zúñiga S, Sola I, Alonso S, Enjuanes L (2004) Sequence motifs involved in the regulation of discontinuous coronavirus subgenomic RNA synthesis. J Virol 78:980–994

that are either amplified as bacterial plasmid DNA or prepared in large amounts by preparative reverse-transcriptase polymerase chain reaction (RT-PCR). The cDNAs are then ligated sequentially, in vitro, to produce a small number of cDNAs which encompass the entire genome. The specific ligation strategy is dictated by the sequence of the coronavirus in question, but a common feature is the use of convenient, naturally occurring or engineered restriction sites, especially if they cleave, for example, interrupted palindromic sequences. It is also necessary to modify the cDNAs which represent the 5' and 3' ends of the coronavirus genome. Normally, a transcription promoter sequence for the bacteriophage T7 RNA polymerase is positioned upstream the coronavirus genome and a (unique) restriction site, followed by the hepatitis δ ribozyme is placed downstream of the poly(A) tail of the coronavirus genome. The terminal cDNA constructs must also have appropriate *Eag*I or *Bsp*120I restriction sites to facilitate cloning into a unique *Not*I restriction site present in the genomic DNA of vaccinia virus, strain v*Not*I/tk.

The Cloning and Propagation of the Coronavirus Genomic cDNA in Vaccinia Virus Vectors. The next stage is to ligate, in vitro, the coronavirus cDNA fragments and the long and short arms of *Not*I-cleaved v*Not*I/tk genomic DNA (Merchlinsky and Moss 1992). This ligation is done in the presence of *Not*I to prevent religation of the vaccinia virus DNA. Subsequently, the ligation reaction is transfected into mammalian cells which have been previously infected with fowlpox virus. Recombinant vaccinia virus, the genome of which includes a full-length copy of the coronavirus genome, is rescued.

Rescue of Recombinant Coronaviruses. Essentially, recombinant coronaviruses are rescued by generating genomic-length RNA transcripts from the coronavirus component of the recombinant vaccinia virus DNA template. These transcripts are then transfected into permissive cells. The transcription reaction is normally done in vitro, but it is also possible to rescue recombinant coronaviruses via the transcription of template DNA in the permissive cell itself. This requires the introduction of non-infectious (i.e. restriction enzyme digested) recombinant vaccinia virus DNA and a source of bacteriophage T7 RNA polymerase, normally a recombinant fowlpox virus, into the permissive cell. Also, as will be described below, we have found that the ability to rescue recombinant coronaviruses is significantly enhanced by (but not dependent on) the directed expression of the coronavirus nucleocapsid protein in the transfected cells.

In addition to these basic concepts, this article will describe how vaccinia virus-mediated homologous recombination can be used to introduce specific mutations into the coronavirus genomic cDNA during its propagation in vaccinia virus. Once an infectious coronavirus cDNA has been obtained, this element of the reverse genetic approach is actually the rate-limiting step. It is, therefore, imperative that a rapid and easy procedure is available. Finally, we shall describe how the coronavirus reverse genetic system has now been extended to the generation of coronavirus replicon RNAs.

2
The Use of Vaccinia Virus as a Vector for Coronavirus cDNA

The first use of vaccinia virus as a cloning vector for full-length coronavirus cDNA was described for the human coronavirus 229E (HCoV 229E) system. Vaccinia virus vectors were chosen for several reasons. First, poxvirus vectors are suitable for the cloning of large cDNA. It has been shown that they have the capacity to accept at least 26 kb of foreign sequence (Smith and Moss 1983), and recombinant vaccinia virus genomes of this size are stable, infectious and replicate in tissue culture to the same titre as non-recombinant virus. Second, vaccinia virus vectors have been developed which are designed for the insertion of foreign DNA by in vitro ligation (Merchlinsky and Moss 1992). This obviates the need for plasmid intermediates carrying the entire cDNA insert. Third, the cloned cDNA insert should be accessible to mutagenesis by vaccinia virus-mediated homologous recombination. Finally, conventional cloning strategies based on procaryotic cloning systems (e.g. plasmid vectors, bacterial artificial chromosomes or bacteriophage lambda vectors) were not applicable to the stable propagation of full-length HCoV 229E cDNA.

2.1
Cloning of Full-Length Coronavirus cDNA into the Vaccinia Virus Genome

As outlined above, the overall strategy to insert full-length coronavirus cDNA fragments into the vaccinia virus genome involves two steps. First, the full-length coronavirus cDNA is assembled by in vitro ligation using multiple cDNAs representing the entire coronavirus genomic RNA. Second, the vaccinia virus v*NotI*/tk genome is used as a cloning vector to insert the full-length cDNA, again by in vitro ligation.

The assembly of full-length coronavirus cDNAs for HCoV 229E, IBV and, recently, MHV-A59 has involved two, three or four cDNA fragments, respectively. The DNA fragments corresponding to the 5'-end of the coronavirus genomes contained the bacteriophage T7 RNA polymerase promoter sequence and, if not encoded at the 5'-end of the coronavirus genome, one or three additional G nucleotides which are required for efficient initiation of the in vitro transcription reaction. The cDNA fragment corresponding to the 3'-end of the coronavirus genome contained a synthetic poly(A) stretch comprised of 20–40 nucleotides (nt), followed by a hepatitis delta ribozyme element and a convenient restriction site that can be used to generate so-called run-off transcripts. In order to insert the full-length cDNAs into a single *NotI* site of the vaccinia virus v*NotI*/tk vector genome, the cDNA fragments corresponding to the 5'- and 3'-genomic termini contained the restriction sites *EagI* or *Bsp*120I. After cleavage, the resulting DNA ends are compatible with *NotI*-cleaved vaccinia virus vector DNA.

The insertion of full-length coronavirus cDNA fragments into the vaccinia virus v*NotI*/tk genome by in vitro ligation required optimization of the procedure. Purified vaccinia virus v*NotI*/tk genomic DNA fragments which had been cleaved with *NotI* were found to be poor substrates for in vitro ligation, most probably because of their large size. In contrast, in vitro assembled full-length coronavirus cDNA fragments which had been cleaved with *EagI* were found to ligate efficiently. Thus, in ligation reactions containing *NotI*-cleaved vaccinia virus vector DNA and coronavirus cDNA inserts, the ligation products were predominantly comprised of multiple insert fragments. Ligation products comprised of vector arms and insert cDNA fragments were not readily detectable. To resolve this problem, we therefore included the *NotI* enzyme in the ligation reaction and, using alkaline phosphatase, dephosphorylated the coronavirus insert DNA fragments. As illustrated in Fig. 1, this strategy resulted in the production of detectable amounts of ligation products comprised of two vaccinia virus vector arms and the coronavirus cDNA insert.

To rescue recombinant vaccinia virus clones containing the full-length coronavirus cDNA the ligation reaction was transfected into CV-1 cells. Because vaccinia virus genomic DNA is not infectious, fowlpox virus has been used as a helper virus (Scheiflinger et al. 1992). Thus CV-1 cells were infected with fowlpox virus before transfection. At 2–3 h after infection/transfection, the cells were collected and transferred with a

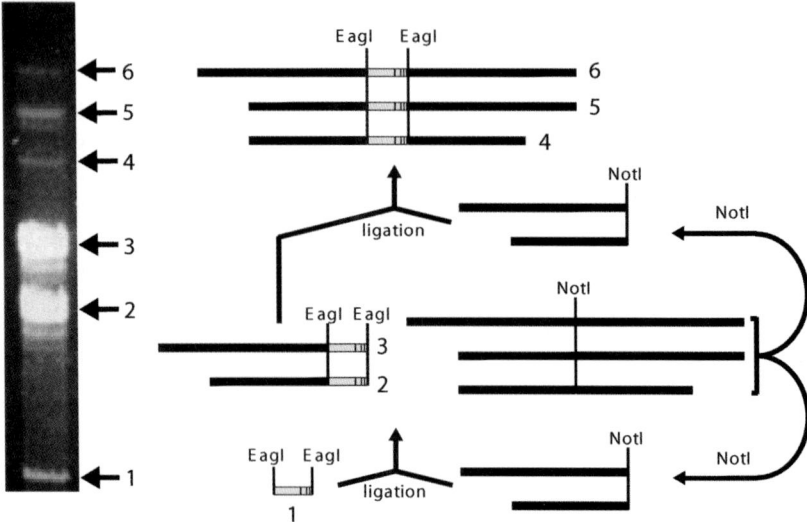

Fig. 1. Cloning of coronavirus cDNA into the vaccinia virus genome. A schematic overview of the optimized ligation reaction using *Eag*I-cleaved and dephosphorylated coronavirus cDNA (*1*) and *Not*I-cleaved vaccinia virus genomic DNA is illustrated. The ligation reactions are carried out at 25°C in the presence of *Not*I enzyme. Also shown is a pulse-field gel electrophoresis analysis of the ligation reaction products. Fragments corresponding to the insert cDNA and ligation products comprised of insert cDNA and vaccinia virus DNA (*2, 3, 4, 5,* and *6*) are indicated

fivefold excess of fresh CV-1 cells into 96-well plates. During a period of 14 days, virus stocks were collected from 96-well plates displaying cytopathic effect. Because fowlpox virus infection of mammalian cells is abortive, the resulting virus stocks contained exclusively vaccinia viruses. Furthermore, the analysis of genomic DNA of rescued vaccinia viruses by Southern blotting showed that a high percentage of the viruses contained the coronavirus cDNA insert (Thiel et al. 2001a, b).

2.2
Mutagenesis of Cloned Coronavirus cDNA

One major advantage of using vaccinia virus as a cloning vector is that the cloned coronavirus cDNA is amenable to site-directed mutagenesis using vaccinia virus-mediated homologous recombination (Ball 1987). We will show one example to demonstrate the ease of using vaccinia virus-mediated recombination to genetically modify coronavirus cDNA inserts.

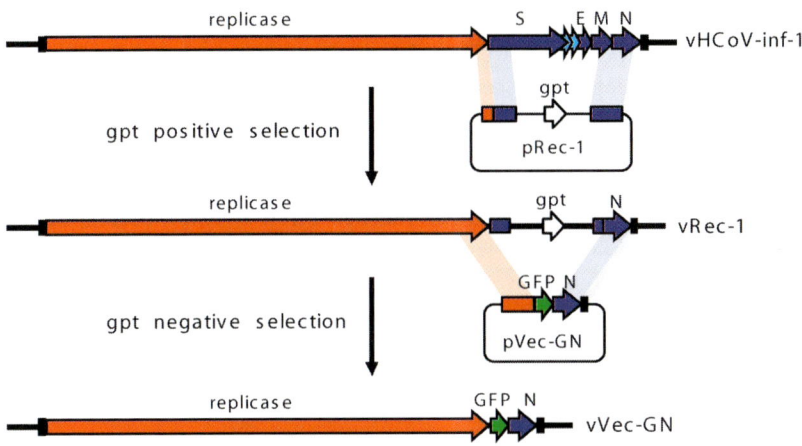

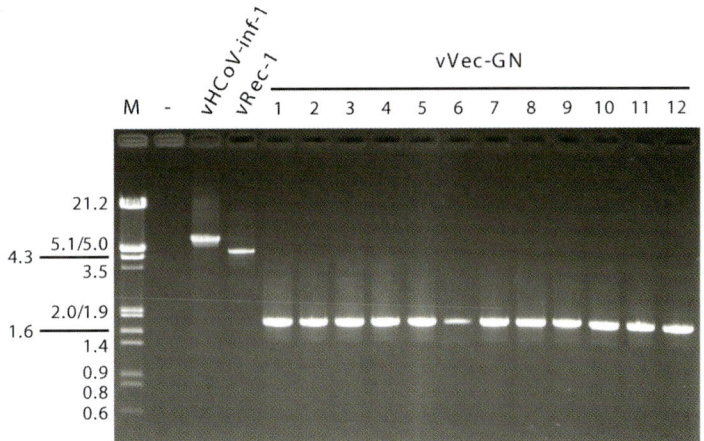

Fig. 2. Mutagenesis of cloned coronavirus cDNA by vaccinia virus-mediated homologous recombination. The generation of the recombinant vaccinia virus vVec-GN containing a modified HCoV 229E genome is illustrated (see text for details). Also shown is a PCR analysis of the region in which the homologous recombination took place within the genomes of the parental vaccinia virus clone vHCoV-inf-1, the intermediate vaccinia virus clone vRec-1 and the recombinant vaccinia virus clone vVec-GN. *Lanes 1–12* show 12 randomly picked recombinant vaccinia virus plaques obtained after the gpt-negative selection, indicating the 100% recovery of desired genotypes

A region corresponding to HCoV nt 20,569–25,653, which lies within the full-length HCoV 229E cDNA insert, has been replaced by the gene encoding the green fluorescent protein (GFP). This results in a recombinant vaccinia virus clone designated vVec-GN. The overall strategy of vVec-GN construction is illustrated in Fig. 2. The procedure is based on using the *E. coli* guanine-phosphoribosyl transferase gene (gpt) as both a positive and a negative selection marker. First, a region corresponding to nt 21,146–24,200 of the HCoV 229E genome was replaced by the *E. coli* gpt gene. To do this, we transfected vHCoV-inf-1-infected CV-1 cells with a plasmid DNA containing the *E. coli* gpt gene located downstream of a vaccinia virus promoter and flanked by HCoV 229E sequences (nt 19,601–21,145 and nt 24,201–25,874) that facilitate recombination. Two days after infection/transfection, a vaccinia virus stock was prepared. To isolate gpt-containing vaccinia viruses, three rounds of plaque purification were done under gpt-positive selection on CV-1 cells in the presence of mycophenolic acid (25 μg/ml), xanthine (250 μg/ml) and hypoxanthine (15 μg/ml). A recombinant vaccinia virus, designated vRec-1, which contained the *E. coli* gpt gene at the expected position, could be easily identified by PCR and Southern blot analysis. In a second step, vRec-1 was used to replace the *E. coli* gpt gene by the GFP gene. CV-1 cells were infected with vRec-1 and transfected with a plasmid DNA encoding the GFP gene flanked by HCoV sequences (nt 19,485–20,568 and nt 25,654–27,273). Again, after 2 days we prepared a vaccinia virus stock from the infected/transfected cells and did three rounds of plaque purification. However, this time we used HeLa-D980R cells and conditions which allow for the selection of vaccinia viruses that have lost the expression of gpt (Kerr and Smith 1991). A PCR analysis of 12 vaccinia virus clones (Fig. 2) demonstrates that, in each case, vaccinia virus-mediated homologous recombination has taken place at the expected position within the cloned HCoV 229E insert DNA. One of the recombinant vaccinia virus clones was subjected to sequencing analysis of the region where the vaccinia virus-mediated recombination had occurred, and the results revealed that vaccinia virus-mediated recombination is precise at the nucleotide level.

2.3
Rescue of Recombinant Coronaviruses

Two strategies have been reported for the rescue of recombinant coronaviruses from full-length cDNA cloned in vaccinia virus vectors. Initially, recombinant HCoV 229E was rescued after transfection of full-length in vitro transcripts of the cloned HCoV 229E cDNA into MRC-5 cells (Thiel et al. 2001a). Alternatively, the rescue of recombinant IBV has been reported by transfecting full-length coronavirus IBV cDNA into chick kidney (CK) cells that had been infected by a recombinant fowlpox virus, rFPV-T7 (Casais et al. 2001). The fowlpox virus mediates the expression of the bacteriophage T7 RNA polymerase. In contrast to the transfection of infectious HCoV 229E RNA, the rescue of recombinant IBV required the directed expression of the IBV nucleocapsid protein [mediated by transfection of a expression plasmid encoding the IBV nucleocapsid (N) protein]. This observation led us to develop a line of BHK cells which express the HCoV 229E N protein, and, indeed, we found that the expression of this protein also facilitates the rescue of recombinant HCoV 229E coronaviruses after the transfection of cells with full-length in vitro transcripts.

2.3.1
Rescue of Recombinant Coronaviruses Using Full-Length In Vitro Transcripts

The overall strategy to recover recombinant human coronavirus from vaccinia virus vHCoV-inf-1 genomic DNA is illustrated in Fig. 3. The full-length HCoV 229E cDNA is cloned downstream of a bacteriophage T7 RNA polymerase promoter, and a *Cla*I restriction endonuclease recognition sequence is located downstream of a synthetic poly(A) sequence, representing the 3' end of the HCoV genome. Genomic vHCoV-inf-1 DNA was prepared from purified recombinant vaccinia virus stocks and cleaved with *Cla*I enzyme. This DNA was then used as template to transcribe, in vitro, a capped RNA corresponding to the HCoV genome with bacteriophage T7 RNA polymerase. When this RNA was transfected into MRC-5 cells by lipofection, cytopathic effects indicative of human coronavirus infection developed throughout the culture after 6–7 days. A virus, designated HCoV-inf-1, was rescued from the tissue culture supernatant, plaque purified and propagated to produce stocks containing approximately 1×10^7 TCID50/ml. Phenotypic analysis revealed that the growth kinetics, cytopathic effect and stability of HCoV-inf-1 were indistinguishable from those of parental virus. Furthermore, Northern hybridization analysis of poly(A)-containing RNA isolated from infected MRC-5 cells demonstrated that the patterns of viral genomic and subgenomic RNAs of HCoV-inf-1 and HCoV 229E were identical. To confirm that, indeed, a recombinant virus had been rescued, the presence of a marker mutation which was introduced into the recombinant HCoV

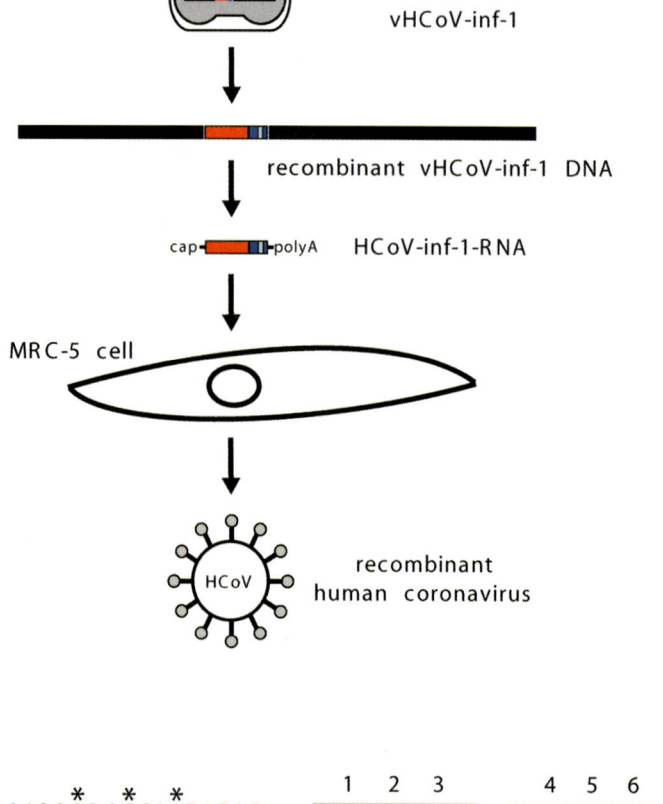
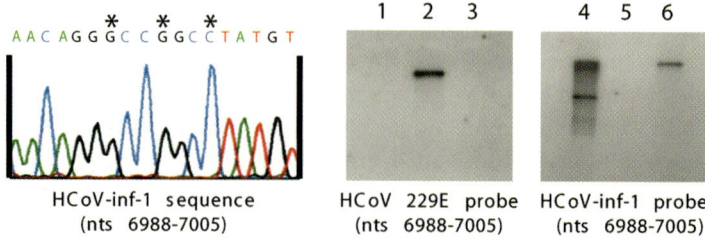

Fig. 3. Rescue of recombinant HCoV 229E from cloned, full-length cDNA. To recover recombinant HCoV 229E, 5′-capped RNA transcripts are produced in vitro with bacteriophage T7 RNA polymerase and vHCoV-inf-1 genomic DNA as template. The transcripts are transfected into MRC-5 cells. The recombinant human coronavirus HCoV-inf-1 contains marker mutations (*) that are evident in the sequence of an RT-PCR product of poly(A)-containing RNA from HCoV-inf-1-infected cells. Also shown is a Northern hybridization analysis of in vitro transcribed HCoV-inf-1 RNA (*lanes 1* and *4*) and poly(A) containing RNA from parental HCoV 229E-infected MRC-5 cells (*lanes 2* and *5*) and HCoV-inf-1-infected MRC-5 cells (*lanes 3* and *6*).

cDNA during the cloning procedure was analysed. As shown in Fig. 3, RT-PCR sequencing analysis of the relevant region of the HCoV-inf-1 genome demonstrates that three nucleotides (nt 6,994, 6,997 and 7,000) were, as predicted, found to be changed compared to the HCoV 229E sequence. These results showed, conclusively, the rescue of recombinant human coronavirus and demonstrated that the coronavirus genomic RNA alone is able to initiate a productive infectious cycle.

2.3.2
Rescue of Recombinant Coronavirus Using Full-Length cDNA

Initially, the rescue of recombinant IBV from cloned cDNA was attempted by generating full-length T7-driven in vitro transcripts by using *Sal*I-restricted vNotI/IBV$_{FL}$ genomic DNA as a template followed by transfection of the RNA into susceptible chick kidney (CK) cells (Casais et al. 2003). Although essentially the same protocol had been used successfully for the rescue of recombinant HCoV 229E, the amounts and purity of full-length synthetic IBV RNA varied and attempts to rescue recombinant IBV from CK cells were not successful. Therefore, an alternative strategy was used (Fig. 4). First, CK cells were infected with rFPV-T7 to provide cytoplasmic bacteriophage T7 RNA polymerase (Britton et al. 1996). At 1 h after vFPV-T7-infection the cells were transfected with *Sal*I- or *Asc*I-restricted vNotI/IBV$_{FL}$ genomic DNA. In addition, a plasmid DNA which mediates the expression of the IBV nucleocapsid (N) protein was co-transfected. When the infected/transfected cells developed cytopathic effects, the supernatant was filtered to remove vFPV-T7 virus and passaged on fresh CK cell monolayers until cytopathic effects characteristic of IBV infection were observed. Further analysis, including Northern blotting, immunofluorescence and finally RT-PCR sequencing to detect marker mutations, confirmed that recombinant IBV had been rescued. This result demonstrates that a reverse genetic system for IBV has been established based on cloning of full-length IBV cDNA

◂——————————————————————

The RNAs were probed with a parental HCoV 229E-specific (*lanes 1–3*) or an HCoV-inf-1-specific radiolabelled oligonucleotide (*lanes 4–6*) corresponding to the genomic nucleotides 6988–7005, respectively. The HCoV 229E-specific probe hybridized to the HCoV 229E genomic RNA but not to recombinant HCoV-inf-1 genomic RNA. In contrast, the HCoV-inf-1-specific probe hybridized to the HCoV-inf-1 genomic RNA and not to the parental HCoV 229E genomic RNA

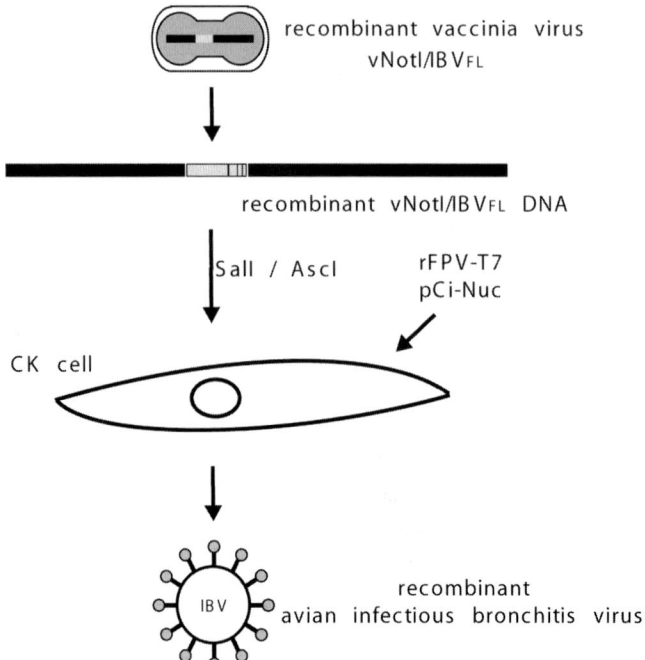

Fig. 4. Rescue of recombinant IBV from cloned, full-length cDNA. To recover recombinant IBV, SalI- or AscI-restricted genomic DNA of the recombinant vaccinia virus vNotI/IBV$_{FL}$ (containing the full-length IBV cDNA) was transfected into chick kidney (CK) cells. The full-length IBV RNA is produced in vivo with a recombinant fowlpox virus rFPV-T7 which expresses the bacteriophage T7 RNA polymerase. In addition, the CK cells are transfected with a plasmid which mediates the expression of the IBV nucleocapsid protein (pCi-Nuc)

in a vaccinia virus vector. It should be noted that this procedure obviated the need to prepare purified vNotI/IBV$_{FL}$ genomic DNA for in vitro transcription and therefore represents an attractive alternative to rescue recombinant coronaviruses from cloned cDNA.

2.3.3
Expression of the Nucleocapsid Protein Facilitates the Rescue of Recombinant Coronaviruses

As described above, the rescue of recombinant IBV using in vivo transcription of full-length IBV RNA was dependent on the co-transfection of a plasmid DNA mediating the expression of the IBV N protein. A similar observation was made for the recovery of recombinant TGEV by Yount et al. (Yount et al. 2000). In contrast, the rescue of recombinant HCoV 229E and MHV-A59 (Coley et al., manuscript in preparation) and the rescue of recombinant TGEV with the BAC system were achieved without the directed expression of N protein. These data indicate that the coronavirus N protein is not absolutely required to establish a productive coronavirus infection with infectious RNA or DNA. However, it clearly facilitates rescue of recombinant viruses. Consequently, we have made two changes in our protocol for the rescue of recombinant coronaviruses with full-length in vitro transcripts.

First, we have used BHK-21 cells as target cells for the delivery of in vitro transcribed coronavirus genomic RNA by electroporation. On the one hand, BHK-21 cells can be efficiently transfected with RNA by electroporation, and at the same time, our data suggest that they are permissive for the production of, at least, HCoV 229E and MHV-A59. Second, to provide directed expression of the coronavirus N protein, we have produced stable BHK-21-derived cell lines which express coronavirus N proteins. An HCoV 229E N protein-expressing cell line, designated BHK-HCoV-N, has been produced for the HCoV 229E reverse genetic system, and an MHV N protein-expressing cell line, designated BHK-MHV-N, has been produced for the MHV-A59 reverse genetic system. Both cell lines are based on the Tet/ON expression system (Gossen et al. 1995) and thus allow the controlled expression of the respective N proteins on induction with doxycyclin. Using the BHK-HCoV-N cell line, we observed that the rescue of recombinant HCoV 229E after electroporation of full-length infectious HCoV RNA is greatly facilitated. The titres of virus in the supernatant of BHK-HCoV-N cells 3 days after transfection were about 1×10^4 TCID50/ml. Moreover, the use of the N-expressing cell line has enabled us to monitor coronavirus RNA synthesis by Northern blot analysis directly in transfected cells. Similar results have been obtained using the BHK-MHV-N cell line for the rescue of recombinant MHV-A59. It should be noted that in addition to the rescue of recombinant coronaviruses, the N-expressing BHK cell lines also facilitate the analysis of coronavirus vector RNAs (see below).

3
Recombinant Coronaviruses

Recombinant coronaviruses generated by the reverse genetic systems described above have been used in the analysis of coronavirus polyprotein processing, the generation of chimeric coronaviruses which might be useful as coronavirus vaccines and the analysis of coronavirus pathogenesis.

3.1
Analysis of HCoV 229E Replicase Polyprotein Processing

The analysis of replication, transcription and polyprotein processing in many positive-strand RNA viruses has been facilitated by the development of reverse genetic systems based on cloned cDNA. The reverse genetic system for coronaviruses enables the generation and analysis of recombinant coronaviruses which carry mutations of choice. Specifically, it is now possible, for the first time, to generate and analyse recombinant coronaviruses which contain genetically modified replicase genes.

The HCoV 229E replicase gene is comprised of two overlapping open reading frames (ORFs), ORF1a and ORF1b, which extend about 20 kb from the 5' end of the genome. It encodes two large polyproteins, pp1a

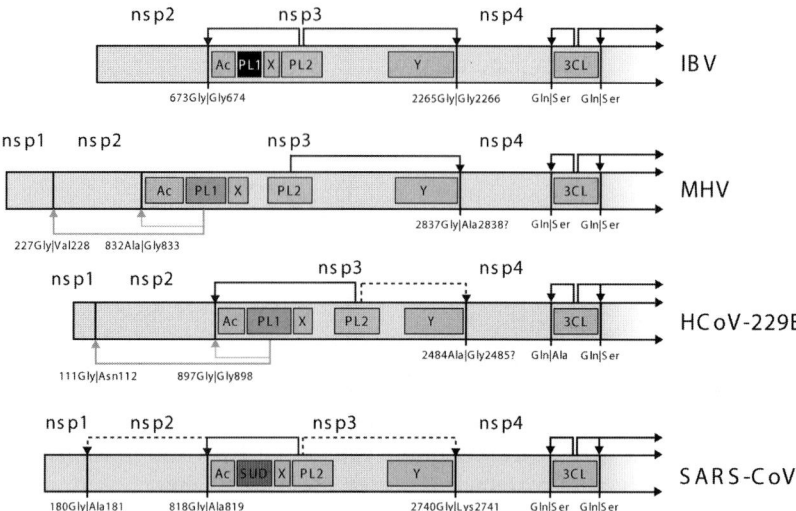

Fig. 5. Proteolytic processing of the amino-proximal regions of coronavirus replicase polyproteins. The amino-proximal regions of IBV, MHV, HCoV 229E and SARS-CoV replicase polyproteins are illustrated. Processing events mediated by the papain-like and main proteinases are indicated (for details see the chapter by Ziebuhr, this volume). Abbreviations: *Ac*, acidic domain; *PL1*, papain-like proteinase 1; *SUD*, SARS-CoV unique domain; *X*, adenosine diphosphate-ribose 1″-phosphatase; *PL2*, papain-like proteinase 2; *Y*, nsp3 C-terminal domain; *3CL*, 3C-like proteinase

and pp1ab, which are extensively processed by the virus main proteinase, M^{pro}, and two accessory papain-like proteinases, PL1^{pro} and PL2^{pro} (Ziebuhr et al. 2000). The N-proximal region of the polyproteins is processed by PL1^{pro} and PL2^{pro} to release three proteins, nsp1 (9 kDa), nsp2 (87 kDa) and nsp3 (177 kDa) and, in addition, the N-terminus of nsp4 (54 kDa) (Fig. 5). In vitro studies of the HCoV 229E polyprotein processing events revealed that PL1^{pro} could cleave between nsp1 and nsp2 (Herold et al. 1998). Furthermore, it has been shown that PL1^{pro} and PL2^{pro} can cleave between nsp2 and nsp3, indicating overlapping substrate specificities. PL2^{pro} cleaves this site efficiently, whereas PL1^{pro}-mediated cleavage is slow and significantly suppressed by proteolytic inactive PL2^{pro} (Ziebuhr et al. 2001).

The in vitro data described above suggest a redundancy of PL^{pro} activities and raised the question of whether both HCoV 229E PL^{pro} activities are indeed required for virus replication. Thus, to gain a better understanding of the physiological roles of coronavirus papain-like proteinases, we have applied our reverse genetic system to the characterization of HCoV 229E PL1^{pro}/PL2^{pro} in virus replication. Based on the recombinant vaccinia virus vHCoV-inf-1, harbouring the full-length HCoV cDNA, we have introduced nucleotide changes resulting in substitutions of the active-site cysteine residues of PL1^{pro} (Cys1054Ala) and PL2^{pro} (Cys1701Ala), respectively. Genomic DNA from the resulting vaccinia virus clones were used as template for T7-driven in vitro transcription to generate full-length HCoV-PL1(−) and HCoV-PL2(−) RNAs, respectively. As shown in Fig. 6 virus-specific RNAs were only detectable in HCoV-PL1(−) RNA-transfected BHK-HCoV-N cells. After transfer of the supernatant to MRC-5 cells only mutant viruses containing the Cys1054Ala nucleotide change (at the PL1^{pro} active site) could be rescued. In contrast, no mutant virus could be rescued if the supernatant of HCoV-PL2(−) RNA-transfected cells was transferred to MRC-5 cells. This result indicates that an active PL1^{pro} enzyme is not required for HCoV 229E RNA synthesis and virus replication. However, the recombinant HCoV-PL1(−) virus displayed reduced growth kinetics, and reversion of mutant virus to the parental sequence at the PL1^{pro} active site (reversion Ala1054Cys) occurred within a few passages on MRC-5 cells. The growth kinetics and RNA synthesis of the reverted virus were indistinguishable from those of the parental HCoV 229E virus. This indicates that although an active PL1^{pro} enzyme is not absolutely required for virus replication it is still beneficial in terms of virus fitness.

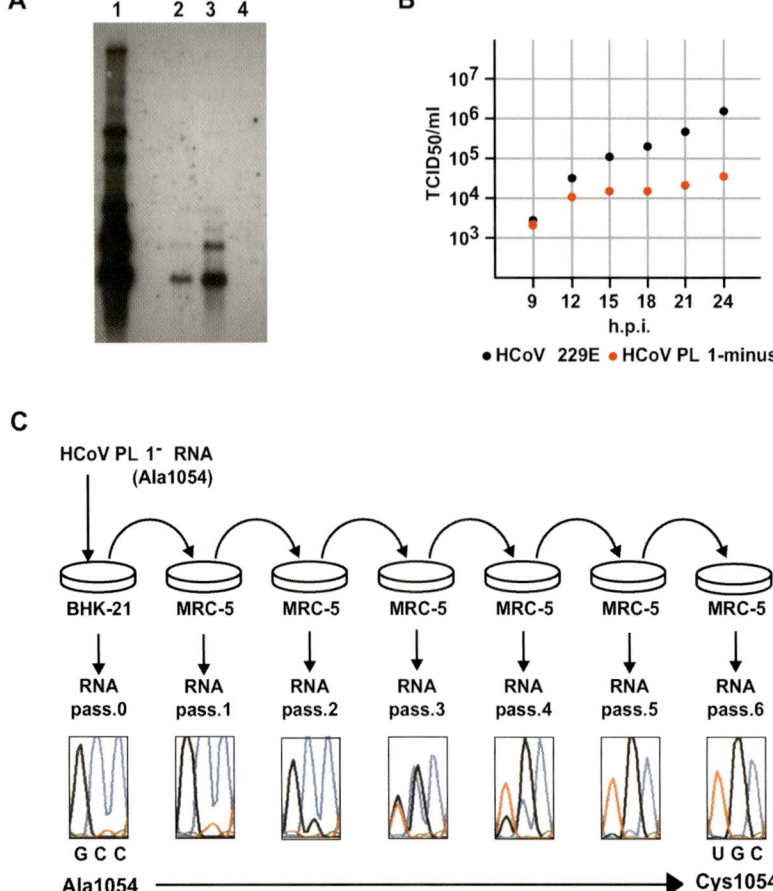

Fig. 6A–C. Rescue of recombinant HCoV 229E ablated for PL1^{pro} activity. **A** Northern blot analysis of poly(A)-containing RNA from BHK-21 cells which have been transfected with in vitro synthesized HCoV-inf-1 RNA (*lane 2*), HCoV-PL1(–) RNA (*lane 3*) or HCoV-PL2(–) RNA (*lane 4*). Poly(A)-containing RNA from HCoV 229E-infected MRC-5 cells was used as a marker. **B** Growth kinetics of parental HCoV 229E and recombinant HCoV-PL1(–) determined after infection of MRC-5 cells (moi=1). **C** The reversion of the PL1^{pro} active site mutation (GCC; Ala1054) of HCoV-PL1(–) to the parental sequence (UGC; Cys1054) during 6 passages on MRC-5 cells is illustrated

3.2
Analysis of IBV Spike Chimeras

IBV is an important veterinary pathogen and, in common with other coronaviruses, the virus surface glycoprotein is thought to be an important determinant of cell tropism. To test this, the vaccinia virus-based coronavirus reverse genetic system has been used to produce a recombinant IBV in which the ectodomain region of the spike gene from IBV M41-CK replaced the corresponding region of the IBV Beaudette genome (Casais et al. 2003). Analysis of the recombinant IBV BeauR-M41(S) showed that it had acquired the same growth characteristics and cell tropism as IBV M41-CK in vitro. These results demonstrate that the IBV spike glycoprotein is a determinant of cell tropism and, importantly, they show that the reverse genetic system can be used to generate recombinant coronaviruses with a precisely modified genome which may be used as vaccine strains.

3.3
Recombinant MHV Is Fully Pathogenic in Mice

MHV is the prototype of the class II coronaviruses, a group that also includes the SARS coronavirus. Moreover, MHV is the most experimentally accessible coronavirus system. For example, MHV-A59 replicates to high titres in cell culture and there exists a collection of temperature-sensitive mutants which are defective in the synthesis of viral RNA (Siddell et al. 2001). These features should facilitate the analysis of coronavirus RNA synthesis and aid in the elucidation of functions associated with coronavirus replicase proteins. Furthermore, the natural host of MHV is the mouse. There is a wealth of genetic and immunological information relating to inbred mouse strains, and there are an increasing number of transgenic mice strains which are ablated or defective in the expression of functional host cell genes, particularly those encoding proteins related to the immunological response to virus infection. MHV infection, therefore, is an ideal tool to study both the innate and adaptive immune responses to viruses. Finally, there are a number of informative animal disease models based on MHV infection, including models of virus-related demyelination and viral hepatitis (Haring and Perlman 2001). These models have provided, and will continue to provide, insights into the pathogenesis of virus infections. Recently, we have established a vaccinia virus-based reverse genetic system for MHV-A59 (Coley et al., manuscript in preparation). Our preliminary results indi-

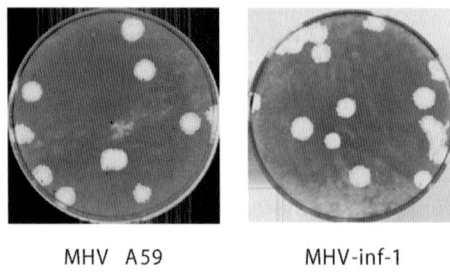

MHV A59   MHV-inf-1

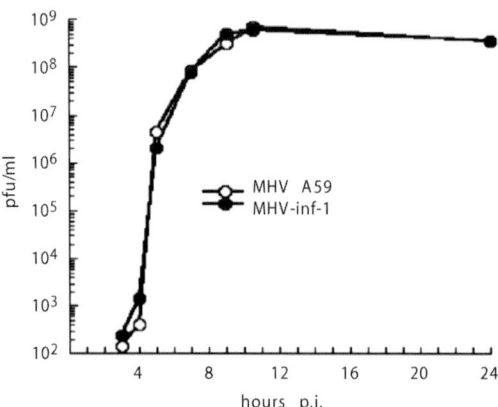

Fig. 7. Plaque morphology and growth kinetics of recombinant MHV-inf-1 generated from cloned full-length cDNA. Recombinant MHV was rescued, plaque purified and propagated by a single passage in murine 17 clone 1 cells. The stock was titrated by plaque assay on 17 clone 1 cells. The photographs show the 10^{-8} dilution of both MHV-A59 (strain OH99) and recombinant MHV-inf-1. The replication kinetics of MHV-A59 and MHV-inf-1 in 17 clone 1 cells were also found to be identical

cate that the recombinant MHV-A59 replicates in cell culture to the same titre and with the same plaque morphology as laboratory-adapted strains of MHV-A59 (Fig. 7). Also, the recombinant virus replicates to comparable titres in mouse tissues, has approximately the same virulence as non-recombinant MHV-A59 and produces the same histopathological changes in the brains and liver of infected mice (Coley et al., manuscript in preparation).

4
Generation of Replicon RNAs

The development of reverse genetic systems for coronaviruses provides an opportunity to carry out an extensive biological characterization of the viral replicative proteins and functions. In addition to the analysis of recombinant coronaviruses, replicative functions can also be studied in non-infectious systems. For example, the concept of using autonomously replicating RNAs (replicon RNAs) has been explored in a number of positive-strand RNA virus systems and has greatly facilitated the functional analysis of viral replication and transcription (Pietschmann and Bartenschlager 2001; Westaway et al. 2003). Furthermore, selectable replicon RNAs that carry a marker for selection in cell culture have been developed in order to generate stable, replicon RNA-containing cell lines. Such cell lines have been used to study replicative functions and are of particular interest if a virus cannot be efficiently propagated in tissue culture, e.g. hepatitis C virus. Because no structural genes and, therefore, no infectious viruses are formed, replicon-based systems also represent an attractive tool for the analysis of replicative functions if the pathogenicity of the virus is a concern.

4.1
Replicase Gene Products Suffice for Discontinuous Transcription

For most positive-strand RNA viruses, the replicase gene and the 5'- and 3'-genomic termini suffice for autonomous replication of the viral RNA. It has been demonstrated that, indeed, the presence of 5'- and 3'-genomic termini of coronavirus genomes is necessary for efficient replication of defective RNAs in helper virus-infected cells (Kim et al. 1993). The availability of reverse genetic systems now allows us to analyse the role of the replicase gene products in coronavirus replication and transcription. Based on the full-length HCoV 229E cDNA, cloned in a vaccinia virus vector, an RNA has been constructed that contains the 5' and 3' ends of the genomic RNA, the entire replicase gene, and a single reporter gene (encoding for GFP) (Fig. 8A; Thiel et al. 2001b). The GFP gene has been cloned downstream of the replicase gene and a transcription regulatory sequence (TRS) to enable the synthesis of a subgenomic mRNA encoding GFP. When this RNA was transfected into BHK-21 cells, a small percentage of cells displayed green fluorescence, indicative for GFP expression. After isolation of poly(A)-containing RNA from BHK-transfected cells an RT-PCR product could be identified which represents the lead-

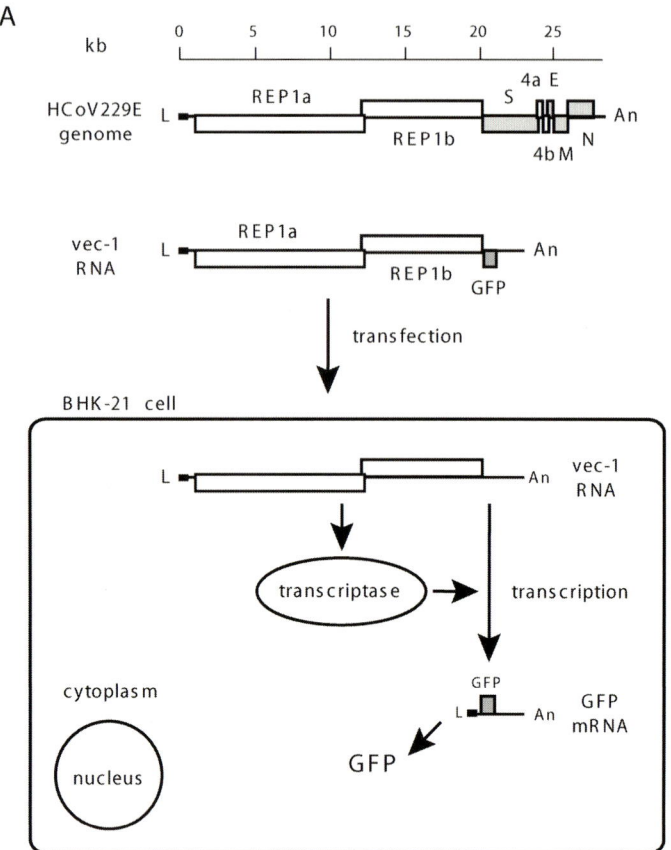
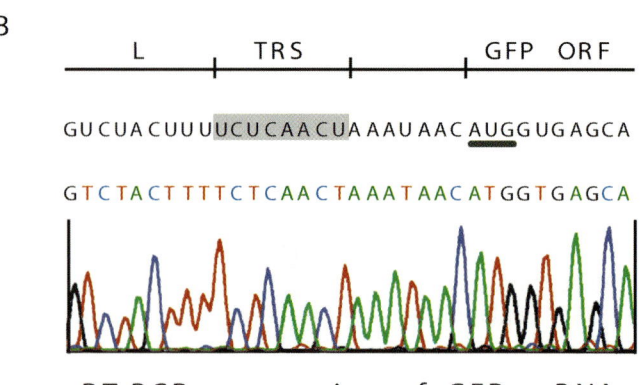

er-body junction of a subgenomic mRNA encoding GFP. Sequence analysis of this RT-PCR product showed that coronavirus-specific, subgenomic mRNA synthesis has occurred and the leader-body fusion has taken place at the expected position (Fig. 8B). These results demonstrate conclusively that the coronavirus replicase gene products are the only viral proteins needed to assemble a functional complex capable of discontinuous transcription. However, it should be noted that these results do not prove that replication of the transfected RNA has occurred, nor do they exclude the possibility that additional viral or host cell proteins may have regulatory roles in coronavirus replication or transcription. It is striking in this respect that only a small percentage (~0.1%) of green fluorescent cells could be observed after RNA transfection into BHK-21 cells. In a similar experiment, HCoV 229E-based vector RNA has been transfected into BHK-21 cells together with a synthetic mRNA encoding for the HCoV 229E N protein. In this case about 3% of green fluorescent cells could be observed (Thiel et al. 2003). This result again indicates that the coronavirus N protein might be involved (directly or indirectly) in the (regulation of) replication and/or transcription of viral RNA. Further studies are needed to address this issue and, clearly, reverse genetic approaches represent valuable tools to study the putative function(s) of coronavirus N proteins.

4.2
Generation of Autonomously Replicating RNAs

Although our analysis of replicon RNAs revealed the basic requirements for coronavirus discontinuous transcription, we were unable to demonstrate replication of coronavirus replicon RNAs in transfected cells. While using the reverse genetic system for the recovery of recombinant coronaviruses and during the development of coronavirus multigene RNA vectors, we have gained evidence that the nucleocapsid protein fa-

Fig. 8A, B. Replicase gene products suffice for coronavirus discontinuous transcription. **A** The structural relationship of the HCoV 229E genome, the in vitro-transcribed HCoV-vec-1 RNA and the intracellular mRNA produced by coronavirus transcriptase-mediated discontinuous transcription is illustrated. **B** The sequence of the transcription regulating sequence (TRS) region of the intracellular GFP mRNA was obtained from an RT-PCR product of poly(A)-containing RNA from HCoV-vec-1-transfected cells. The sequences corresponding to the HCoV 229E leader (*L*), the TRS region and the first 10 nt of the GFP-ORF are shown

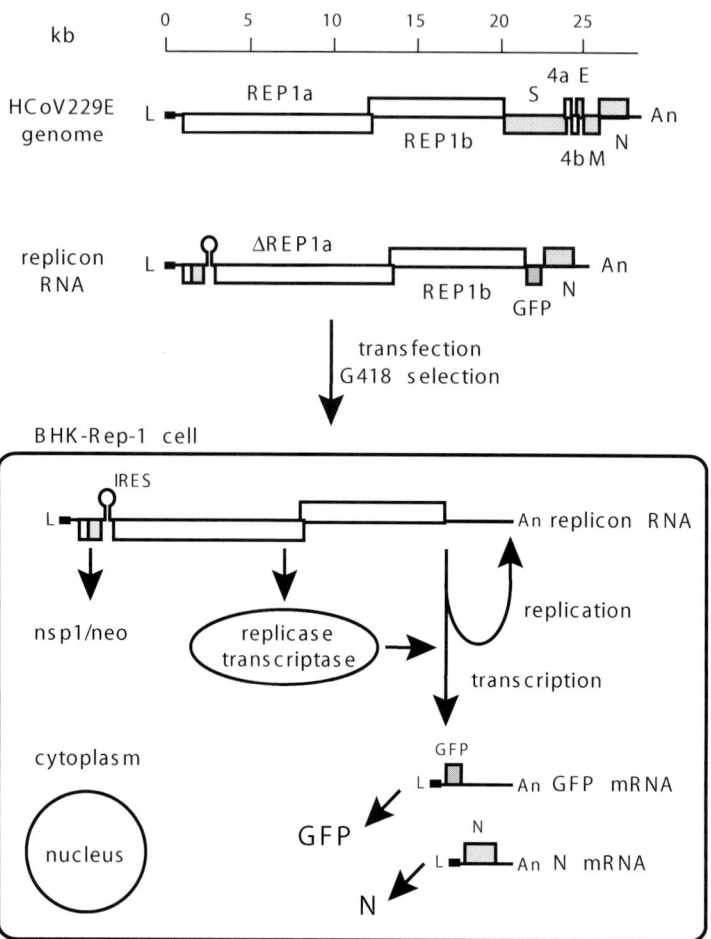

Fig. 9. Generation of stable cell lines containing autonomously replicating coronavirus RNA. The structural relationship of the HCoV 229E genome, the in vitro-transcribed replicon RNA and the intracellular mRNAs (GFP and N mRNAs) produced by coronavirus replicase/transcriptase-mediated discontinuous transcription is illustrated. The insertion of the neomycin resistance gene at the replicase nsp1/nsp2 junction allows the selection of stable cell lines with G418

cilitates the rescue of recombinant coronaviruses and, furthermore, increases the number of cells that contain an active transcription complex after electroporation of vector RNAs. Consequently, we constructed an RNA that contains the 5′ and 3′ ends of the HCoV 229E genome, the

HCoV 229E replicase gene, the GFP reporter gene and the HCoV 229E nucleocapsid gene. In addition, to provide the basis for the selection of stable cell clones that contain autonomously replicating RNA, we have introduced a selectable marker gene into the replicase gene conferring resistance to neomycin (Fig. 9).

After transfection of this selectable replicon RNA into BHK-21 cells we were, indeed, able to select multiple cell clones containing an autonomously replicating RNA. These cells have been shown to stably maintain the replicating RNA and, moreover, display green fluorescence due to the replicon RNA-mediated GFP expression (Hertzig et al. 2004). This is the first example of coronavirus-derived RNAs which can be selected for replication in cell culture. The selectable replicon RNA, in combination with GFP reporter gene expression, provides an excellent basis to analyse the function of coronavirus replicase inhibitors in cell culture. Thus it is now possible to test the effects of antiviral compounds on coronavirus replication by simply seeding out the replicon RNA-containing cells and assaying for reporter gene expression levels. Decreasing reporter gene expression will indicate putative antiviral activity of a particular compound, which can then be tested for specificity and efficacy. Given the striking similarities of replicative enzymes and functions amongst coronaviruses, the currently available replicon RNA, based on HCoV 229E, can already be used to screen for suitable drugs which target coronavirus infections, including SARS. In the long term, a SARS-CoV replicon RNA can be used to screen for SARS-specific inhibitors. The SARS-CoV replicon system will provide a versatile platform technology which allows the development of a high-throughput antiviral screening assay based on reporter gene expression. Furthermore, it circumvents biosafety concerns associated with SARS-CoV, because it represents a rapid, convenient and safe assay for SARS-CoV replication without the need to grow infectious SARS-CoV.

5
Development of Coronavirus-Based Multigene Vectors

The molecular biology of coronaviruses and the specific features of the human coronavirus 229E (HCoV 229E) system indicate that HCoV 229E-based vaccine vectors have the potential to become a new class of viral vaccines. First, the receptor for HCoV 229E, human aminopeptidase N (hAPN or CD13) is expressed on human dendritic cells (DC) and macrophages, indicating that targeting of HCoV 229E-based vectors to professional antigen-presenting cells can be achieved by receptor-mediated transduction. Second, coronaviruses display a unique transcription mechanism resulting in the synthesis of multiple subgenomic mRNAs encoding mainly structural proteins. Because the structural genes can be replaced by multiple heterologous genes, these vectors represent safe, non-infectious vector RNAs. Third, it has been shown that HCoV 229E multigene vectors can be packaged to virus-like particles (VLPs) if the structural proteins are expressed *in trans*. Most interestingly, VLPs containing HCoV 229E-based vector RNA have the ability to transduce human DC and to mediate heterologous gene expression in these cells (Thiel et al. 2003). Thus the expression of multiple antigens in combination with specific DC tropism represents an unprecedented potential to induce both T cell and antibody responses against multiple antigens. Finally, coronavirus infections are mainly associated with respiratory and enteric diseases and natural transmission of coronaviruses occurs via mucosal surfaces. HCoV 229E infections are mainly encountered in children, and re-infection occurs frequently in adults. It is therefore unlikely that preexisting immunity against HCoV 229E will have a significant impact on the vaccination efficiency if HCoV 229E-based vectors are used in humans.

5.1
Multigene Expression Using Coronavirus-Based Vectors

With the reverse genetic systems available, it is now possible to make use of the unique characteristics of coronavirus transcription to develop coronavirus expression vectors. The rationale of expressing heterologous genes with coronavirus-mediated transcription is to insert a transcriptional cassette, comprised of a TRS located upstream of the gene of interest, into a coronavirus genome, minigenome or vector RNA. We have explored a vector RNA-based strategy using the HCoV 229E reverse genetic system and could show for human coronavirus vector RNAs that a region of at least 5.7 kb is dispensable for discontinuous transcription (Thiel et al. 2001b). This region contained all structural genes, and therefore our vector RNAs are not infectious. For the construction of coronavirus-based non-infectious, multigene vectors, we consider that about one-third of the genome, or up to 9 kb, could be replaced by multiple transcriptional cassettes. Indeed, we could demonstrate that it is possible to construct a human coronavirus vector RNA capable of mediating the expression of multiple heterologous proteins, namely the chloramphenicol-acetyltransferase, the firefly luciferase and GFP

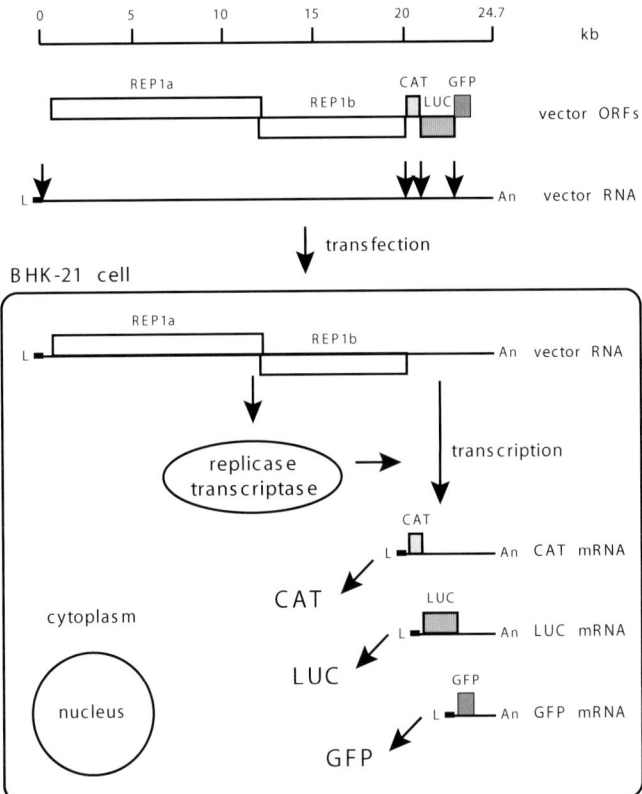

Fig. 10. Multigene RNA vectors based on coronavirus transcription. The structural relationship of HCoV-based vector ORFs, TRS elements (*arrows*) and the intracellular mRNAs produced by coronavirus-mediated transcription is illustrated, together with the intracellular translation products (i.e. the replicase/transcriptase, CAT, LUC and GFP)

(Fig. 10). These results indicate that coronavirus-based vector systems might be useful for heterologous gene expression, especially for longer and multiple genes.

5.2
Coronavirus-Based Vectors as Potential Vaccines

An important consideration for viral vaccine vectors is their potential for efficient delivery of their genetic material to specific target cells. For example, targeting of viral vaccine vectors to dendritic cells (DC) is highly desirable in order to optimize vaccine efficacy. In this respect, it is important to note that the HCoV 229E receptor, human aminopeptidase N (hAPN or CD13), is expressed at high levels on human DC (Summers et al. 2001). This implies that HCoV 229E-based VLPs could be used to efficiently (receptor-mediated uptake) transduce these cells. Indeed, it has been demonstrated that HCoV 229E-based VLPs can be used to transduce immature and mature human DC and that vector-mediated heterologous gene expression can be achieved in human DC (Thiel et al. 2003). Multigene vectors, based on HCoV 229E, represents a particularly promising tool to genetically deliver multiple genes such as tumour or HIV antigens and immunostimulatory cytokines to human DC.

Despite the remarkable potential of human coronavirus vectors as vaccines, the efficacy of coronavirus multigene vectors has not yet been demonstrated in vivo. For obvious reasons, a small animal model is desirable to address this issue. We have, therefore, established a reverse genetic system for MHV (see Sect. 3.3) which allows construction of MHV vector RNAs resembling their HCoV 229E counterparts. Murine DC can be infected with MHV, indicating that the analysis of recombinant MHV vectors in the context of a murine animal model may well serve as a

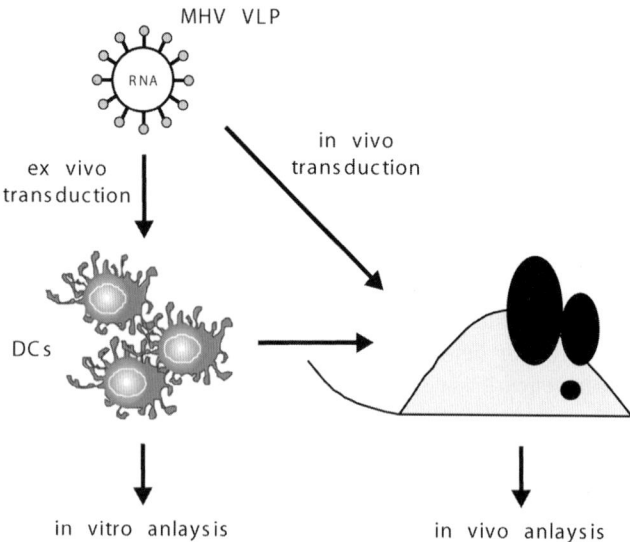

Fig. 11. A murine model to assess the efficacy of coronavirus-based multigene vaccine vectors

paradigm for the development of coronavirus vaccine vectors (Fig. 11). In addition, the murine model will allow the use of inbred and transgenic mice and a variety of established immunological techniques which are indispensable for the analysis of vector-induced immune responses.

An important prerequisite to study of the efficacy of coronavirus vaccine vectors is the availability of VLPs that can be produced to high titres. Therefore, packaging cell lines must be established which mediate the expression of coronavirus structural proteins *in trans*. It has been shown that recombinant MHV defective-interfering particles can be produced in the absence of helper virus if the structural proteins are expressed *in trans* (Bos et al. 1996). Furthermore, a replication-competent but propagation-deficient TGEV vector RNA which lacks the E gene can be packaged when the E protein is expressed under the control of the cytomegalovirus promoter or by using alphavirus-based expression systems (Curtis et al. 2002; Ortego et al. 2002). Once an efficient coronavirus packaging strategy has been established, the efficacy of coronavirus multigene vaccine vectors can be assessed in vivo with the murine animal model.

6
Discussion

In the past, the study of coronavirus genetics was broadly restricted to the analysis of temperature-sensitive (*ts*) mutants, the analysis of defective RNA templates which depend on replicase proteins provided by a helper virus and the analysis of mutant and chimeric viruses generated by targeted recombination. Each of these methods produced valuable information, but, in one way or another, each has its own limitations. For example, the generation of *ts* mutants is essentially a random process and a great deal of effort must be expended to produce a comprehensive collection of mutants representing the possible complementation groups or cistrons encoded in the coronavirus genome. Similarly, targeted recombination is a powerful tool for the generation of site-specific coronavirus mutants and chimeras. However, for technical reasons, it is restricted to the 3' one-third of the genome, encoding the structural and accessory proteins. So the development of reverse genetic approaches which do not have these limitations should provide a tremendous impetus to the study of coronavirus replication and biology. We predict that the new reverse genetic systems, including the one described here, will be used very quickly to study the structure and function of coronavirus replicase genes and genes that are non-essential for replication in cell culture but clearly provide a selective advantage in vivo. We also predict that the reverse genetic systems will lead to the development of sophisticated RNA replicon systems which will be used as vectors for the delivery of heterologous genes in gene therapy and as biosafe diagnostic tools for the identification of, for example, coronavirus replicase inhibitors.

References

Almazan F, Gonzalez JM, Penzes Z, Izeta A, Calvo E, Plana-Duran J, Enjuanes L (2000) Engineering the largest RNA virus genome as an infectious bacterial artificial chromosome. Proc Natl Acad Sci USA 97:5516–5521

Ball LA (1987) High-frequency homologous recombination in vaccinia virus DNA. J Virol 61:1788–1795

Bos EC, Luytjes W, van der Meulen HV, Koerten HK, Spaan WJ (1996) The production of recombinant infectious DI-particles of a murine coronavirus in the absence of helper virus. Virology 218:52–60

Britton P, Green P, Kottier S, Mawditt KL, Penzes Z, Cavanagh D, Skinner MA (1996) Expression of bacteriophage T7 RNA polymerase in avian and mammalian cells by a recombinant fowlpox virus. J Gen Virol 77:963–7

Casais R, Dove B, Cavanagh D, Britton P (2003) Recombinant avian infectious bronchitis virus expressing a heterologous spike gene demonstrates that the spike protein is a determinant of cell tropism. J Virol 77:9084–9089

Casais R, Thiel V, Siddell SG, Cavanagh D, Britton P (2001) Reverse genetics system for the avian coronavirus infectious bronchitis virus. J Virol 75:12359–12369

Curtis KM, Yount B, Baric RS (2002) Heterologous gene expression from transmissible gastroenteritis virus replicon particles. J Virol 76:1422–1434

Gossen M, Freundlieb S, Bender G, Muller G, Hillen W, Bujard H (1995) Transcriptional activation by tetracyclines in mammalian cells. Science 268:1766–1769

Haring J, Perlman S (2001) Mouse hepatitis virus. Curr Opin Microbiol 4:462–466

Herold J, Gorbalenya AE, Thiel V, Schelle B, Siddell SG (1998) Proteolytic processing at the amino terminus of human coronavirus 229E gene 1-encoded polyproteins: identification of a papain-like proteinase and its substrate. J Virol 72:910–918

Hertzig T, Scandella E, Schelle B, Ziebuhr J, Siddell SG, Ludewig B, Thiel V (2004) Rapid identification of coronavirus replicase inhibitors using a selectable replicon RNA. J Gen Virol 85:1717–1725

Kerr SM, Smith GL (1991) Vaccinia virus DNA ligase is nonessential for virus replication: recovery of plasmids from virus-infected cells. Virology 180:625–632

Kim YN, Jeong YS, Makino S (1993) Analysis of *cis*-acting sequences essential for coronavirus defective interfering RNA replication. Virology 197:53–63

Merchlinsky M, Moss B (1992) Introduction of foreign DNA into the vaccinia virus genome by in vitro ligation: recombination-independent selectable cloning vectors. Virology 190:522–526

Ortego J, Escors D, Laude H, Enjuanes L (2002) Generation of a replication-competent, propagation-deficient virus vector based on the transmissible gastroenteritis coronavirus genome. J Virol 76:11518–11529

Pietschmann T, Bartenschlager R (2001) The hepatitis C virus replicon system and its application to molecular studies. Curr Opin Drug Discov Devel 4:657–664

Scheiflinger F, Dorner F, Falkner FG (1992) Construction of chimeric vaccinia viruses by molecular cloning and packaging. Proc Natl Acad Sci USA 89:9977–9981

Siddell S, Sawicki D, Meyer Y, Thiel V, Sawicki S (2001) Identification of the mutations responsible for the phenotype of three MHV RNA-negative ts mutants. Adv Exp Med Biol 494:453–458

Smith GL, Moss B (1983) Infectious poxvirus vectors have capacity for at least 25000 base pairs of foreign DNA. Gene Ther 25:21–28

Summers KL, Hock BD, McKenzie JL, Hart DN (2001) Phenotypic characterization of five dendritic cell subsets in human tonsils. Am J Pathol 159:285–295

Thiel V, Herold J, Schelle B, Siddell SG (2001a) Infectious RNA transcribed in vitro from a cDNA copy of the human coronavirus genome cloned in vaccinia virus. J Gen Virol 82:1273–1281

Thiel V, Herold J, Schelle B, Siddell SG (2001b) Viral replicase gene products suffice for coronavirus discontinuous transcription. J Virol 75:6676–6681

Thiel V, Karl N, Schelle B, Disterer P, Klagge I, Siddell SG (2003) Multigene RNA vector based on coronavirus transcription. J Virol 77:9790–9798

Westaway EG, Mackenzie JM, Khromykh AA (2003) Kunjin RNA replication and applications of Kunjin replicons. Adv Virus Res 59:99–140

Yount B, Curtis KM, Baric RS (2000) Strategy for systematic assembly of large RNA and DNA genomes: transmissible gastroenteritis virus model. J Virol 74:10600–10611

Ziebuhr J, Snijder EJ, Gaorbalenya AE (2000) Virus-encoded proteinases and proteolytic processing in the *Nidovirales*. J Gen Virol 81:853–879

Ziebuhr J, Thiel V, Gorbalenya AE (2001) The autocatalytic release of a putative RNA virus transcription factor from its polyprotein precursor involves two paralogous papain-like proteases that cleave the same peptide bond. J Biol Chem 276:33220–33232

Development of Mouse Hepatitis Virus and SARS-CoV Infectious cDNA Constructs

R. S. Baric (✉) · A. C. Sims

Department of Epidemiology, University of North Carolina at Chapel Hill, Chapel Hill, NC 27599-7400, USA
rbaric@email.unc.edu

1	Introduction	230
2	The Coronavirus Genome	230
3	Systematic Approaches to Assembling Coronavirus cDNAs from a Panel of Contiguous Subclones	231
4	Assembling MHV Infectious cDNAs	235
4.1	Applications in Genomics	240
4.2	Engineering MHV Genomes	241
5	SARS-CoV Infectious Clone	245
6	Future Applications	246
	References	248

Abstract The genomes of transmissible gastroenteritis virus (TGEV) and mouse hepatitis virus (MHV) have been generated with a novel construction strategy that allows for the assembly of very large RNA and DNA genomes from a panel of contiguous cDNA subclones. Recombinant viruses generated from these methods contained the appropriate marker mutations and replicated as efficiently as wild-type virus. The MHV cloning strategy can also be used to generate recombinant viruses that contain foreign genes or mutations at virtually any given nucleotide. MHV molecular viruses were engineered to express green fluorescent protein (GFP), demonstrating the feasibility of the systematic assembly approach to create recombinant viruses expressing foreign genes. The systematic assembly approach was used to develop an infectious clone of the newly identified human coronavirus, the serve acute respiratory syndrome virus (SARS-CoV). Our cloning and assembly strategy generated an infectious clone within 2 months of identification of the causative agent of SARS, providing a critical tool to study coronavirus pathogenesis and replication. The availability of coronavirus infectious cDNAs heralds a new era in coronavirus genetics and genomic applications, especially within the replicase proteins whose functions in replication and pathogenesis are virtually unknown.

1
Introduction

Molecular analysis of the structure and function of RNA virus genomes has been profoundly advanced by the availability of full-length cDNA clones, the source of infectious RNA transcripts that replicate efficiently when introduced into permissive cell lines (Boyer and Haenni 1994). Coronaviruses contain the largest single-stranded, positive-polarity RNA genome of about 30 kb (Cavanagh et al. 1997; de Vries et al. 1997; Eleouet et al. 1995). Until recently, coronavirus genetic analysis has been limited to analysis of temperature-sensitive (ts) mutants (Fu and Baric 1992, 1994; Lai and Cavanagh 1997; Schaad and Baric 1994; Stalcup et al. 1998), defective interfering (DI) RNAs (Izeta et al. 1999; Narayanan and Makino 2001; Repass and Makino 1998; Williams et al. 1999), and recombinant viruses generated by targeted recombination (Fischer et al. 1997; Hsue and Masters 1999; Kuo et al. 2000). Among these, targeted recombination is the seminal approach developed to systematically assess the function of individual mutations in the 3'-most ~10 kb of the MHV genome. Methods to assemble an MHV full-length infectious construct have been hampered by the large size of the genome, the regions of chromosomal instability, and the inability to synthesize full-length transcripts (Almazán et al. 2000; Masters 1999; Yount et al. 2000). This is especially problematic within the group 2 coronavirus replicase, where several regions of chromosomal toxicity and instability have hampered the development of infectious cDNAs. Full-length infectious constructs will allow for the systematic dissection of the structure and function of each viral gene, the phenotypic consequences of gene rearrangement on virus replication and pathogenesis, the development of coronavirus heterologous gene expression systems, and a clearer understanding of the transcription and replication strategy of the *Coronaviridae*. In this report, we review strategies for building coronavirus infectious cDNAs by using mouse hepatitis virus strain A59 as a model.

2
The Coronavirus Genome

The coronavirus genome, a single-stranded RNA, is the largest viral RNA genome known to exist in nature (27.6–31.3 kb). Genomic RNAs have a 5' terminal cap and a 3' terminal poly (A) tail. In addition, a leader sequence of 65–98 nucleotides and a 200- to 400-base pair untranslated region are located at the 5' terminus, whereas a 200- to 500-base pair untranslated region is located at the 3' terminus. The 5' most two-thirds of the genome encodes the replicase gene in two open reading frames (ORFs), 1a and 1b, the latter of which is expressed by ribosomal frameshifting (Almazán et al. 2000; Eleouet et al. 1995). Like many other positive-sense RNA viruses, the coronavirus replicase is translated as a large precursor polyprotein that is processed by viral proteinases, giving rise to ~15 replicase proteins. The functions of most of the coronavirus replicase proteins are unknown. However, based on nucleotide sequence homology and empirical studies, identifiable functions include two papainlike cysteine proteases, a chymotrypsin-like 3C protease, a cysteine-rich growth factor-related protein, an RNA-dependent RNA polymerase, a nucleoside triphosphate (NTP)-binding/helicase domain, and a zinc-finger nucleic acid-binding domain (Enjuanes et al. 2000a; Penzes et al. 2001; Siddell 1995). Most of the replicase gene products colocalize with replication complexes at sites of RNA synthesis on internal membranes. However, a spectrum of genetically informative mutations have not been systematically targeted to any of these replicase proteins, so we have little insight into the organization of the replicase complex and the location of functional motifs, which regulate transcription, replication, and RNA recombination. Because of the extremely rich milieu of molecular reagents that are available against the replicase proteins, the availability of a molecular clone of MHV allows for the first time a systematic genetic analysis of gene 1 function in coronavirus replication.

3
Systematic Approaches to Assembling Coronavirus cDNAs from a Panel of Contiguous Subclones

Coronavirologists have seized on several different strategies to build infectious cDNA clones. However, all were primarily designed to circumvent problems associated with the large size of the coronavirus genome, regions of chromosomal instability, and other problems associated with the production of full-length infectious transcripts (Almazán et al. 2000; Masters 1999; Yount et al. 2000). Our solution was to assemble infectious cDNAs from a panel of contiguous subclones that spanned the entire length of the TGEV and MHV genomes. Each subclone was flanked by unique restriction sites with characteristics that allow for the systematic and precise assembly of a full-length cDNA with in vitro ligation. For this strategy to be efficient, restricted subclone fragments had to be incapable of self-concatemer formation and not spuriously assemble with other noncontiguous subclones.

Conventional class II restriction enzymes, such as *Eco*RI, leave identical sticky ends that assemble with similarly cut DNA in the presence of DNA ligase (Pingoud and Jeltsch 2001; Sambrook et al. 1989). Because these enzymes leave identical compatible ends, digested fragments randomly self-assemble into large concatamers and, therefore, they are poor choices for assembling large intact genomes or chromosomes. However, a second group of class II restriction enzymes (i.e., *Bgl*I, *Bst*XI, *Sfi*I) also recognize a symmetrical sequence but leave random sticky ends 1–4 nucleotides in length, and consequently, restrict assembly cascades along specific pathways (Table 1). For example, the type II restriction enzyme, *Bgl*I, recognizes the symmetrical sequence **GCCNNNN↓NGGC** and cleaves a random DNA sequence on average every ~4,096 base pairs. Because 64 different 3-nucleotide overhangs can be generated, DNA frag-

Table 1. Selected restriction enzymes used in assembly of recombinant full-length genomes

Restriction enzyme[a]	Recognition site	No. of variable sticky end	Average cutting frequency[b]	Actual frequency of compatible ends[b]
*Bgl*I	GCCNNNN↓NGGC CGGN↑NNNNCCG	3 nt/64 potential ends	~4,096 nt	~261,344 nt
*Bst*XI	CCANNNNN↓NTGG GGTN↑NNNNNACC	4 nt/256 potential ends	~4,096 nt	~1,045,376 nt
*Sfi*I	GGCCNNNN↓NGGCC CCGGN↑NNNNCCGG	3 nt/64 potential ends	~65,536 nt	~4,194,304 nt
*Sap*I	GCTCTTCN↓NNNN CGAGAAGNNNNN↑	3 nt/64 potential ends	~16,385 nt (in either strand)	~1,048,640 nt*
*Aar*I	CACCTGCNNNN↓NNNN GTGGACGNNNNNNNN↑	4 nt/256 potential ends	~16,385 nt (in either strand)	~4,194,304 nt*
*Esp*3I (*Bsm*BI)	CGTCTCN↓NNNN GCAGAGNNNNN↑	4 nt/256 potential ends	~4,096 nt (in either strand)	~1,048,576 nt*

[a] Other enzymes leaving many different overhangs: *Bsm*FI, *Ecl*HkI, *Fok*I, *Mbo*II, *Tth*IIII, *Ahd*I, *Drd*I, *Bsp*MI, *Bsm*AI, *Bcg*I, *Bmr*I, *Bpm*I, *Bsa*I, *Bse*I, *Ear*I, *Pfi*MI, *Bst*V2, *Vpa*K32I, *Abe*I, *Ppi*I.

[b] Assuming a totally random DNA sequence; *asymmetric cutters like *Sap*I, *Aar*I and *Esp*3I can have recognition sites in either strand of DNA so actual site frequency is ~1/2 of indicated values and can be engineered as "no-see-um" (Yount et al. 2002).

ments will only assemble with the appropriate 3-nucleotide complementary overhang generated at an identical *Bgl*I restriction site. As a result, identical ends are generated every ~264,000 base pairs, providing a powerful means for the construction of very large DNA and RNA genomes. Consonant with these findings, the type IIS restriction enzyme, *Esp*3I, recognizes an asymmetric sequence and makes a staggered cut 1 and 5 nucleotides downstream of the recognition sequence, leaving 256, mostly asymmetrical, 4-nucleotide overhangs (GCTCTCN↓NNNN). As identical *Esp*3I sites are generated every ~1,000,000 base pairs or so in a random DNA sequence, most restricted fragments usually do not self-assemble (Yount et al. 2002). Rather, specific recursive assembly pathways can be designed that hypothetically allow assembly of >1 million base pair DNA genomes (~2^{256} fragments) (Table 1). We took advantage of several unique properties inherent in type II restriction enzymes to build coronavirus infectious cDNAs.

Initially, we isolated five cDNA subclones spanning the entire TGEV genome (designated TGEV A, B, C, D/E, and F) by RT-PCR using primers that introduced unique *Bgl*I restriction sites at the 5′ and 3′ ends of each fragment without altering the amino acid coding sequences of the virus (Table 2). The TGEV A, C, DE, and F clones were stable in plasmid DNAs in *Escherichia coli*. The B fragment, however, was unstable, containing deletions or insertions in the wild-type sequence at a region of instability in the TGEV genome noted by other investigators (Almazán et al. 2000; Eleouet et al. 1995). To prevent fragment instability, we used primer-mediated mutagenesis to bisect the B fragment at the unstable site with an adjoining *Bst*XI (CCA**TTCAC**↓TTGG) site, resulting in TGEV B1 and TGEV B2 amplicons (Fig. 1; Table 2). It is likely that sequences

Table 2. Design of TGEV junction sequences

Restriction site junction	Location	Junction
5′-GCCT**GTT**↓TGGC-3′ 3′-CGGA↑**CAA**ACCG-5′	*Bgl*I, nt 6,159	A-B1
5′-CCA**TTCAC**↓TTGG-3′ 3′-GGTA↑**AGTG**AACC-5′	*Bst*XI, nt 9,949	B1-B2
5′-GCCG**CAT**↓TGGC-3′ 3′-CGGC↑**GTA**GCCG-5′	*Bgl*I, nt 11,355	B2-C
5′-GCCT**TCT**↓TGGC-3′ 3′-CGGA↑**AGA**ACCG-5′	*Bgl*I, nt 16,595	C-D/E1
5′-GCC**GTGC**↓AGGC-3′ 3′-CGGC↑**ACGT**CCG-5′	*Bgl*I, nt 23,487	D/E1-F

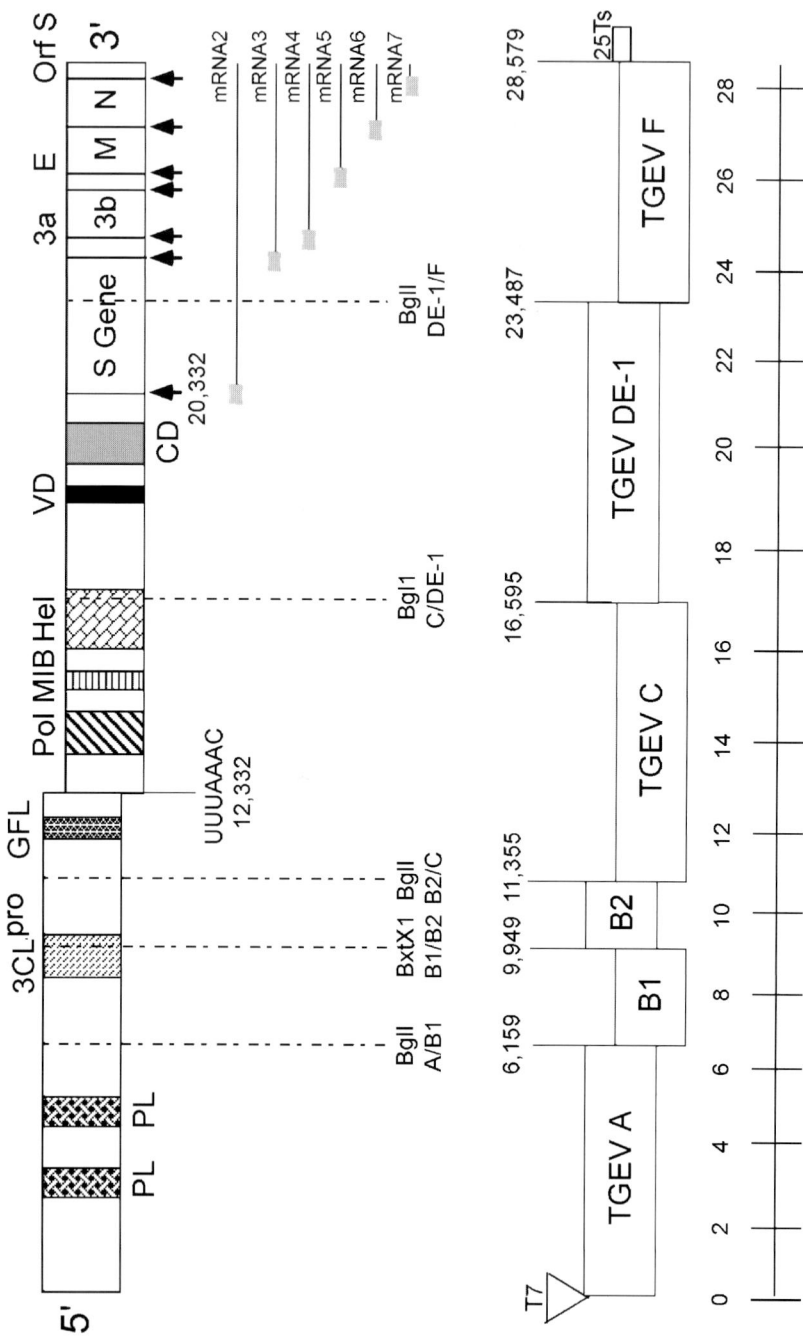

(9600–9950) in and around the TGEV 3C like protease (3CL^{pro}) motif are either bactericidal or unstable in microbial vectors (Almazán et al. 2000; Yount et al. 2000). The resulting 6 fragments, TGEV A, B1, B2, C, D/E, and F, were ligated in vitro to generate a full-length cDNA of the TGEV genome (Fig. 1). Molecularly cloned viruses were indistinguishable from wild type and contained the marker mutations and unique *Bgl*I and *Bst*XI junction sequences used in the assembly of the infectious construct (Yount et al. 2000).

4
Assembling MHV Infectious cDNAs

One potential problem with the original approach was that several "silent" mutations were inserted to introduce the unique *Bgl*I sites into the TGEV component clones. To circumvent this problem, a variation of the systematic assembly approach was used to build the group II coronavirus, mouse hepatitis virus (MHV) infectious cDNA (Yount et al. 2002). The enzyme *Esp*3I recognizes an asymmetrical site and cleaves external to the recognition sequence, allowing for traditional and "no-see-um" cloning applications (Fig. 2, Table 1). With traditional approaches, *Esp*3I sites can be oriented to reform the recognition site after ligation of two MHV cDNAs, leaving the restriction site within the genomes of recombinant viruses. However, the *Esp*3I recognition site is asymmetrical, so a simple reverse orientation allows for the insertion of an *Esp*3I recognition sequence on the ends of two adjacent clones with the cleavage site derived from virtually any 4-nucleotide sequence combination dictated by the virus sequence. On cleavage and ligation with the adjoining fragment, the *Esp*3I sites are lost from the final ligation products, leaving a

Fig. 1. Strategy for the systematic assembly of TGEV full-length cDNA. The TGEV genome is a positive-sense, single-stranded RNA of about 28.5 kb. Six independent subclones (*A*, *B1*, *B2*, *C*, *DE*, and *F*) that span the entire length of the genome were isolated by RT-PCR using primer pairs that introduced unique *Not*I, *Bgl*I, and/or *Bst*XI restriction sites at each end. On ligation, the intact viral genome is generated as a cDNA. A unique T7 start site and a 25 poly(T) tail allow for in vitro transcription of full-length, capped, polyadenylated transcripts (Yount et al. 2000). *PL*, papainlike protease; *3CL^{pro}*, 3CL protease; *GFL*, growth factor like; *pol*, polymerase motif; *MIB*, metal binding motif; *hel*, helicase motif; *VD/CD*, variable or conserved domains

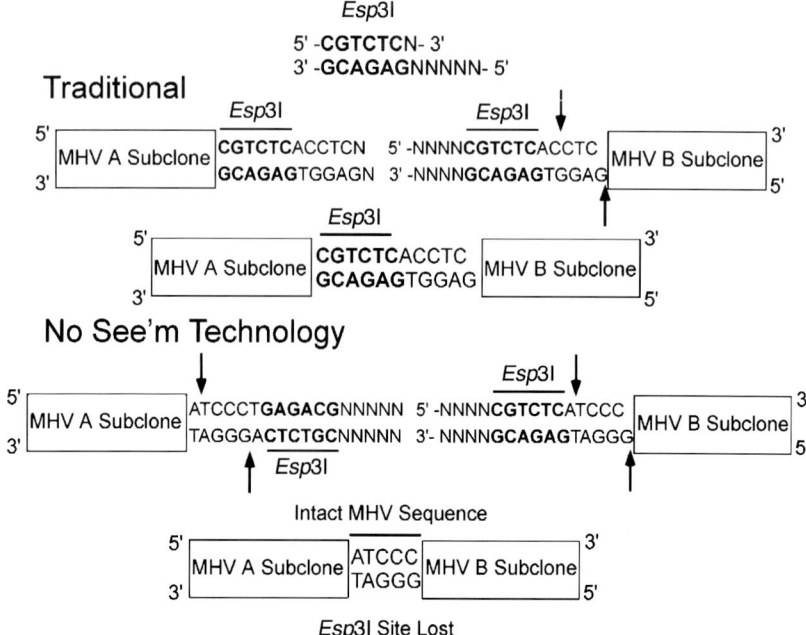

Fig. 2. Use of *Esp*3I in the traditional and "no-see-um" approaches. The traditional approach to the use of *Esp*3I involves the ligation of two fragments containing identical *Esp*3I restriction sites, resulting in a ligation product with an intact *Esp*3I site remaining. In the "no-see-um" approach, a simple reverse orientation of the restriction sites allows for the specific removal of the *Esp*3I site from the two fragments, resulting in a ligation product lacking the engineered restriction site. The use of the "no-see-um" technology allows for the assembly of large DNAs from smaller subclones without the incorporation of unique restriction sites into the genome. (Yount et al. 2002)

seamless junction compiled from the exact MHV-A59 sequence. Because of this property, unique junctions can be inserted at virtually any position between two component clones without mutating the viral genome sequence. Additionally, a large number of other restriction enzymes share this property (e.g., *Sap*I, *Aar*I), expanding the utility of the "no-see-um" technology (Table 1).

During the isolation of the MHV component clones, it was also necessary to remove three preexisting *Esp*3I sites located throughout the MHV ORF1 sequence (Bonilla et al. 1994). Mutations inserted to ablate these sites were used as marker mutations to distinguish molecularly cloned and wild-type virus. We then isolated seven consensus cDNAs that spanned the entire length of the MHV-A59 genome in the same manner as the TGEV infectious construct (Fig. 3). This was necessary because the MHV-A59 genome contains several major regions of sequence toxicity in microbial cloning vectors, most of which map between ~10 and 15 kb in the MHV ORF 1a/ORF 1b polyprotein and an unstable region mapping ~5.0 kb in ORF 1a. As described for the TGEV B fragment, cDNAs were isolated after intersecting the toxic domains and separating them into independent subclones. However, many subclones were still unstable in traditional PUC-based cloning vectors (e.g., pGem, TopoII) even when maintained at low temperature. Consequently, we used pSMART cloning vectors (Lucigen), which lack a promoter and indicator gene and contain transcriptional and translational terminators surrounding the cloning site. Instability appears to be associated with expression, as this entire domain (nucleotides 9,555–15,754) is also stable in ycast vectors (pYES2.1 Topo TA Cloning Kit from Invitrogen) that maintain tight regulation over foreign gene expression (Yount et al., unpublished results). Full-length MHV-A59 cDNA was systematically assembled through the simultaneous in vitro ligation of a series of seven subgenomic cDNAs (Yount et al. 2002). In the future, it may be possible to construct larger subgenomic fragments spanning the entire genome by using the pSMART cloning vectors, thereby simplifying the assembly strategy, although we have not tested this directly.

The TGEV and MHV A fragments contain a T7 promoter, whereas the TGEV F and MHV G fragments terminate in a poly(T) tract at the 3' end, allowing for in vitro T7 transcription of infectious capped, polyadenylated transcripts. The poly(A) tails generated from these transcripts are 25 nucleotides in length, which appears sufficient for transcript infectivity. At this time, we do not know the minimal number of 3' poly(A) residues necessary for transcript infectivity or whether a 5' methylated cap is essential. Electroporation of the genomic-length RNAs resulted in the production of recombinant MHV virus with growth characteristics identical to those of the wild-type viruses (Yount et al. 2000, 2002). Importantly, the molecularly cloned viruses contained marker mutations engineered into the component clones. Inclusion of nucleocapsid(N)-encoding transcripts enhanced the infectivity of full-length MHV and TGEV transcripts. In MHV, N transcripts enhanced the infectivity of full-length MHV-A59 transcripts by 10- to 15-fold as evidenced by increased viral antigen expression and virus titers at 25 h postinfection (Yount et al. 2002). It is unclear whether MHV N transcripts, N protein, or both are essential for increased virus yields after electroporation, or

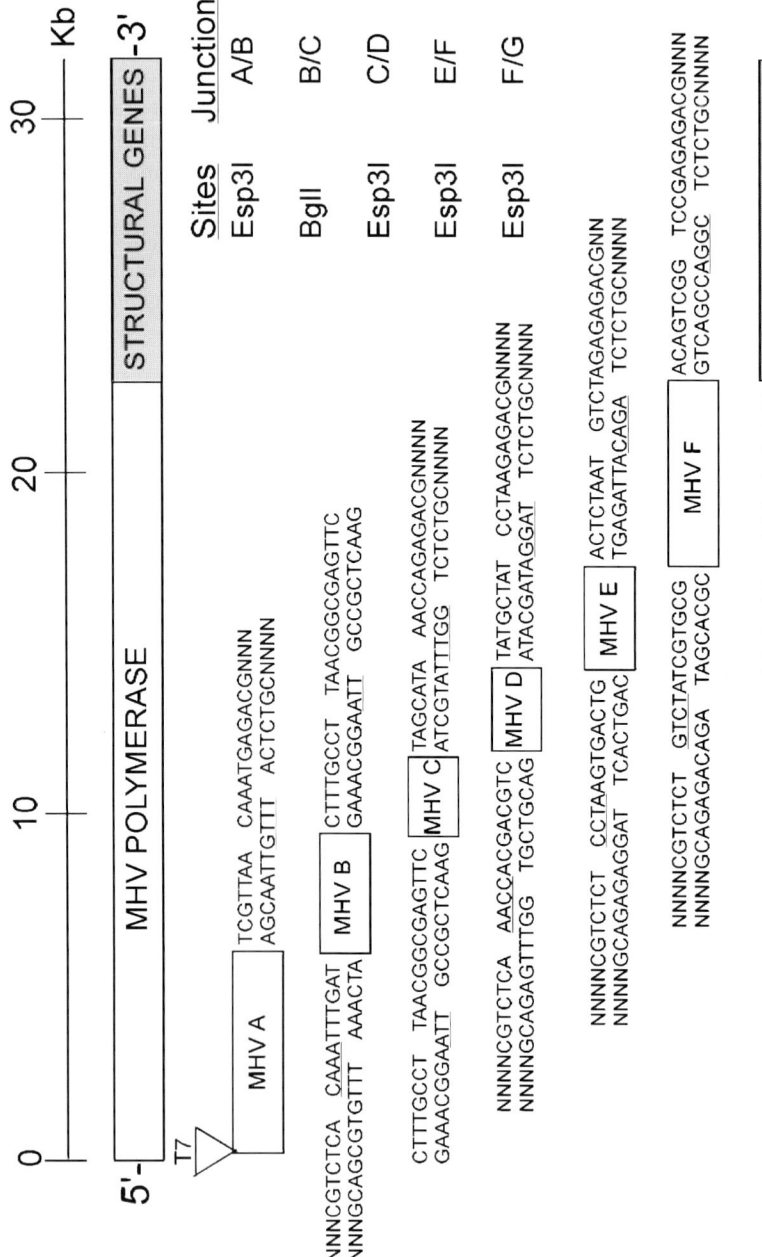

whether this effect would be observed with transcripts encoding unrelated genes. Coronaviruses have been demonstrated to package low concentrations of subgenomic mRNAs, especially N transcripts, and several studies have suggested that N may function in transcription and replication and are tightly associated with the replication complex. With IBV, but not TGEV or HCoV-229E, N transcripts are absolutely essential for full-length transcript infectivity (Casais et al. 2001). With HCoV-229E, other groups have shown that the N gene is not required for subgenomic transcription (Thiel et al. 2001). Clearly, additional studies are needed to evaluate the role of N protein in RNA transcript infectivity.

The MHV cDNA cassettes can be ligated systematically as described for TGEV or simultaneously. Although numerous incomplete assembly intermediates were evident, our demonstration that simultaneous ligation of seven cDNAs will result in full-length cDNA will simplify the complexity of the assembly strategy. At this time, there is no evidence to indicate that this approach might introduce spurious mutations or genome rearrangements from aberrant assembly cascades. However, it is possible that such variants might arise after RNA transfection, as a consequence of high-frequency MHV RNA recombination between incomplete and genome-length transcripts. It is likely that such variants would be replication impaired and rapidly out-competed by wild-type virus. A second limitation is that the yield of full-length cDNA product is reduced, resulting in less robust transfection efficiencies compared with the more traditional systematic assembly method. At this time, the MHV approach suffers from the large number of component clones (seven), which increase the complexity of the system and reduce the yield of full-length cDNA product after in vitro ligation. If the large number of toxic domains in the MHV genome is duplicated in other group II coronaviruses, this will likely interfere with the development of other infec-

Fig. 3. Systematic assembly strategy for the construction of MHV-A59 full-length cDNA. The MHV genome is a positive-sense, single-stranded RNA of ~31.5 kb. Seven independent subclones (*A*, *B*, *C*, *D*, *E*, *F*, and *G*) that span the entire MHV genome were isolated by RT-PCR. Unique *Bgl*I and *Esp*3I restriction sites, located at the 5′ and 3′ ends of each subclone, were used to assemble a full-length cDNA. A unique T7 start site was inserted at the 5′ end of the MHV A fragment and a 25 poly(T) tail was inserted at the 3′ end of the MHV F fragment, allowing for in vitro transcription of full-length, capped, poly-adenylated transcripts. Note: *Esp*3I sites are lost in the assembly process. (Yount et al. 2002)

tious cDNAs as well. Topics of future research include: (1) Can group II coronavirus cDNAs be stabilized as full-length constructs in bacterial artificial chromosomes or poxvirus vectors as has been reported with TGEV, IBV, and HCoV 229E? (2) How does N function to enhance infectivity of full-length transcripts? (3) How can we enhance yields or the infectivity of coronavirus infectious cDNAs and transcripts and allow for critical review of the consequences of lethal mutations? (4) Can we reduce the number of component clones needed to assemble group II coronavirus infectious cDNAs?

4.1
Applications in Genomics

Our assembly strategy for coronavirus infectious constructs is simple and straightforward, although the synthesis of full-length transcripts is technically challenging. In contrast to infectious clones of other positive-strand viruses, our TGEV and MHV constructs must be assembled de novo and do not exist intact in bacterial or viral vectors. This does not restrict the method's applicability for reverse genetic applications. Rather, it allows for rapid genetic manipulation of independent subclones, which minimizes the introduction of spurious mutations elsewhere in the genome during recombinant DNA manipulation. Theoretical limits of our method may exceed several million base pairs of DNA and will likely surmount the cloning capacity of bacterial (BAC) and eukaryotic artificial chromosome vectors (Grimes and Cooke 1998). Our systematic assembly method should also be appropriate for constructing full-length infectious clones of other large RNA viruses, including coronaviruses (27–32 kb), toroviruses (24–27 kb), and filoviruses like Marburg (19 kb) (de Vries et al. 1997; Peters et al. 1996). Viral genomes that are unstable in prokaryotic vectors can also be cloned by these methods (Boyer and Haenni 1994; Rice et al. 1989). Moreover, the technique should allow the systematic assembly of full-length infectious dsDNA genomes of adenoviruses, herpesviruses, and perhaps other large DNA viruses that promise to be powerful tools in vaccination, gene transfer, and gene therapy (Smith and Enquist 2000; van Zijl et al. 1988). Recently, genome sequences from a large number of prokaryotic and eukaryotic organisms have been obtained, providing significant insight into gene organization, structure, and function (Cho et al. 1999; Hutchison et al. 1999) (TIGR homepage http://www.tigr.org). Using this strategy, it may be possible to reconstruct a minimal microbial genome from the bottom up. However, problems associated with isolating large DNA fragments and the introduction of large DNA genomes into environments that permit replication will likely be significant hurdles. Nevertheless, our assembly strategy may provide a means to analyze the function of large blocks of DNA, such as pathogenesis islands, or to engineer chromosomes that contain large gene cassettes of interest (Cho et al. 1999).

4.2
Engineering MHV Genomes

Coronaviruses provide a unique system for the incorporation and expression of one or more foreign genes (Enjuanes and Van der Zeijst 1995). Coronavirus genes rarely overlap, simplifying the design and expression of foreign genes from downstream intergenic sequences (IS) start sites. Integration of the coronavirus RNA genome into the host cell chromosome is unlikely (Lai and Cavanagh 1997). Additionally, recombinant viruses or replicon particles could be readily targeted to other mucosal surfaces in swine or to other species by simple replacements in the S glycoprotein gene, which has been shown to determine tissue- and species tropism (Ballesteros et al. 1997; Delmas et al. 1992; Kuo et al. 2000; Leparc-Goffart et al. 1998; Sánchez et al. 1999; Tresnan et al. 1996). Furthermore, coronaviruses infect a number of different species, including human, porcine, bovine, canine, and feline, and are available for the development of expression systems (Sánchez et al. 1992). Additionally, the coronavirus helical ribonucleocapsid structure may further relax the packaging constraints of the virus, as compared to icosahedral structures (Enjuanes and Van der Zeijst 1995; Lai and Cavanagh 1997; Risco et al. 1996). Selected questions that remain unanswered include: (1) What is the coding capacity of coronavirus based expression systems? (2) What is the minimal genome required for efficient replication? (3) Can high-titer coronavirus replicon particles be obtained for vaccine applications? (4) What are the minimal sequence requirements for subgenomic transcription? (5) How many foreign genes can be coordinately regulated without impeding virus replication or immunogenicity? (6) What are the efficacy, stability, and safety of the recombinant coronaviruses in natural settings? Clearly, these vaccine-related topics will provide fruitful avenues of investigation over the next decade and will greatly enhance our understanding of the mechanics of coronavirus transcription, replication, assembly and release, and pathogenesis.

The future development of vaccines and expression vectors are particularly intriguing applications of our TGEV and MHV infectious clones. Importantly, at least two TGEV downstream ORFs encode luxury func-

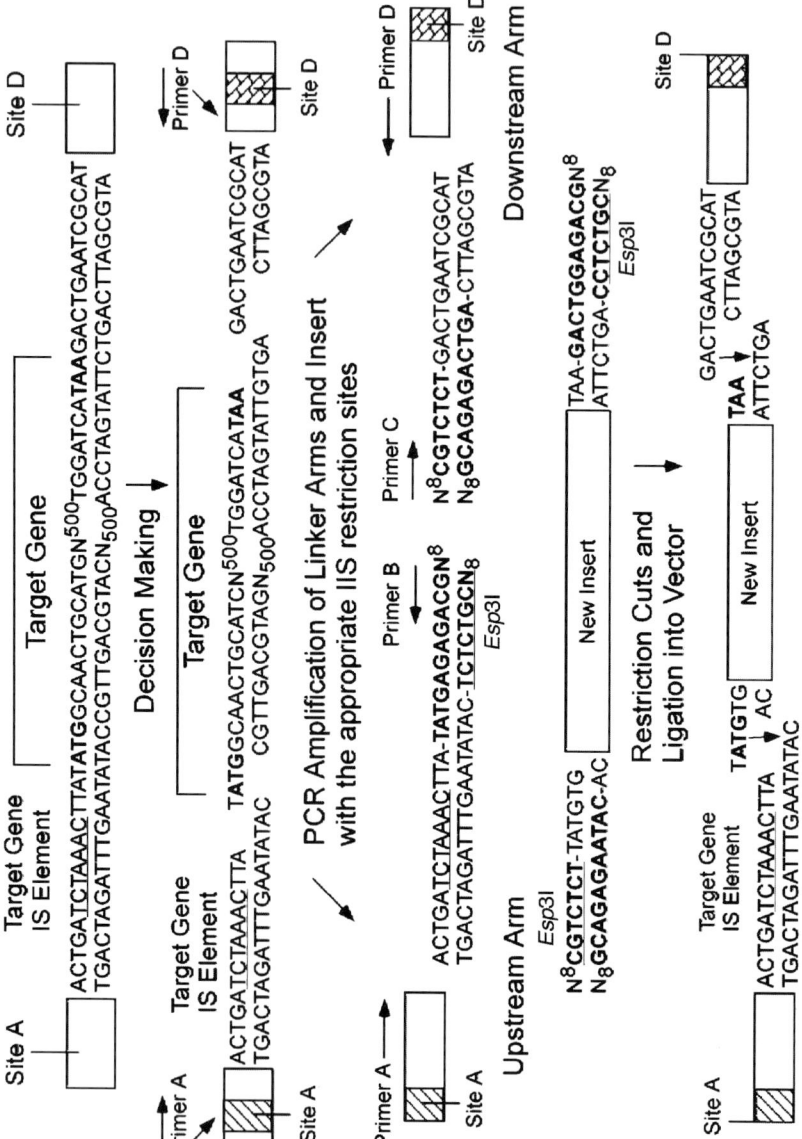

Fig. 4. Rapid mutagenesis of the MHV infectious cDNA with Class IIS restriction endonucleases. Seamless insertion of foreign genes into the coronavirus genome can be accomplished with Class IIS restriction enzymes. In this case, a target gene is systematically removed and replaced by a new gene (*new insert*). Using a primer with overlaps a unique upstream (*Site A*) restriction site, the upstream arm amplicon is

tions (ORF 3a and 3b) that may be deleted from the viral genome without impacting infectivity (Curtis et al. 2002; Laude et al. 1990; McGoldrick et al. 1999; Wesley et al. 1991). We have developed a rapid approach that allows seamless insertion of foreign sequences into virtually any nucleotide position in the MHV genome, based on class IIS restriction endonucleases (Fig. 4). In this approach, flanking sequences around the target domain are amplified as separate arms linked by unique class IIS restriction site oriented as described in Fig. 3. A third amplicon encoding the payload sequence of interest is isolated and flanked by similar class IIS sites. After restriction digestion and ligation, the foreign sequences are inserted into the backbone sequence at any given nucleotide, leaving no evidence of the restriction sites that were used to "sew" the new sequences into the MHV backbone. We have successfully expressed GFP from the ORF 3a locus of TGEV (Curtis et al. 2002) and ORF 4 of MHV (Fig. 5) (manuscript in preparation), demonstrating the feasibility of the method and the use of TGEV and MHV as expression vectors. In the case with TGEV, GFP expression was stable for at least 10 passages. In addition, we have removed the ORF 3a and replaced it with GP5 of PRRSV to create icTGEV PRRSV GP5 recombinant viruses (Curtis KM and Baric RS, unpublished data). Recombinant viruses expressed the PRRSV GP5 glycoprotein as evidenced by indirect immunofluorescence assay (IFA) and RT-PCR using primer pairs within the TGEV leader and PRRSV GP5 gene (data not shown). Recently, expression of the reporter gene β-glucuronidase (GUS) and PRRSV ORF 5 from a TGEV-derived minigenome was demonstrated (Alonso et al. 2002). Importantly, strong humoral immune responses against GUS and PRRSV ORF5 were generated in swine with these vectors, demonstrating the feasibility of coronavirus-based vectors for future vaccine development.

◄───

amplified with a second primer (*Site B*) containing a *Esp*3I recognition at the 5' end of the nonsense strand of DNA by PCR. A similar approach is used to amplify the downstream arm (*Site C* and *D* primers). The insert DNA is amplified with primer pairs containing compatible C and D *Esp*3I sites. After PCR amplification and restriction digestion, the new insert can be inserted into the viral genome without evidence of the restriction sites used in the assembly cascade. A large number of class IIS restriction enzymes greatly enhances the plasticity of the approach

Fig. 5a, b. Recombinant MHV-A59 expressing GFP. With standard molecular techniques, ORF 4 was removed and the gene encoding GFP inserted downstream of the ORF 4 IS (**a**). DBT cells were infected with wild-type MHV-A59 (*A* and *C*) or icMHV-A59 GFP (*B* and *D*) and subsequently analyzed for CPE by light microscopy (*A* and *B*) and GFP expression by fluorescent microscopy (*C* and *D*) (**b**)

5
SARS-CoV Infectious Clone

Rapid response and control of exigent emerging pathogens require an approach to quickly generate full-length cDNAs from which molecularly cloned viruses are rescued, allowing for genetic manipulation of the genome. Identification of the first human coronavirus to cause considerable morbidity and mortality worldwide provided the first template to test the rapidity of our systematic assembly strategy (Drosten et al. 2003; Ksiazek et al. 2003). Development of novel vaccine candidates and therapeutics requires a better understanding of viral pathogenesis, a process greatly facilitated by the availability of an infectious clone. A systematic assembly strategy based on the TGEV infectious clone was employed to create an infectious construct of the SARS-CoV, within ~2 months of the identification and isolation of genomic SARS-CoV RNA (Yount et al. 2003). Consensus clones were assembled from sibling clones of each SARS-CoV fragment by taking advantage of the special properties of asymmetric type IIS restriction enzymes. Within 9 weeks, infectious clone SARS-CoV was isolated that was phenotypically indistinguishable from wild-type SARS-CoV strains.

The SARS-CoV genome was cloned as six contiguous subclones that could be systematically linked by unique *BglI* restriction endonuclease sites (Fig. 6). Two *BglI* junctions were derived from sites encoded within the SARS-CoV genome at nt 4,373 (A/B junction) and nt 12,065 (C/D junction). A third *BglI* site at nt 1,577 was removed, and new *BglI* sites were inserted by the introduction of silent mutations into the SARS-CoV sequence at nt 8,700 (B/C junction), nt 18,916 (D/E junction) and nt 24,040 (E/F junction). The resulting cDNAs include SARS A (nt 1–4,436), SARS B (nt 4,344–8,712), SARS C (nt 8,695–12,070), SARS D (nt 12,055–18,924), SARS E (nt 18,907–24,051), and SARS F (nt 24,030–29,736) subclones. The SARS A subclone also contains a T7 promoter, and the SARS F subclone terminates in 21Ts, allowing synthesis of capped, polyadenylated transcripts. SARS-CoV infectious clone virus was assembled, transcribed and transfected as described previously, and recombinant viruses contained the marker mutations inserted into the infectious clone. Recombinant viruses produced a mild pneumonia on x-ray in macaques similar to wild-type viruses and replicated to similar titers in the mouse model (unpublished observation). These data suggest that recombinant viruses recapitulated the pathogenesis of wild type in animal models, allowing for the identification of pathogenesis determinants and developing attenuated viruses as candidate live and killed vaccines.

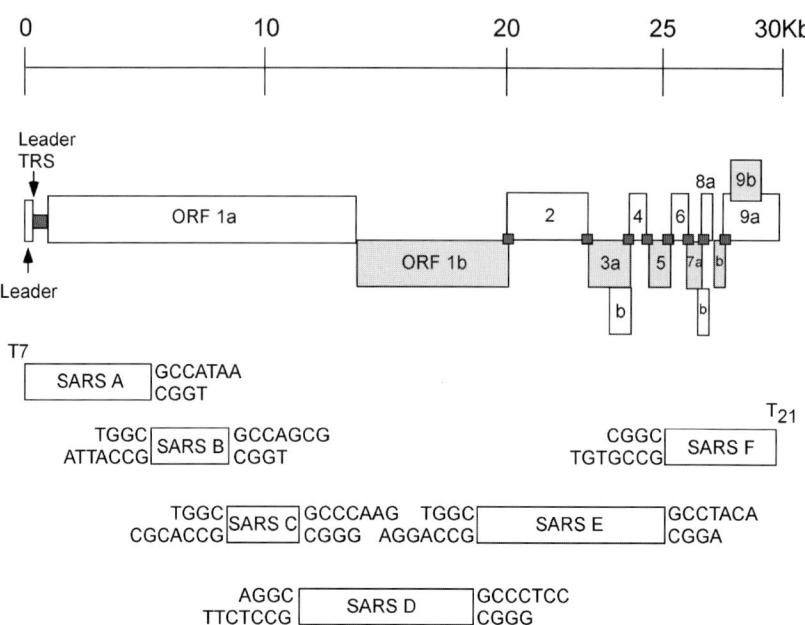

Fig. 6. Systematic assembly strategy for the SARS-CoV infectious clone. The SARS-CoV genome is about 30 kb in length and contains ~14 open reading frames (ORFs). The predicted functions of the group specific ORFs (*ORF 3a/b, ORF 6, ORF 7a/b, ORF 8a/b, ORF 9b*) are unknown. *Dark gray squares* indicate highly conserved consensus sequence sites that function in subgenomic RNA synthesis. Six independent subclones (*A, B, C, D, E*, and *F*) that span the entire SARS-CoV genome were isolated by RT-PCR (genome fragments are not shown to scale). The A fragment spans nt 1–4436, the B fragment nt 4344–8712, the C fragment nt 8695–12,070, the D fragment nt 12,055–18,924, the E fragment 18,907–24,051, and the F fragment nt 24,030–29,736. Unique *Bgl*I restriction sites located at the 5′ and 3′ ends of each subclone were used to assemble a full-length cDNA. A unique T7 start site was inserted at the 5′ end of the SARS-CoV A fragment, and a 21 poly(T) tail was inserted at the 3′ end of the SARS-CoV G fragment, allowing for in vitro transcription of full-length, capped, polyadenylated transcripts

6
Future Applications

The availability of infectious cDNA clones will undoubtedly have a profound effect on the field of coronavirology. These new tools will facilitate basic studies and allow for more precise analyses of the molecular mechanisms of viral replication, including the definition of RNA elements important for RNA replication, subgenomic RNA transcription, and genomic RNA packaging. In addition, studies of gene function will be enhanced by the availability of infectious cDNA clones by allowing for the construction of recombinant viruses and/or replicons containing mutations and the analysis of their effects on viral replication and assembly. MHV has long been used as a premiere model to study coronavirus assembly and release, replication, transcription, entry, and pathogenesis. The availability of MHV and SARS-CoV infectious cDNA clones will complement the existing targeted recombination approaches by providing a tool for the mutagenesis of the replicase gene, which encode a large number of cleavage products that have not been fully characterized. The structure and function of the ~20-kb MHV replicase domain will likely remain a fertile area of research for the next decade and reveal novel protein functions that participate and regulate discontinuous transcription and high-frequency RNA recombination. Although large panels of reagents are available for analyzing replicase protein expression, processing, and subcellular localization, a spectrum of genetically informative mutations have not been systematically targeted to any of these replicase proteins. Given the complexity and size of the coronavirus replicase gene, the number of potential mutants that can be generated is enormous and will likely require bioinformatic approaches for building and testing specific hypotheses. For example, the ORF1a C-terminal MHV p15 protein is highly conserved among group I through III coronaviruses and contains a large number of conserved cysteine residues and predicted phosphorylation, myristilation, and glycosylation sites (prosite, unpublished) (Fig. 7). The original sequence report of p15 also suggested possible similarities to growth factor-like proteins (Lee et al. 1991). Recent studies with an IBV homolog suggest that p15 exists as a dimer and accumulates on stimulation with epidermal growth factor, providing some evidence that the protein might be involved in the growth factor signaling pathway (Ng and Liu 2002). A single amino acid mutation has been identified in p15 of the temperature sensitive mutant, LA6, an MHV-A59 mutant with a defect in RNA synthesis at nonpermissive temperature (Siddell et al. 2001). The availability of infectious cDNAs allows, for the first time, a systematic mutagenesis approach for studying the function of specific structural features within this and other replicase proteins.

Coupled with the capacity to isolate large panels of mutants in each of the replicase proteins, selected questions include: (1) Are each of the PL1^{pro}, PL2^{pro}, and 3CL^{pro} cleavage sites necessary for MHV replication? (2) Are the PL1^{pro}, PL2^{pro}, or 3CL^{pro} proteases essential for replication? (3) Are any replicase proteins nonessential? (4) Is replicase gene order

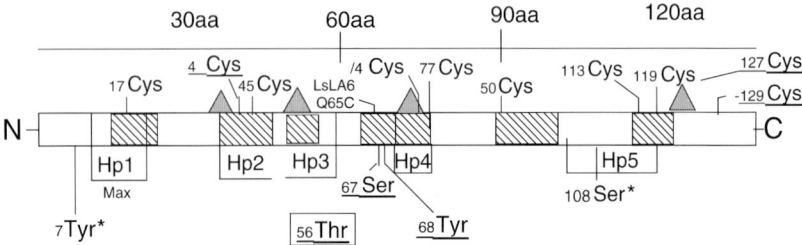

Fig. 7. Potential sites of mutagenesis within the C-terminal Orf1a p15 replicase protein. The MHV p15 replicase protein is highly conserved among all *Coronaviridae* (*hatched domains*), contains several hydrophobic domains (*Hp1–5*) and several potential sites for myristylation (*gray triangles*), and 10 highly conserved cysteine residues (*Cys*). Several sites for phosphorylation and glycosylation are predicted with prosite analysis, although it is unclear whether p15 is phosphorylated or glycosylated

critical? (5) Are replicase proteins interchangeable between the group 1 and/or group 2 coronaviruses? (6) How do replication complexes form on membranes? (7) What replicase complexes regulate discontinuous transcription and synthesis of genome-length and subgenomic-length mRNAs and negative-strand RNAs? (8) What are the *cis*-acting sequence elements required for genomic RNA packaging and replication? (9) What are the structure-function relationships within and between various replicase proteins and/or RNAs? (10) What are the functions of the group-specific ORFs, and how do they influence pathogenesis? The next decade of research may well be defined as the golden age of coronavirus genetics.

References

Almazán F, González JM, Pénzes Z, Izeta A, Calvo E, Plana-Durán J, Enjuanes L (2000) Engineering the largest RNA virus genome as an infectious bacterial artificial chromosome. Proc Natl Acad Sci USA 97:5516–5521

Alonso S, Sola I, Teifke J, Reimann I, Izeta A, Balach M, Plana-Durán J, Moormann RJM, Enjuanes L (2002) In vitro and in vivo expression of foreign genes by transmissible gastroenteritis coronavirus-derived minigenomes. J Gen Virol 83:567–579

Ballesteros ML, Sánchez CM, Enjuanes L (1997) Two amino acid changes at the N-terminus of transmissible gastroenteritis coronavirus spike protein result in the loss of enteric tropism. Virology 227:378–388

Bonilla PJ, Gorbalenya AE, Weiss SR (1994) Mouse hepatitis virus strain A59 RNA polymerase gene ORF 1a: heterogeneity among MHV strains. Virology 198:736–740

Boyer JC, Haenni AL (1994) Infectious transcripts and cDNA clones of RNA viruses. Virology 198:415–426

Casais R, Thiel V, Siddell SG, Cavanagh D, Britton P (2001) Reverse genetics system for the avian coronavirus infectious bronchitis virus. J Virol 75:12359–12369

Cavanagh D, Brian DA, Britton P, Enjuanes L, Horzinek MC, Lai MMC, Laude H, Plagemann PGW, Siddell S, Spaan W, Talbot PJ (1997) *Nidovirales*: a new order comprising *Coronaviridae* and *Arteriviridae*. Arch Virol 142:629–635

Cho MK, Magnus D, Caplan AL, McGee D, Ethics of Genomics Group (1999) GENETICS:Ethical Considerations in Synthesizing a Minimal Genome. Science 286:2087–2090

Curtis KM, Yount B, Baric RS (2002) Heterologous gene expression from transmissible gastroenteritis virus replicon particles. J Virol 76:1422–1434

Delmas B, Gelfi J, L'Haridon R, Vogel LK, Norén O, Laude H (1992) Aminopeptidase N is a major receptor for the enteropathogenic coronavirus TGEV. Nature 357:417–420

de Vries AAF, Horzinek MC, Rottier PJM, de Groot RJ (1997) The genome organization of the *Nidovirales*: similarities and differences between arteri-, toro-, and coronaviruses. Semin Virol 8:33–47

Drosten C, Günther S, Preiser W, van der Werf S, Brodt H-R, Becker S, Rabenau H, Panning M, Kolesnikova L, Fouchier RAM, Berger A, Burguiere A-M, Cinatl J, Eickmann M, Escriou N, Grywna K, Kramme S, Manuguerra J-C, Muller S, Rickerts W, Sturmer MV, S., Klenk H-D, Osterhaus ADME (2003) Identification of a novel coronavirus in patients with severe acute respiratory syndrome. N Engl J Med 348:1967–1976

Eleouet JF, Rasschaert D, Lambert P, Levy L, Vende P, Laude H (1995) Complete sequence (20 kilobasee not been fully characterized. The structure and function of the ~20-kb MHV replicase domain will likely remain a fertile area of research for the next decade and reveal novel protein functions that participate and regulate discontinuous transcription and high-frequency RNA recombination. Although large panels of reagents are available for analyzing replicase protein expression, processing, and subcellular localization, a spectrum of genetically informative mutations have not been systematically targeted to any of these replicase proteins. Given the complexity and size of the coronavirus replicase gene, the number of potential mutants that can be generated is enormous and will likely require bioinformatic approaches for building and testing specific hypotheses. For example, the ORF1a C-terminal MHV p15 protein is highly conserved among group I through III coronaviruses and contains a large number of conserved cysteine residues and predicted phosphorylation, myristylation, and glycosylation sites (prosite, spect of coronavirus transcription. J Virol 71:5148–5160

Fu K, Baric RS (1994) Map locations of mouse hepatitis virus temperature-sensitive mutants: confirmation of variable rates of recombination. J Virol 68:7458 7466

Fu KS, Baric RS (1992) Evidence for variable rates of recombination in the MHV genome. Virology 189:88–102

Grimes B, Cooke H (1998) Engineering mammalian chromosomes. Hum Mol Genet 7:1635–1640

Hsue B, Masters PS (1999) Insertion of a new transcriptional unit into the genome of mouse hepatitis virus. J Virol 73:6128–6135

Hutchison CA III, Peterson SN, Gill SR, Cline RT, White O, Fraser CM, Smith HO, Venter JC (1999) Global transposon mutagenesis and a minimal mycoplasma genome. Science 286:2165–2169

Izeta A, Smerdou C, Alonso S, Penzes Z, Méndez A, Plana-Durán J, Enjuanes L (1999) Replication and packaging of transmissible gastroenteritis coronavirus-derived synthetic minigenomes. J Virol 73:1535–1545

Ksiazek TG, Erdman D, Goldsmith C, Zaki S, Peret T, Emery S, Tong S, Urbani C, Comer JA, Lim W, Rollin PE, Dowell S, Ling A-E, Humphrey C, Shieh W-J, Guarner J, Paddock CD, Rota P, Fields B, DeRisi J, Yang J-Y, Cox N, Hughes J, LeDuc JW, Bellini WJ, Anderson LJ (2003) A novel coronavirus associated with severe acute respiratory syndrome. N Engl J Med 348:1953–1966

Kuo L, Godeke G-J, Raamsman MJB, Masters PS, Rottier PJM (2000) Retargeting of coronavirus by substitution of the spike glycoprotein ectodomain: crossing the host cell species barrier. J Virol 74:1393–1406

Lai MMC, Cavanagh D (1997) The molecular biology of coronaviruses. Adv Virus Res 48:1-100

Laude H, Rasschaert D, Delmas B, Godet M, Gelfi J, Bernard C (1990) Molecular biology of transmissible gastroenteritis virus. Vet Microbiol 23:147–154

Lee HJ, Shieh CK, Gorbalenya AE, Koonin EV, Lamonica N, Tuler J, Bagdzhadzhyan A, Lai MMC (1991) The complete sequence (22 kilobases) of murine coronavirus gene-1 encoding the putative proteases and RNA polymerase. Virology 180:567–582

Leparc-Goffart I, Hingley ST, Chua MM, Phillips J, Lavi E, Weiss SR (1998) Targeted recombination within the spike gene of murine coronavirus mouse hepatitis virus-A59: Q159 is a determinant of hepatotropism. J Virol 72:9628–9636

Masters PS (1999) Reverse genetics of the largest RNA viruses. Adv Virus Res 53:245–264

McGoldrick A, Lowings JP, Paton DJ (1999) Characterisation of a recent virulent transmissible gastroenteritis virus from Britain with a deleted ORF 3a. Arch Virol 144:763–770

Narayanan K, Makino S (2001) Cooperation of an RNA packaging signal and a viral envelope protein in coronavirus RNA packaging. J Virol 75:9059–9067

Ng LFP, Liu DX (2002) Membrane association and dimerization of a cysteine-rich, 16-kilodalton polypeptide released from the C-terminal region of the coronavirus infectious bronchitis virus 1a polyprotein. J Virol 76:6257–6267

Penzes Z, González JM, Calvo E, Izeta A, Smerdou C, Mendez A, Sánchez CM, Sola I, Almazán F, Enjuanes L (2001) Complete genome sequence of transmissible gastroenteritis coronavirus PUR46-MAD clone and evolution of the Purdue virus cluster. Virus Genes 23:105–118

Peters CJ, Sanchez A, Rollin PE, Ksiazek TG, Murphy FA (1996) Filoviridae: Marburg and Ebola Viruses. In: Fields BN, Knipe DM, Howley PM, Chanock RM, Melnick JL, Monath TP, Roizman B and Straus SE (eds) Field's Virology. Lippincott Williams and Wilkens, Philadelphia, pp 1161–1176

Pingoud A, Jeltsch A (2001) Structure and function of type II restriction endonucleases. Nucl Acids Res 29:3705–3727

Repass JF, Makino S (1998) Importance of the positive-strand RNA secondary structure of a murine coronavirus defective interfering RNA internal replication signal in positive-strand RNA synthesis. J Virol 72:7926–7933

Rice CM, Grakoui A, Galler R, Chambers TJ (1989) Transcription of infectious yellow fever RNA from full-length cDNA templates produced by in vitro ligation. New Biol 1:285–296

Risco C, Antón IM, Enjuanes L, Carrascosa JL (1996) The transmissible gastroenteritis coronavirus contains a spherical core shell consisting of M and N proteins. J Virol 70:4773–4777

Sambrook J, Fritsch EF, Maniatis T (1989) Molecular cloning: A laboratory manual, 2nd edn. Cold Spring Harbor Laboratory, Cold Spring Harbor, New York

Sánchez CM, Gebauer F, Suñé C, Méndez A, Dopazo J, Enjuanes L (1992) Genetic evolution and tropism of transmissible gastroenteritis coronaviruses. Virology 190:92–105

Sánchez CM, Izeta A, Sánchez-Morgado JM, Alonso S, Sola I, Balasch M, Plana-Durán J, Enjuanes L (1999) Targeted recombination demonstrates that the spike gene of transmissible gastroenteritis coronavirus is a determinant of its enteric tropism and virulence. J Virol 73:7607–7618

Schaad M, Baric RS (1994) Genetics of mouse hepatitis virus transcription: evidence that subgenomic negative strands are functional templates. J Virol 68:8169–8179

Siddell SG (1995) The Coronaviridae: an introduction. In: Siddell SG (ed) The Coronaviridae. Plenum Press, New York The Viruses, pp 1–10

Siddell SG, Sawicki D, Meyer Y, Thiel V, Sawicki S (2001) Identification of the mutations responsible for the phenotype of three MHV RNA-negative ts mutants. Adv Exp Med Biol 494:453–458

Smith GA, Enquist LW (2000) A self-recombining bacterial artificial chromosome and its application for analysis of herpesvirus pathogenesis. Proc Natl Acad Sci USA 97:4873–4878

Stalcup RP, Baric RS, Leibowitz JL (1998) Genetic complementation among three panels of mouse hepatitis virus gene 1 mutants. Virology 241:112–121

Thiel V, Herold J, Schelle B, Siddell SG (2001) Viral replicase gene products suffice for coronavirus discontinuous transcription. J Virol 75:6676–6681

Tresnan DB, Levis R, Holmes KV (1996) Feline aminopeptidase N serves as a receptor for feline, canine, porcine, and human coronaviruses in serogroup I. J Virol 70:8669–8674

van Zijl M, Quint W, Briaire J, de Rover T, Gielkens A, Berns A (1988) Regeneration of herpesviruses from molecularly cloned subgenomic fragments. J Virol 62:2191–2195

Wesley RD, Woods RD, Cheung AK (1991) Genetic analysis of porcine respiratory coronavirus, an attenuated variant of transmissible gastroenteritis virus. J Virol 65:3369–3373

Williams GD, Chang R-Y, Brian DA (1999) A phylogenetically conserved hairpin-type 3′ untranslated region pseudoknot functions in coronavirus RNA replication. J Virol 73:8349–8355

Yount B, Curtis KM, Baric RS (2000) Strategy for systematic assembly of large RNA and DNA genomes: the transmissible gastroenteritis virus model. J Virol 74:10600–10611

Yount B, Denison MR, Weiss SR, Baric RS (2002) Systematic assembly of a full length infectious cDNA of mouse hepatitis virus stain A59. J Virol 76:11065–11078

Yount B, Curtis KM, Fritz EA, Hensley LE, Jahrling PB, Prentice E, Denison MR, Geisbert TW, Baric RS (2003) Reverse genetics with a full-length infectious cDNA of severe acute respiratory syndrome coronavirus. Proc Natl Acad Sci USA 100:12995–13000

Subject Index

accessory proteinase 64
acidic domain 61, 62, 67
aconitase 15, 116
– cytoplasmic 116
– mitochondrial 115
adenosine diphosphate-ribose 63
– 1"-phosphatase 61
ADRP 63, 64, 67
Alb4 138–142, 146
alphavirus 15, 81
Arteriviridae 35, 165
arterivirus 58, 64, 68, 168, 172, 185
– helicase 77, 82
– replication complex 82
3CL^{pro} autocatalytic release 73

BAC 240
– system 211
BCoV (bovine coronavirus) 3, 17, 103, 108, 141, 146, 148
BHK-21 211, 213, 217, 219
bovine coronavirus (BCoV) 103, 141, 146, 148

canine coronavirus (CCoV) 136, 165
carcinoembryonic antigen-related cell adhesion molecules (CEACAM) 186
CAT (chloramphenicol acetyltransferase) 179
catalytic
– dyad 67
– residues 69
CCoV (canine coronavirus) 165
cDNA clone 162
CEACAM (carcinoembryonic antigen-related cell adhesion molecules) 186

charperone protein 98
chick kidney (CK) 207, 209
chloramphenicol acetyltransferase (CAT) 179, 222
chloromethyl ketone inhibitor 71
chymotrypsin 3C-like protease (3CLP) 65, 97, 101
cis-acting
– leader RNA 99
– replication sequences 14
– RNA elements 8, 12
– signal 117
cloning capacity 183, 184
3CLP (chymotrypsin 3C-like protease) 61, 68, 72, 101, 105, 102, 106, 107, 203
core sequence (CS) 170
Coronaviridae 35
coronavirology 246
coronavirus 58
– 5'UTR 9
– gene order 19
– genome 2, 230
– transcription 34, 42
CPD 63, 64
crystal structure 70, 74
cyclic phosphodiesterase 63
cycloheximide 42, 48
cysteine proteinase 65
cytomegalovirus (CMV) promoter 166
cytoplasmic aconitase (c-aconitase) 116
cytoskeletal protein 98

dactinomycin 33, 41
dentritic cell 221, 223

DI RNA 9, 12, 17, 21, 38, 112, 140–142, 148
discontinuous transcription 38, 40, 47, 171, 219
domain III 69, 75, 76
double membrane vesicles (DMVs) 81
double-stranded (ds) RNA 76
dsDNA 240
duplex unwinding activity 77, 105

EAV (equine arteritis virus) 39, 106
ectodomain region 215
elongation factor 98
endocytic carrier vesicle (ECV) 81
endoribonuclease (XendoU) 63
endosomal membrane 81
equine arteritis virus (EAV) 106
Escherichia coli 206, 233
eukaryotic iniation factor (eIF) 98, 115
evolution 68
ExoN (exonuclease) 61, 63, 64

feline
– coronavirus (FCoV) 165
– interfectious peritonitis virus (FIPV) 143–145, 150, 151
flavivirus 15
fMHV 143–145
fowlpox virus 203, 207

G–D–D signature 62
gene expression 184, 186, 188
GFP (green fluorescent protein) 168, 182, 206, 217, 220, 223, 244
– reporter gene 221
β-glucuronidase (GUS) 178, 243
glycoprotein S
  32, 186, 241
glycosylation 247
Golgi membrane 12
green fluorescent protein (GFP) 168, 206
GUS (β-glucuronidase) 243
– gene 185

HCoV (human coronavirus) 104
– -229E 18, 178, 180, 185, 205, 208, 239
– infected cells 106
HE 179
helical ribonucleosid structure 241
helicase 58, 61, 63, 65, 76, 97, 102, 104
helper-dependent expression system 178
hemagglutinin-esterase (HE) 109, 179
hepato-encephalopathy 32
heterogeneous nuclear ribonucleoprotein (hnRNP) 98
hnRNP (heterogeneous nuclear ribonucleoprotein) 98
– A1 99, 110, 112, 113, 118
– I 112
host range-based selection 142–145, 147, 153
human
– aminopeptidase N (hAPN) 187, 221, 224
– coronavirus (HCoV) 104
hydrophobic domain 82, 108

IBV (infectious bronchitis coronavirus) 10, 12, 17, 103, 106, 135, 136, 142, 152, 179, 182, 200, 207, 209
– spike chimeras 215
IFA (immunofluorescence assay) 243
IG (intergenic) sequence 97, 110
IGS 36, 39
immunofluorescence assay (IFA) 243
infectious bronchitis virus (IBV) 135, 136, 142, 152, 200
interferon-γ (IFN-γ) 179
intergenic (IG) sequence 97
internal ribosome entry site (IRES) 112
IRES (internal ribosome entry site) 112

leader proteinase of food and mouth disease virus (FMDV) 67
luciferase (LUC) gene 183, 184, 222
lysosomal membrane 81

Subject Index

M protein   82
m-aconitase   115
main proteinase   65, 69
membrane compartment   80
membrane-associated protein (MP)   101, 108
methyltransferase   61, 63
MHV (mouse hepatitis virus)   8, 12, 32, 44, 96, 103, 135, 137, 138, 140, 231, 235, 243
– A59   37, 41, 49, 52, 200, 203, 211, 211, 216, 236
– defective interfering (DI) RNA   96
– genome   241
– polymerase gene products   100
– replication complex   11
– subgenomic minus strand   37
minigenome   174, 175, 178, 184
minus strand
– synthesis   37, 41
– discontinous synthesis   35
mitochondrial aconitase   115
Mononegavirales   171, 185
mouse hepatitis virus (MHV)   135, 137, 138, 140
MP (membrane-associated protein)   101, 108
multivesicular body (MVB)   81
myristylation   247

N (nucleocapsid)
– gene   20
– mRNA   220
– protein   9, 16, 18, 33, 52, 82, 107, 108, 209
negative strand RNA synthesis   171
Nidovirales   33, 34, 58, 173
nidovirus   168
– family   64
nsp
– -1   82
– -2   82
– -3   67, 82
– -4   82
– -5   69, 78, 81
– -6   82
– -8   82
– -10   78, 79
– -12   63, 78–81
– -13   63, 77, 80–82
– -14   63, 64
– -15   63
– -16   63
NTPase activity   77
nucleocapsid (N) protein   107
nucleoside triphosphate (NTP)   76, 231

2'-O-MT   64
ORF (open reading frame)   7, 97, 100, 212, 231
– -1   33
– -1b   40
– 3a   169
oxyanion hole   72

PABP   99, 114, 117, 118
packaging
– cell line   225
– signal   17, 189
papain-like
– cystein protease (PLP)   61, 65, 101
– protease   97
– proteinase   68, 213
pathogenesis   149, 150
PEDV   11
phosphorylation   247
picornavirus   15, 65, 69
– 3C proteinase   69, 74
PL1^{pro} 66, 67
PL2^{pro} 66
PLP (papain-like cystein protease)   101, 102, 106
plus-stranded RNA virus   2, 36
polarity   77
– 5' to 3' polarity   77
poliovirus   21, 96
polymerase   38, 40, 97
– initiation   47
– protein   97
polyprotein processing   65
porcine respiratory and reproductive syndrome virus (PRSSV)   164

porcine respiratory coronavirus (PRCV)   164
positive-strand synthesis   97
potyvirus   69
poxvirus vector   202
pp (polyprotein)
– 1a   33, 50, 59
– 1ab   33, 50, 59
PRCV (porcine respiratory coronavirus)   164
proofreading   64
protease   105
proteinase
– 3C-like main proteinase   61, 65, 70
– inhibitor   64
proteolytic processing   60, 64, 212
PRRSV GP5   243
pseudoknot   7, 10, 14, 101, 116
pSMART   237
PTB   99, 112, 118
– PTB-binding nucleotides   113
puromycin   42

RdRp   36, 58, 63, 65, 78, 102
recombinant
– coronavirus   206, 211
– mouse hepatitis (MHV-A59)   200
recombination   64
replicase   60
– complex   79
– gene   59
– polyprotein   107
– protein   97
– subcellular localization   79
replication   51, 96, 109
– complex   82
replicative/transcriptive intermediate   48–50
reverse genetics   173, 180
RF   44, 46
rFPV-T7   209
RI   43, 44, 46
ribonucleoprotein (RNP)   109
ribosomal frameshifting   59, 63, 102, 168
ribosome binding   47

RNA
– +RNA virus   58
– 5' RNA cap structure   78
– 5'-triphosohatase   43, 78
– interference (RNAi)   118
– polymerase   61, 78
– processing   63, 65, 98
– pseudoknot structure   59
– replication   99, 114, 217
– replicon RNA   217, 220
– RNA-binding domain (RBD)   110
– synthesis   33, 35, 103, 111, 134, 141, 148, 149, 153
RNase (ribonuclease)   36, 44, 45
ronivirus   58, 64

S gene   177
S protein   72, 73, 163
SARS   215
– coronavirus (SARS-CoV)   3, 4, 34, 40, 164, 186, 245
– – 3C-like main proteinase   76
– – infectious clone   246
– – PL2pro   67
– – replicon   221
– – unique domain   61
serine proteinase-like structure   65
sgmRNA (subgenomic RNA)   7, 16, 18, 21, 173
single-stranded RNA   48
slippery sequence   10
small nucleolar (sno) RNA   63
SNEMN   19, 20
subgenomic mRNA (sgmRNA)   38, 173

temperature sensitive (ts) mutant   75, 78, 101, 225, 230
TGEV (transmissible gastroenteritis virus)   10–12, 17, 19, 36, 103, 135, 142, 148, 150, 152, 163, 231, 235, 239, 243
– rTGEV-Δ7   169
TM1-3   73, 82
tobacco vein mottling virus (TVMV)   104
torovirus   58, 64, 168
transmembrane domain   61

Subject Index

trans activity   73
– leader RNA   99
transcription   51
– factor-like domain   68
transmissible gastroenteritis virus (TGEV)   103, 135, 142, 148, 150, 152
tRNA splicing   63
TRS (transcription-regulating sequence)   39, 97, 169
TRS-L   175, 177
TVMV (tobacco vein mottling virus)   104

3'-UTR   111
5'-UTR   9, 15

vaccine   136, 151, 188, 241
vaccinia virus   105, 167, 200
– vector   201
Venezuelan equine encephalitis (VEE)   183
viral hepatitis   215
virion assembly   146
virus vector   162, 170, 178–184, 187–180
virus-like particle (VLP)   222

XendoU (endoribonuclease)   63, 64

Y domain   61

zinc-binding domain (ZBD)   68, 77

Current Topics in Microbiology and Immunology

Volumes published since 1989 (and still available)

Vol. 244: **Daëron, Marc; Vivier, Eric (Eds.):** Immunoreceptor Tyrosine-Based Inhibition Motifs. 1999. 20 figs. VIII, 179 pp. ISBN 3-540-65789-4

Vol. 245/I: **Justement, Louis B.; Siminovitch, Katherine A. (Eds.):** Signal Transduction and the Coordination of B Lymphocyte Development and Function I. 2000. 22 figs. XVI, 274 pp. ISBN 3-540-66002-X

Vol. 245/II: **Justement, Louis B.; Siminovitch, Katherine A. (Eds.):** Signal Transduction on the Coordination of B Lymphocyte Development and Function II. 2000. 13 figs. XV, 172 pp. ISBN 3-540-66003-8

Vol. 246: **Melchers, Fritz; Potter, Michael (Eds.):** Mechanisms of B Cell Neoplasia 1998. 1999. 111 figs. XXIX, 415 pp. ISBN 3-540-65759-2

Vol. 247: **Wagner, Hermann (Ed.):** Immunobiology of Bacterial CpG-DNA. 2000. 34 figs. IX, 246 pp. ISBN 3-540-66400-9

Vol. 248: **du Pasquier, Louis; Litman, Gary W. (Eds.):** Origin and Evolution of the Vertebrate Immune System. 2000. 81 figs. IX, 324 pp. ISBN 3-540-66414-9

Vol. 249: **Jones, Peter A.; Vogt, Peter K. (Eds.):** DNA Methylation and Cancer. 2000. 16 figs. IX, 169 pp. ISBN 3-540-66608-7

Vol. 250: **Aktories, Klaus; Wilkins, Tracy, D. (Eds.):** Clostridium difficile. 2000. 20 figs. IX, 143 pp. ISBN 3-540-67291-5

Vol. 251: **Melchers, Fritz (Ed.):** Lymphoid Organogenesis. 2000. 62 figs. XII, 215 pp. ISBN 3-540-67569-8

Vol. 252: **Potter, Michael; Melchers, Fritz (Eds.):** B1 Lymphocytes in B Cell Neoplasia. 2000. XIII, 326 pp. ISBN 3-540-67567-1

Vol. 253: **Gosztonyi, Georg (Ed.):** The Mechanisms of Neuronal Damage in Virus Infections of the Nervous System. 2001. approx. XVI, 270 pp. ISBN 3-540-67617-1

Vol. 254: **Privalsky, Martin L. (Ed.):** Transcriptional Corepressors. 2001. 25 figs. XIV, 190 pp. ISBN 3-540-67569-8

Vol. 255: **Hirai, Kanji (Ed.):** Marek's Disease. 2001. 22 figs. XII, 294 pp. ISBN 3-540-67798-4

Vol. 256: **Schmaljohn, Connie S.; Nichol, Stuart T. (Eds.):** Hantaviruses. 2001, 24 figs. XI, 196 pp. ISBN 3-540-41045-7

Vol. 257: **van der Goot, Gisou (Ed.):** Pore-Forming Toxins, 2001. 19 figs. IX, 166 pp. ISBN 3-540-41386-3

Vol. 258: **Takada, Kenzo (Ed.):** Epstein-Barr Virus and Human Cancer. 2001. 38 figs. IX, 233 pp. ISBN 3-540-41506-8

Vol. 259: **Hauber, Joachim, Vogt, Peter K. (Eds.):** Nuclear Export of Viral RNAs. 2001. 19 figs. IX, 131 pp. ISBN 3-540-41278-6

Vol. 260: **Burton, Didier R. (Ed.):** Antibodies in Viral Infection. 2001. 51 figs. IX, 309 pp. ISBN 3-540-41611-0

Vol. 261: **Trono, Didier (Ed.):** Lentiviral Vectors. 2002. 32 figs. X, 258 pp. ISBN 3-540-42190-4

Vol. 262: **Oldstone, Michael B.A. (Ed.):** Arenaviruses I. 2002, 30 figs. XVIII, 197 pp. ISBN 3-540-42244-7

Vol. 263: **Oldstone, Michael B. A. (Ed.):** Arenaviruses II. 2002, 49 figs. XVIII, 268 pp. ISBN 3-540-42705-8

Vol. 264/I: **Hacker, Jörg; Kaper, James B. (Eds.):** Pathogenicity Islands and the Evolution of Microbes. 2002. 34 figs. XVIII, 232 pp. ISBN 3-540-42681-7

Vol. 264/II: **Hacker, Jörg; Kaper, James B. (Eds.):** Pathogenicity Islands and the Evolution of Microbes. 2002. 24 figs. XVIII, 228 pp. ISBN 3-540-42682-5

Vol. 265: **Dietzschold, Bernhard; Richt, Jürgen A. (Eds.):** Protective and Pathological Immune Responses in the CNS. 2002. 21 figs. X, 278 pp. ISBN 3-540-42668-X

Vol. 266: **Cooper, Koproski (Eds.):** The Interface Between Innate and Acquired Immunity, 2002, 15 figs. XIV, 116 pp. ISBN 3-540-42894-1

Vol. 267: **Mackenzie, John S.; Barrett, Alan D. T.; Deubel, Vincent (Eds.):** Japanese Encephalitis and West Nile Viruses. 2002. 66 figs. X, 418 pp. ISBN 3-540-42783-X

Vol. 268: **Zwickl, Peter; Baumeister, Wolfgang (Eds.):** The Proteasome-Ubiquitin Protein Degradation Pathway. 2002, 17 figs. X, 213 pp. ISBN 3-540-43096-2

Vol. 269: **Koszinowski, Ulrich H.; Hengel, Hartmut (Eds.):** Viral Proteins Counteracting Host Defenses. 2002, 47 figs. XII, 325 pp. ISBN 3-540-43261-2

Vol. 270: **Beutler, Bruce; Wagner, Hermann (Eds.):** Toll-Like Receptor Family Members and Their Ligands. 2002, 31 figs. X, 192 pp. ISBN 3-540-43560-3

Vol. 271: **Koehler, Theresa M. (Ed.):** Anthrax. 2002, 14 figs. X, 169 pp. ISBN 3-540-43497-6

Vol. 272: **Doerfler, Walter; Böhm, Petra (Eds.):** Adenoviruses: Model and Vectors in Virus-Host Interactions. Virion and Structure, Viral Replication, Host Cell Interactions. 2003, 63 figs., approx. 280 pp. ISBN 3-540-00154-9

Vol. 273: **Doerfler, Walter; Böhm, Petra (Eds.):** Adenoviruses: Model and Vectors in Virus-Host Interactions. Immune System, Oncogenesis, Gene Therapy. 2004, 35 figs., approx. 280 pp. ISBN 3-540-06851-1

Vol. 274: **Workman, Jerry L. (Ed.):** Protein Complexes that Modify Chromatin. 2003, 38 figs., XII, 296 pp. ISBN 3-540-44208-1

Vol. 275: **Fan, Hung (Ed.):** Jaagsiekte Sheep Retrovirus and Lung Cancer. 2003, 63 figs., XII, 252 pp. ISBN 3-540-44096-3

Vol. 276: **Steinkasserer, Alexander (Ed.):** Dendritic Cells and Virus Infection. 2003, 24 figs., X, 296 pp. ISBN 3-540-44290-1

Vol. 277: **Rethwilm, Axel (Ed.):** Foamy Viruses. 2003, 40 figs., X, 214 pp. ISBN 3-540-44388-6

Vol. 278: **Salomon, Daniel R.; Wilson, Carolyn (Eds.):** Xenotransplantation. 2003, 22 figs., IX, 254 pp.ISBN 3-540-00210-3

Vol. 279: **Thomas, George; Sabatini, David; Hall, Michael N. (Eds.):** TOR. 2004, 49 figs., X, 364 pp.ISBN 3-540-00534-X

Vol. 280: **Heber-Katz, Ellen (Ed.):** Regeneration: Stem Cells and Beyond. 2004, 42 figs., XII, 194 pp.ISBN 3-540-02238-4

Vol. 281: **Young, John A. T. (Ed.):** Cellular Factors Involved in Early Steps of Retroviral Replication. 2003, 21 figs., IX, 240 pp. ISBN 3-540-00844-6

Vol. 282: **Stenmark, Harald (Ed.):** Phosphoinositides in Subcellular Targeting and Enzyme Activation. 2003, 20 figs., X, 210 pp. ISBN 3-540-00950-7

Vol. 283: **Kawaoka, Yoshihiro (Ed.):** Biology of Negative Strand RNA Viruses: The Power of Reverse Genetics. 2004, 24 figs., IX, 350 pp. ISBN 3-540-40661-1

Vol. 284: **Harris, David (Ed.):** Mad Cow Disease and Related Spongiform Encephalopathies. 2004, 34 figs., IX, 219 pp. ISBN 3-540-20107-6

Vol. 285: **Marsh, Mark (Ed.):** Membrane Trafficking in Viral Replication. 2004, 19 figs., IX, 259 pp. ISBN 3-540-21430-5

Vol. 286: **Madshus, Inger H. (Ed.):** Signalling from Internalized Growth Factor Receptors. 2004, 19 figs., IX, 187 pp. ISBN 3-540-21038-5

Vol. 287: **Enjuanes, Luis (Ed.):** Coronavirus Replication and Reverse Genetics. 2005, 49 figs., XI, 257 pp. ISBN 3-540-21494-1

Printing: Saladruck, Berlin
Binding: Stein+Lehmann, Berlin